Cryptic Species

Morphological Stasis, Circumscription, and Hidden Diversity

Cryptic species are organisms that look identical but represent distinct evolutionary lineages. They are an emerging trend in organismal biology across all groups, from flatworms, insects, amphibians, and primates, to vascular plants. This book critically evaluates the phenomenon of cryptic species and demonstrates how they can play a valuable role in improving our understanding of evolution, in particular of morphological stasis. It also explores how the recognition of cryptic species is intrinsically linked to the so-called species problem, the lack of a unifying species concept in biology, and suggests alternative approaches. Bringing together a range of perspectives from practising taxonomists, the book presents case studies of cryptic species across a range of animal and plant groups. It will be an invaluable text for all biologists interested in species and their delimitation, definition, and purpose, including undergraduate and graduate students and researchers.

ALEXANDRE K. MONRO is a research leader at the Royal Botanic Gardens, Kew, London. He is a taxonomist who has specialised in nettles and the generation of biological collections and their application to taxonomy and conservation. He has co-authored 5 field guides, more than 90 peer-reviewed scientific papers, and made more than 16,000 biological collections. He is a council member of the Systematics Association.

SIMON J. MAYO is an honorary research associate at the Royal Botanic Gardens, Kew, London. He is a taxonomist specialising in aroids and Brazilian flora. He has authored 112 peer-reviewed papers, 24 flora accounts, and 7 books on Araceae, Brazilian flora, monocot evolution, and computational botany. His current focus is on algorithmic delimitation and the cognitive basis of taxonomic species.

The Systematics Association Special Volume Series

SERIES EDITOR

GAVIN BROAD

Department of Life Sciences, The Natural History Museum, London, UK

The Systematics Association promotes all aspects of systematic biology by organizing conferences and workshops on key themes in systematics, running annual lecture series, publishing books and a newsletter, and awarding grants in support of systematics research. Membership of the Association is open globally to professionals and amateurs with an interest in any branch of biology, including palaeobiology. Members are entitled to attend conferences at discounted rates, to apply for grants and to receive the newsletter and mailed information; they also receive a generous discount on the purchase of all volumes produced by the Association.

The first of the Systematics Association's publications The New Systematics (1940) was a classic work edited by its then-president Sir Julian Huxley. Since then, more than 70 volumes have been published, often in rapidly expanding areas of science where a modern synthesis is required.

The Association encourages researchers to organize symposia that result in multi-authored volumes. In 1997 the Association organized the first of its international Biennial Conferences. This and subsequent Biennial Conferences, which are designed to provide for systematists of all kinds, included themed symposia that resulted in further publications. The Association also publishes volumes that are not specifically linked to meetings, and encourages new publications (including textbooks) in a broad range of systematics topics.

More information about the Systematics Association and its publications can be found at our website: www.systass.org.

Previous Systematics Association publications are listed after the index for this volume.

Systematics Association Special Volumes published by Cambridge University Press:

78. *Climate Change, Ecology and Systematics* (2011)
    Trevor R. Hodkinson, Michael B Jones, Stephen Waldren and John A. N. Parnell
79. *Biogeography of Microscopic Organisms: Is Everything Small Everywhere?* (2011)
    Diego Fontaneto
80. *Flowers on the Tree of Life* (2011)
    Livia Wanntorp and Louis Ronse De Craene
81. *Evolution of Plant-Pollinator Relationships* (2011)
    Sébastien Patiny
82. *Biotic Evolution and Environmental Change in Southeast Asia* (2012)
    David J. Gower, Kenneth G. Johnson, James E. Richardson, Brian R. Rosen, Lukas Rüber and Suzanne T. Williams
83. *Early Events in Monocot Evolution* (2013)
    Paul Wilkin and Simon J. Mayo
84. *Descriptive Taxonomy: The Foundation of Biodiversity Research* (2014)
    Mark F. Watson, Chris Lyal and Colin Pendry
85. *Next Generation Systematics* (2016)
    Peter D. Olson, Joseph Hughes and James A. Cotton
86. *The Future of Phylogenetic Systematics: The Legacy of Willi Hennig* (2016)
    David Williams, Michael Schmitt, and Quentin Wheeler
87. *Interactions in the Marine Benthos: Global Patterns and Processes* (2019)
    Stephen J. Hawkins, Katrin Bohn, Louise B. Firth and Gray A. Williams
88. *Cladistics: A Guide to Biological Classification* 3rd edition (2020)
    David M. Williams and Malte C. Ebach
89. *Cryptic Species: Morphological Stasis, Circumscription, and Hidden Diversity* (2022)
    Alexandre K. Monro and Simon J. Mayo

The Systematics Association Special Volume 89

Cryptic Species

Morphological Stasis, Circumscription, and Hidden Diversity

Edited by

Alexandre K. Monro

Royal Botanic Gardens, Kew, London, UK

Simon J. Mayo

Royal Botanic Gardens, Kew, London, UK

THE Systematics ASSOCIATION

Cambridge University Press

CAMBRIDGE
UNIVERSITY PRESS

University Printing House, Cambridge CB2 8BS, United Kingdom

One Liberty Plaza, 20th Floor, New York, NY 10006, USA

477 Williamstown Road, Port Melbourne, VIC 3207, Australia

314–321, 3rd Floor, Plot 3, Splendor Forum, Jasola District Centre, New Delhi – 110025, India

103 Penang Road, #05-06/07, Visioncrest Commercial, Singapore 238467

Cambridge University Press is part of the University of Cambridge.

It furthers the University's mission by disseminating knowledge in the pursuit of education, learning, and research at the highest international levels of excellence.

www.cambridge.org

Information on this title: www.cambridge.org/9781316513644

DOI: 10.1017/9781009070553

First published 2022

A catalogue record for this publication is available from the British Library.

Library of Congress Cataloging-in-Publication Data
Names: Monro, Alexandre K., 1968- editor. | Mayo, S. J., editor.
Title: Cryptic species : morphological stasis, circumscription, and hidden diversity / edited by Alexandre K. Monro, Royal Botanic Gardens, Kew, Simon J. Mayo, Royal Botanic Gardens, Kew.
Description: First edition. | Cambridge, United Kingdom ; New York, NY : Cambridge University Press, [2022] | Series: Systematics Association special volume series ; no. 89 | Includes bibliographical references and index.
Identifiers: LCCN 2022013575 (print) | LCCN 2022013576 (ebook) | ISBN 9781316513644 (hardback) | ISBN 9781009074506 (paperback) | ISBN 9781009070553 (epub)
Subjects: LCSH: Species. | Biology–Classification. | BISAC: SCIENCE / Life Sciences / Evolution
Classification: LCC QH380 .C79 2022 (print) | LCC QH380 (ebook) | DDC 576.8/6–dc23/eng/20220510
LC record available at https://lccn.loc.gov/2022013575
LC ebook record available at https://lccn.loc.gov/2022013576

ISBN 978-1-316-51364-4 Hardback

Additional resources for this publication at www.cambridge.org/9781316513644.

Contents

Contributors

Marta Álvarez-Presas Department of Genetics, Microbiology and Statistics, Institute of Biodiversity Research (IRBio), University of Barcelona, Spain; School of Biological Sciences, University of Bristol, Bristol, UK

Richard M. Bateman Royal Botanic Gardens, Kew, London, UK

José Cerca Natural History Museum, University of Oslo, Oslo, Norway; Department of Natural History, NTNU University Museum, Norwegian University of Science and Technology, Trondheim, Norway

Cene Fišer University of Ljubljana, Biotechnical Faculty, Ljubljana, Slovenia

David J. Gower Department of Life Sciences, The Natural History Museum, London, UK; Island Biodiversity and Conservation Centre, University of Seychelles, Anse Royale, Mahé, Seychelles

Tatiana Korshunova Koltzov Institute of Developmental Biology, Moscow, Russia

Klemen Koselj University of Ljubljana, Biotechnical Faculty, Ljubljana, Slovenia; Max Planck Institute for Ornithology, Acoustic and Functional Ecology, Seewiesen, Germany

Jim Labisko Department of Genetics, Evolution and Environment, Division of Biosciences, University College London, London, UK; Department of Life Sciences, The Natural History Museum, London, UK; Island Biodiversity and Conservation Centre, University of Seychelles, Anse Royale, Mahé, Seychelles

Matt Lavin Department of Plant Sciences and Plant Pathology, Montana State University, Bozeman, MT, USA

Simon T. Maddock School of Biology, Chemistry and Forensic Science, Faculty of Science and Engineering, University of Wolverhampton, Wolverhampton, UK; Department of Life Sciences, The Natural History Museum, London, UK; Island Biodiversity and Conservation Centre, University of Seychelles, Anse Royale, Mahé, Seychelles

Alexander Martynov Zoological Museum of the Moscow State University, Moscow, Russia

Eduardo Mateos Department of Evolutionary Biology, Ecology and Environmental Sciences, University of Barcelona, Spain

Simon J. Mayo Royal Botanic Gardens, Kew, London, UK

Alexandre K. Monro Royal Botanic Gardens, Kew, London, UK

Pablo Muñoz-Rodríguez Department of Plant Sciences, University of Oxford, Oxford, UK

R. Toby Pennington Geography, University of Exeter, Exeter, UK; Royal Botanic Garden Edinburgh, Edinburgh, UK

Marta Riutort Department of Genetics, Microbiology and Statistics, Institute of Biodiversity Research (IRBio), University of Barcelona, Spain

Sara Rocha CINBIO, Universidade de Vigo, Vigo, Spain

Robert W. Scotland Department of Plant Sciences, University of Oxford, Oxford, UK

Ronald Sluys Naturalis Biodiversity Center, Leiden, The Netherlands

Torsten H. Struck Natural History Museum, University of Oslo, Oslo, Norway

Paul H. Williams Department of Life Sciences, The Natural History Museum, London, UK

John R. I. Wood Department of Plant Sciences, University of Oxford, Oxford, UK

1

Introduction

Alexandre K. Monro

Introduction

This Systematics Association Special Volume is the result of a symposium titled, 'Cryptic Taxa – Artefact of Classification or Evolutionary Phenomena?' held on 17 June as part of the association's 10th Biennial Meeting in 2019. The symposium comprised five presentations, Torsten Struck, Paul Williams, Matt Lavin, Mark Wilkinson and Jim Labisko. For the purposes of this volume, we also invited contributions by Cene Fišer and Klemen Koselj, Alexander Martynov and Tatiana Korshunova, Simon Mayo, Richard Bateman, Marta Álvarez-Presas, and Pablo Muñoz, with the aim of providing a broader perspective on the subject, not only with respect to theory and practice but also with respect to the organisms that they work on.

My motivation for organising the meeting was scepticism. Scepticism that stemmed from the feeling that what was being observed were species whose evolutionary history had resulted in strong genetic partitioning, and that in the absence of a universally accepted species concept, it was arbitrary to designate the rank of species based on DNA alone for organisms where other sources of observations are available. Second, as a taxonomist tasked with producing field guides, identification keys, and identifying biological collections in herbaria, I did not welcome the prospect of taxa that are impossible to identify without access to a DNA laboratory and funds, the latter being difficult to access even in world-leading biological collections institutions.

The first time that I encountered the concept of cryptic species was in the mid-1990s, at which time they were novel and controversial. In the last decade, however, with the wide availability of DNA sequence observations and improved techniques for extracting DNA from biological collections, the description of cryptic species is becoming commonplace (Figure 1.1), including of hitherto well-known species: For example, the Tapanuli Orangutan (*Pongo tapanuliensis,* Nater et al. 2017), the Baltic flounder (*Platichthys solemdali,* Momigliano et al. 2018), or the Kabomani Tapir (*Tapirus kabomani*, Cozzuol et al.

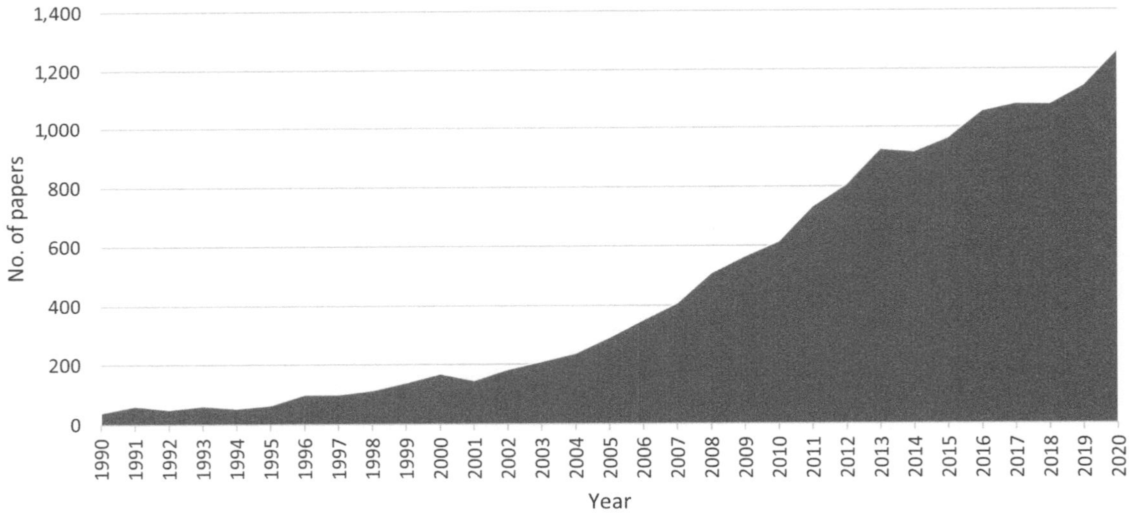

Figure 1.1 Frequency of papers with 'cryptic' and 'species' in the title (1990–2020).
Source: Scopus search 'TITLE-ABS-KEY (cryptic AND & AND species)', 2021 (undertaken 14 June 2021).

2013); but also amongst many less well-known groups of organisms, for example, sponges (Xavier et al. 2010), marine interstitial ghost-worms (Cerca et al. 2020), copepods (Fišer et al. 2015), roundworms (Armenteros et al. 2014), flatworms (Álvarez-Presas et al. 2015; Leria et al. 2020), malaria parasites (alveolates, Bensch et al. 2004), sea-slugs (Korshunova et al. 2020), rotifers (Gabaldón et al. 2015; Mills et al. 2017), cod icefish (Dornburg et al. 2016), lizards (Leavitt et al. 2007), ferns (Bauret et al. 2017), mosses (McDaniel and Shaw 2003), fungi (Muggia et al. 2015), and bumblebees (Williams et al. 2016).

Preparing for the symposium, I began to realise that the notion of – and process of discovery for – cryptic species touches the heart of several major debates in biology, including 'what are species?', 'how should we recognise them?', the notion of punctuated equilibria, and that of morphological stasis in the fossil record. In addition, in the midst of a biodiversity crisis (Koh et al. 2004) the phenomenon of cryptic species indicates that there may be a greater diversity of evolutionary lineages in need of conservation than has been suggested by morphology alone (Funk et al. 2012; Chapters 8–11), implying the need for a more nuanced approach to species conservation (Carroll et al. 2014).

Rather than simply being a distracting artefact of new sequencing technologies, phylogenetic techniques, and opportunism, any consideration of the notion of cryptic species exposes a fundamental and well-documented weakness of contemporary systematic biology: that we do not yet have the conceptual framework or the quality and breadth of observations to be able to say what a species is, and, as a result, to assert crypsis in relation to one. Striving to resolve some of the debates around cryptic species might not only provide the tools and framework to answer some major questions in biology but also make taxon delimitation and the documentation of diversity a more rigorous and useful scientific undertaking.

This book is organised to present overviews of cryptic (sibling) species in the context of species delimitation and the taxonomic method (Chapters 2–4), followed by reviews of cryptic species concepts and their value to evolutionary biology (Chapters 5–8) and then some case examples from diverse groups of organisms (Chapters 3, 5, 9–11).

1.1 Were There Cryptic Species before Darwin?: Cryptic Species and the Concepts of Species

Cryptic species are, logically, by-products of the application of a species concept to group a set of individuals into units referred to as 'species'. The notion of species as units of diversity predates classical Candollean, Linnean, and Aristotelian attempts to classify diversity and can be found in all human cultures (Atran 1998; Berlin 1973, 1992; Berlin et al. 1974; Bulmer et al. 1968; Coyne 2004; Diamond 1966; Ludwig 2017; Majnep and Bulmer 1977; Mayr 1963; Slater 2015). It also likely occurs, at a biological level at least, amongst non-humans (Poelstra et al. 2014; Robinson et al. 2015).

The term 'species' originated in the fourteenth century (Online Etymological Dictionary 1993). It denotes, 'appearance, form, kind' (Oxford University Press 1993) and as such is congruent with morphological species concepts. The notion of species as real entities ('natural things') that existed in nature rather than defined by humans dates back at least to Locke (Locke 1689, see also Mayo (Chapter 2)). The notion that species are the product of an evolutionary process is most closely associated with Darwin, who emphasised such a relationship in the title of his epoch-defining work, *The Origin of Species* (Darwin 1859). Since Darwin first linked the phenomenon of species to that of evolution, most systematic biologists equate species with separately evolving lineages – equivalent to branches of the 'Tree of Life' (e.g. De Queiroz 2007; Padial and De la Riva 2021), with the logical consequence that the basis and process of species delimitation centres on assigning individuals to a phylogenetic lineage (see Chapter 2 for context). Freudenstein et al. (2016) and Chapter 8 argue, however, that lineage divergence alone is not sufficient to delimit species. Templeton (1989), under his 'Cohesion Species Concept', rather than focussing on isolation or divergence, applies explicitly evolutionary criteria to define species as, 'the most inclusive group of organisms having the potential for genetic and/or demographic exchangeability' (Templeton 1989: 181).

For practical and academic reasons (see Chapter 2), there are now probably as many ways to assign an individual to a species as there are taxonomists doing so, a situation referred to as the 'species problem' (Mayden 1997). Compounding this, even if there was agreement on the criteria for delimiting species, we would rarely have the resources to do so confidently, either with respect to the number of populations sampled or with respect to the observations made from each. The reality is that the incidental evidence (Padial and De la Riva 2021) or operational criteria (De Queiroz 2007) used in the delimitation of the vast majority of species comprise just two classes of observations, morphological and/or molecular (DNA sequence), from a very small sample of individuals (Chapters 3, 4, and 9). Morphological evidence formed the basis of species delimitation for all groups of organisms for over 250 years, albeit mostly from the very small, arbitrary, and biased sample of characters preserved in biological collections and mostly interpreted outside of any explicit hypothesis of homology or species. For the last 25 years, the sequence of nucleotides in DNA has provided an independent class of observations for which increasingly robust statistical analyses have been developed, incorporating complex mathematical inferences for evolutionary phenomena (e.g. coalescent, Bayesian, substitution models), to delimit putative evolutionary lineages. DNA sequence observation, however, also suffers from very small sample sizes, both with respect to the proportion of populations sampled and, to a lesser and varying extent, to the proportion of the genome sampled. As implied by allopatry, geography, sometimes in association with ecological niche, is also a source of observations for the delimitation of species and subspecies (Darwin 1859; Jordan 1905; Rensch 1938) across many species concepts and frequently underpins the decision to delimit new taxa. With the exception of the rank of subspecies, however, geographical observations are rarely applied formally for the purposes of species delimitation but are generally held to be confirming factors (Davis and Heywood 1963) or 'soft characters' (see Chapter 5).

Morphological, DNA sequence, and geographical observations are just three out of an increasing number that could be available for taxon delimitation. For example, developmental (ontogenetic), physiological, transcriptomic, proteomic, behavioural, ecological niche, ecological network, immunological, biochemical, and holobiome could all provide observation useful in testing hypotheses of species, but they have been largely ignored for the process of species delimitation (Chapters 3 and 4). Both Bateman (Chapter 3) and Martynov and Korshunova (Chapter 4) suggest that no species would likely be considered cryptic were there adequate sampling of populations and morphology together with the inclusion of additional sources of observations, such as ontogenetic (Chapter 4), chemical, electrical, magnetic, sensory, ecological (Chapters 7 and 8), or were morphological characters to be observed and evaluated adequately (Chapter 3) from an effective sample of populations (Chapters 3 and 4).

Within the context of crypsis, discordance between DNA sequence and morphological estimates of divergence can result in two phenomena, (1) that of cryptic species, whereby DNA sequence observations suggest lineage divergence equivalent to a distinct species but morphological observations do not, or (2) of polymorphic species, where morphologically distinct species are suggested by DNA sequence observations to represent a single lineage (Chapter 8; Dexter et al. 2010). This latter group of species has been the focus of far less research.

It is the lack of congruence between the, arguably superficial (Chapter 3) sampling of morphological and DNA sequence observations and their use as incidental evidence that has fuelled a renewed interest in and description of cryptic species. Basically, DNA sequence observations are suggesting greater or lesser lineage divergence than morphological observations and where greater, then this is being used to propose morphologically cryptic lineages at the rank of species. This lack of congruence could be attributed to the identification of early diverging lineages, equivalent to De Queiroz's 'gray zone' of speciation (De Queiroz 2007: Fig. 1; Chapter 7). Struck and Cerca (Chapter 6) and Muñoz et al. (Chapter 5) suggest, however, that this is not the usual case, with crypsis being identified in lineages up to 140 million years old (Chapter 6). In order to prevent early-diverging lineages from being designated as cryptic species, Struck (Struck et al. 2018a, b; Chapter 6) proposes that one should explicitly show that the species are morphologically more similar to each other than would be expected given the time that has passed since their last common ancestor. This is something that is possible to establish, with some degree of error, using DNA sequence observations and/or fossils.

Given the limitations of sample size and bias in the taxonomic process, it could be argued that the current state of knowledge on species can best be described as superficial or tokenistic, as suggested by Bateman (Chapter 3). As a result, we do not have the necessary observations to formulate or apply universal species concepts. More useful is a less mechanistic definition such as Templeton's Cohesion Species Concept (Templeton 1989), or the flexibility to delimit species which conspecific genetic samples resolve as paraphyletic (Freudenstein et al. 2016; Muñoz-Rodríguez et al. 2019; Pennington and Lavin 2016; Chapter 8) and a more rigorous circumscription (see Chapter 3), applied with the recognition of the limitations of our sampling and methods, may, therefore, be more useful for the purposes of exploring evolution, but also for the establishment of a stable classification of life on earth. Within such a context, cryptic species could be viewed as nodes (Chapters 3, 5–8) for which there is evidence of lineage divergence but not of morphological change.

1.2 Cryptic Species, Morphological Stasis: Artefacts of Taxonomic Method

The fact that the definition of cryptic species is problematic does not mean that the phenomena it highlights are not important. In fact, it may be that the use of the term 'cryptic species' within a jungle of problematic species concepts has prevented a key phenomenon, morphological stasis, from getting the research focus it should have. With hindsight, the great debate over the tempo of evolution (punctuated equilibria, gradualism) occurred prematurely, prior to the 'molecular revolution' that has enabled the pairing of the palaeontological perspective of morphology with evaluations of lineage divergence.

There are several reasons for the discordance between DNA sequences and morphological observations. Some of these can be considered experimental error, whereby prior to the application of DNA sequence observations, the delimitation of a species was based on too few morphological observations. For example, the bumblebee, *Bombus kluanensis* (Williams et al. 2016; Chapter 8), was initially recognised from a coalescent analysis of a small subsample of the mitochondrial genome (COX1, Williams et al. 2019). This triggered a morphological re-evaluation of the biological collections that recovered diagnostic morphological character states. Another example is the Tapanuli Orangutan (*Pongo tapanuliensis*, Nater et al. 2017), for which cranio-mandibular and dental characters (albeit from a single individual) were identified following analyses of whole mitochondrial genomes. In both cases, the species turned out not to be cryptic, as morphological differences were observable. They had just not been detected earlier.

There are, however, many cases where morphological crypsis is confirmed and five evolutionary processes can be proposed to account for these (Chapters 6, 8–10): (I) recent lineage divergence that has not yet resulted in morphological divergence, (II) parallel or (III) convergent morphological evolution, (IV) morphological stasis, or (V) introgression. Evidence for all five has been observed for cryptic species (Chapters 6 and 7). These are all, however, distinct, testable, evolutionary phenomena that are not best served by being combined or obscured under the term 'cryptic species' (Chapter 6). Of these phenomena, recent lineage divergence and convergence have been the subject of substantial research effort by evolutionary and population biologists. Parallelism, effectively representing convergence within closely related lineages, has also been the subject of some research from speleo- (Gross 2016; Khalik et al. 2020; Powers et al. 2020) and hydrothermal vent biologists (Yuan et al. 2020).

Morphological stasis, however, remains relatively little studied outside palaeontology (Gingerich 2019), despite being a major feature of the paleontological record (Gould 2002; Stanley 1979) and presumably of evolution. In addition, the explanations for stasis have been controversial (Davis et al. 2014), with both genetic-developmental constraints and stabilising selection being invoked (Charlesworth and Lande 1982; Davis et al. 2014; Estes and Arnold 2007; Raff 1996; Smith 1981). Understanding the causes and implications of morphological stasis in evolution could therefore provide a productive research focus for which cryptic species would be key study organisms/scenarios. It is for this reason that the definition, terminology, and methods used in the recognition of such taxa are important.

Beyond morphological stasis, cryptic species are probably best referred to using terms that highlight the evolutionary phenomena more clearly, such as 'convergence', 'parallelism', or a term explicitly indicative of lineage divergence, such as 'ochlospecies' (Chapter 8). Struck's proposal – that the term 'cryptic species' should be restricted to morphological stasis, defined as lineages morphologically more similar to each other than one would be expected given time since lineage divergence – is useful and pragmatic as highlighted earlier in the chapter.

Delimiting species solely on lineage divergence is pragmatic where there is an abundance of DNA sequence observations and a paucity of other observations. It does, however, invoke operational criteria as definitional concepts, a practice subject to substantial epistemological criticism (De Queiroz 2007). Perhaps more importantly, incongruence between morphological and lineage divergence, whilst identifying important evolutionary phenomena, should not automatically be translated into taxonomic actions. Rather, the identification of incongruence between DNA sequence and morphological observations should be the starting point for hypothesis-testing and the generation of observations from additional sources. For example, the use of geographical, ontogenetic, physiological, behavioural, ecological, and chemical observations. Where these observations corroborate the DNA sequence observations then there is sense in recognising the metapopulation or lineage using a taxonomic rank.

It could be argued that adopting such an approach effectively weights morphological observations over DNA sequence ones. This I think is the reality of a taxonomy designed by and for people, and for a multiplicity of uses.

1.3 Taxonomy Is Not Just About Documenting Evolution

Taxonomy concerns the construction of classifications in general. Here we refer to that long-term enterprise undertaken by biological scientists that results in a classification and identification system founded on species taxa, named according to international codes of nomenclature. This framework, which is intended to encompass all organic life, serves the needs of a range of audiences and applications. Biological classification takes place within a constrained resource, both with respect to observations but also to the number of people delivering it and the narrow window of time in which it is taking place. Scientific practitioners are heavily influenced by evolutionary theory and for many this demands an assumed link between species as units of both diversity and the evolutionary process, but for other users, classification serves as a tool and surrogate for predicting properties (traits) related to usefulness, for measuring biological diversity, for predicting and mitigating the impacts of human activities, and for developing and testing theories about the history of life on earth (evolutionary biology, biogeography) and a shared understanding of the living world (aesthetic).

Evolutionary relationships have provided a robust framework for doing so and they are largely reflected in the classification of life. The aim of taxonomy is not, however, to reveal the footprint of evolution, but rather to use evolutionary relationships to provide a robust and stable classification and so make the universe of biodiversity accessible to all. A classification needs to meet the requirements for a well-documented and wide range of uses and users. It is the diversity of these, from local farmers, pharmacists, amateur naturalists, archaeologists, anthropologists, ecologists, environmental scientists, physicists to systematic and evolutionary biologists, that places an emphasis on morphometrics, broad predictiveness, and ease of diagnosis and makes taxonomy a fundamentally pragmatic undertaking. It is because of this need for accessibility and ease of diagnosis, the fact that the foundations of post-Linnean, 'Candollean' (or 'natural', see Chapter 2) taxonomy were built on morphological observations, and the importance of integrating fossils, that morphology remains key to species recognition. Whilst DNA sequence observations enable the assigning of individuals to phylogenetic lineages and so provide a major tool for species identification and delimitation, they are limited in their accessibility to a relatively small and wealthy group of academics, and commercial and government agencies. They also still rely on referencing a nomenclatural system wedded to (and so do not function outside of) morphology-based classifications.

1.4 How Best to Document Cryptic Species/Morphological Stasis in Nomenclature

To summarise, the term cryptic species is problematic on two levels. (1) It assumes a lineage-dominated view of species definition and delimitation, which results from a decision to apply operational criteria as definitional concepts. (2) Depending on the definition used (e.g. Struck et al. 2018b; a; Chapter 6), it conflates and obscures noteworthy evolutionary phenomena, the most important of which is probably morphological stasis. Those phenomena are not best served by the *status quo*, whereby 'cryptic' species are described under new binomials that offer no indication of the sister cryptic species(s) and which cannot be diagnosed without the infrastructure and resources to generate DNA sequence observations.

For multicellular and many unicellular organisms, trinomials may be a more useful vehicle for naming cryptic taxa as they flag the relationship between cryptic 'sister' species. In the case of the International Code of Nomenclature for algae, fungi, and plants (Turland et al. 2018), the International Code of Nomenclature of Prokaryotes (Anon 2019), and the International Code of Zoological Nomenclature (Ride et al. 1999) there is the availability of subspecies as a rank, which would fit well within a heuristic framework and cohesive species concept. Within lineage-focussed concepts this may be problematic as there may be a perceived implication of incomplete lineage divergence. In the case of viruses, while the International Code of Virus Classification and Nomenclature does not provide a framework for subspecific ranks, which are devolved to specialist groups (International Committee on Taxonomy of Viruses 2005), there are groups of organisms for which nomenclatural codes do not permit trinomials.

Other notations are possible. Hybrid plant species (nothotaxa) are indicated by placing a multiplication sign before the species epithet (ICBN H.3A.1. 2018). For example,

Verbascum × *schiedeanum* W. D. J. Koch indicates that the taxon is a hybrid. It is conceivable that a similar notation could be used to indicate cryptic status, although the use of the letter 'c' would need to be spaced in a way to avoid orthographic confusion. This would also require the concerted modification of nomenclatural codes, which in turn would require broad consensus, expressed by the votes of the systematics communities.

In the absence of a universal species concept, and given the heuristic nature of species delimitation and recognition, the use of subspecific rank is probably the best way to document morphological stasis.

The symposium and the papers that emerged from it and are presented here show clearly how the topic of 'species' remains central to biodiversity sciences and the subject of wide-ranging and lively debate. In almost every paper there is a call for change, either of direction or for the inclusion of new developments, and their focus ranges from abandoning species altogether (Chapter 4) to highlighting the fact that there is still no accessible reference system for the 300 years-worth of accumulated knowledge of species' delimitation (Chapter 2): our representation of the biological universe is still a chaotic torso. Other authors highlight the need for international cooperation as the only meaningful basis for generating such a representation – a collective effort that requires long-term institutional investment – and that the methodology of monograph production requires a favourable institutional (and political) framework.

Taxonomists need to remember that species, as well as being the products of an evolutionary process, are also conventions on which this language and scientific facts are built (Fleck 1935). The issue of species as units of biological diversity, therefore, goes well beyond the relatively simple problem of scientific definition (Lherminier 2015), because at its root what is involved in the notion of species is a key part of our mental language that we all need for understanding our living world. Now, more than ever, this is a language in which everybody has a stake, as we experience the mass extinction of biodiversity and the loss of the ecosystem services that are, at least in part, derived from them (Haines-Young and Potschin 2010).

Acknowledgements

I would like to thank all those who contributed chapters to this book and to the Systematics Association for their Conference Organization Grant. I would especially like to thank my co-editor Simon Mayo for his hard work on the manuscripts, his invaluable comments and suggestions on this Introduction, and his stimulating conversations and encyclopaedic knowledge of the history of taxon delimitation and of the so-often overlooked relevant German literature of the late nineteenth and early twentieth centuries. I would also like to thank Raquel Negrao for help with the indexing of the book and Simon Mayo, Richard Bateman, and Tom Wells for stimulating conversations on the subjects of species concepts and taxonomy.

References

Álvarez-Presas, M., Amaral, S. V., Carbayo, F., Leal-Zanchet, A. M., and Riutort, M. (2015) Focus on the details: Morphological evidence supports new cryptic land flatworm (Platyhelminthes) species revealed with molecules. *Organisms Diversity & Evolution* 15: 379–403. https://doi.org/10.1007/s13127-014-0197-z

Anon (2019) International Code of Nomenclature of Prokaryotes. *International Journal of Systematic and Evolutionary Microbiology* 69: S1–S111. https://doi.org/10.1099/ijsem.0.000778

Armenteros, M., Ruiz-Abierno, A., and Decraemer, W. (2014) Taxonomy of Stilbonematinae (Nematoda: Desmodoridae): Description of two new and three known species and phylogenetic relationships within the family. *Zoological Journal of the Linnean Society* 171: 1–21. https://doi.org/10.1111/zoj.12126

Atran, S. (1998) Folk biology and the anthropology of science: Cognitive universals and cultural particulars. *Behavioral and Brain Sciences* 21: 547–569. https://doi.org/10.1017/S0140525X98001277

Bauret, L., Gaudeul, M., Sundue, M. A. A. et al. (2017) Madagascar sheds new light on the molecular systematics and biogeography of grammitid ferns: New unexpected lineages and numerous long-distance dispersal events. *Molecular Phylogenetics and Evolution* 111: 1–17. https://doi.org/10.1016/j.ympev.2017.03.005

Bensch, S., Péarez-Tris, J., Waldenströum, J., and Hellgren, O. (2004) Linkage between nuclear and mitochondrial DNA sequences in avian malaria parasites: Multiple cases of cryptic speciation? *Evolution* 58: 1617–1621. https://doi.org/10.1111/j.0014-3820.2004.tb01742.x

Berlin, B. (1973) Folk systematics in relation to biological classification and nomenclature. *Annual Reviews* 4: 259–271.

(1992) *Ethnobiological Classification: Principles of Categorization of Plants and Animals in Traditional Societies.* Princeton University Press, Princeton, NJ.

Berlin, B., Breedlove, D. E., and Raven, P. H. (1974) *Principles of Tzeltal Plant Classification.* Academic Press, New York and London.

Bulmer, R. N. H., Menzies, J. I., and Parker, F. (1968) Kalam classification of birds and reptiles. *Journal of the Polynesian Society* 84: 267–308.

Carroll, S. P., Jørgensen, S. P., Kinnison, M. T. et al. (2014) Applying evolutionary biology to address global challenges. *Science* 346: 313–323. https://doi.org/10.1126/science.1245993

Cerca, J. Meyer, C., Purschke, G., and Struck, T. H. (2020) Delimitation of cryptic species drastically reduces the geographical ranges of marine interstitial ghost-worms (Stygocapitella; Annelida, Sedentaria). *Molecular Phylogenetics and Evolution* 143: 106663. https://doi.org/10.1016/j.ympev.2019.106663

Charlesworth, B. and Lande, R. (1982) Morphological stasis and developmental constraint: No problem for Neo-Darwinism. *Nature* 296: 610–610. https://doi.org/10.1038/296610a0

Coyne, J. A. and Orr, O. H. (2004) *Speciation.* Sinauer Associates, Sunderland, MA.

Cozzuol, M. A., Clozato, C. L., Holanda, E. C. et al. (2013) A new species of tapir from the Amazon. *Journal of Mammalogy* 94: 1331–1345. https://doi.org/10.1644/12-MAMM-A-169.1

Darwin, C. (1859) *On the Origin of Species by Means of Natural Selection, or the Preservation of Favoured Races in the Struggle for Life.* John Murray, London.

Davis, C. C., Schaefer, H., Xi, Z. et al. (2014) Long-term morphological stasis maintained by a plant-pollinator mutualism. *Proceedings of the National Academy of Sciences* 111: 5914–5919. https://doi.org/10.1073/pnas.1403157111

Davis, P. H. and Heywood, V. H. (1963) *Principles of Angiosperm Taxonomy*. Oliver & Boyd, Edinburgh.

Dexter, K. G., Pennington, T. D., and Cunningham, C. W. (2010) Using DNA to assess errors in tropical tree identifications: How often are ecologists wrong and when does it matter? *Ecological Monographs* 80: 267–286. https://doi.org/10.1890/09-0267.1

Diamond, J. M. (1966) Zoological classification system of a primitive people. *Science* 151: 1102–1104. https://doi.org/10.1126/science.151.3714.1102

Dornburg, A., Federman, S., Eytan, R. I., and Near, T.J. (2016) Cryptic species diversity in sub-Antarctic islands: A case study of Lepidonotothen. *Molecular Phylogenetics and Evolution* 104: 32–43. https://doi.org/10.1016/j.ympev.2016.07.013

Estes, S. and Arnold, S. J. (2007) Resolving the paradox of stasis: Models with stabilizing selection explain evolutionary divergence on all timescales. *The American Naturalist* 169: 227–244. https://doi.org/10.1086/510633

Fišer, Ž. et al. (2015) Morphologically cryptic Amphipod species are "ecological clones" at regional but not at local scale: A case study of four Niphargus species. *PLoS ONE* 10: e0134384. https://doi.org/10.1371/journal.pone.0134384

Fleck, L. (1980 [1935]) *Enstehung und Entwicklung einer wissenschaftlichen Tatsache: Einführung in die Lehre vom Denkstil und Denkkollektiv*. Suhrkamp, Frankfurt am Main.

Freudenstein, J. V., Broe, M. B., Folk, R. A., and Sinn, B. T. (2016) Biodiversity and the species concept: Lineages are not enough. *Systematic Biology* 66: 644–656. https://doi.org/10.1093/sysbio/syw098

Funk, W. C., Caminer, M., and Ron, S. R. (2012) High levels of cryptic species diversity uncovered in Amazonian frogs. *Proceedings of the Royal Society B: Biological Sciences* 279: 1806–1814. https://doi.org/10.1098/rspb.2011.1653

Gabaldón, C., Serra, M., Carmona, M. J., and Montero-Pau, J. (2015) Life-history traits, abiotic environment and coexistence: The case of two cryptic rotifer species. *Journal of Experimental Marine Biology and Ecology* 465: 142–152. https://doi.org/10.1016/j.jembe.2015.01.016

Gingerich, P. D. (2019) *Rates of Evolution: A Quantitative Synthesis*. Cambridge University Press, Cambridge.

Gould, S. J. (2002) *The Structure of Evolutionary Theory*. Harvard University Press, Cambridge, MA.

Gross, J. B. (2016) *Convergence and Parallelism in Astyanax Cave-Dwelling Fish: Evolutionary Biology*. Springer, Cham, IL.

Haines-Young, R. and Potschin, M. (2010) The links between biodiversity, ecosystem services and human well-being. In: D. G. Raffaelli and C. L. J. Frid (eds.) *Ecosystem Ecology: A New Synthesis*. BES Ecological Reviews Series, Cambridge University Press, Cambridge, pp. 110–139.

International Committee on Taxonomy of Viruses (2005) The international code of virus classification and nomenclature of ICTV. *Virus Taxonomy: Eighth Report of the International Committee on Taxonomy of Viruses*: 1209–1214. Available from: https://talk.ictvonline.org/information/w/ictv-information/383/ictv-code

Jordan, K. (1905) Der Gegensatz zwischen geographischer und nichtgeogeographischer Variation. *Zeitschrift für wissenschaftliche Zoologie* 83: 15–210.

Khalik, M. Z., Bozkurt, E., and Schilthuizen, M. (2020) Morphological parallelism of sympatric cave-dwelling microsnails of the genus Georissa at Mount Silabur, Borneo (Gastropoda, Neritimorpha, Hydrocenidae). *Journal of Zoological Systematics and Evolutionary Research* 58: 648–661. https://doi.org/10.1111/jzs.12352

Koh, L. P., Dunn, R. R., Sodhi, N. S. et al. (2004) Species coextinctions and the biodiversity crisis. *Science* 305: 1632–1634. https://doi.org/10.1126/science.1101101

Korshunova, T., Fletcher, K., Picton, B. et al. (2020) The Emperor's Cadlina, hidden diversity and gill cavity evolution: New insights for the taxonomy and phylogeny of dorid nudibranchs (Mollusca: Gastropoda). *Zoological Journal of the Linnean Society* 189: 762–827. https://doi.org/10.1093/zoolinnean/zlz126

Leavitt, D. H., Bezy, R. L., Crandall, K. A. et al. (2007) Multi-locus DNA sequence data reveal a history of deep cryptic vicariance and habitat-driven convergence in the desert night lizard Xantusia vigilis species complex (Squamata: Xantusiidae). *Molecular Ecology* 16: 4455–4481. https://doi.org/10.1111/j.1365-294X.2007.03496.x

Leria, L., Vila-Farré, M., Álvarez-Presas, M. et al. (2020) Cryptic species delineation in freshwater planarians of the genus Dugesia (Platyhelminthes, Tricladida): Extreme intraindividual genetic diversity, morphological stasis, and karyological variability. *Molecular Phylogenetics and Evolution* 143: 106496. https://doi.org/10.1016/j.ympev.2019.05.010

Lherminier, P. (2015) La valeur de l'espèce. *La Pensée N°* 383: 75–85. https://doi.org/10.3917/lp.383.0075

Locke, J. (1689) *An Essay Concerning Human Understanding.* 1975 ed. P. H. Nidditch (ed.) Clarendon Press, Oxford.

Ludwig, D. (2017) Indigenous and scientific kinds. *The British Journal for the Philosophy of Science* 68: 187–212. https://doi.org/10.1093/bjps/axv031

Majnep, I. S. and Bulmer, R. N. H. (1977) *Birds of My Kalam Country.* Auckland University Press, Oxford University Press, Auckland and Oxford.

Mayden, R. L. (1997) A hierarchy of species concepts: The denouement in the saga of the species problem. In: H. A. D. and M. R. W. M. F. Claridge (ed.) *The Systematics Association Special Volume Series, Species: The Units of Diversity.* Chapman & Hall, London, pp. 381–423.

Mayr, E. (1963) *Animal Species and Evolution.* Belknap Press, Harvard University Press, Cambridge, MA.

McDaniel, S. F. and Shaw, A. J. (2003) Phylogeographic structure and cryptic speciation in the trans-Antarctic moss Pyrrhobryum minioides. *Evolution* 57: 205–215. https://doi.org/10.1111/j.0014-3820.2003.tb00256.x

Mills, S., Alcántara-Rodríguez, J. A., Ciros-Pérez, J. et al. (2017) Fifteen species in one: Deciphering the Brachionus plicatilis species complex (Rotifera, Monogononta) through DNA taxonomy. *Hydrobiologia* 796: 39–58. https://doi.org/10.1007/s10750-016-2725-7

Momigliano, P., Denys, G. P. J., Jokinen, H., and Merilä, J. (2018) Platichthys solemdali sp. nov. (Actinopterygii, Pleuronectiformes): A New flounder species from the Baltic Sea. *Frontiers in Marine Science* 5. https://doi.org/10.3389/fmars.2018.00225

Muggia, L., Kocourkova, J., and Knudsen, K. (2015) Disentangling the complex of Lichenothelia species from rock communities in the desert. *Mycologia* 107: 1233–1253. https://doi.org/10.3852/15-021

Muñoz-Rodríguez, P., Carruthers, T., Wood, J. R. I. et al. (2019) A taxonomic monograph of Ipomoea integrated across phylogenetic scales. *Nature Plants* 5: 1136–1144. https://doi.org/10.1038/s41477-019-0535-4

Nater, A., Mattle-Greminger, M. P., Nurcahyo, A. et al. (2017) Morphometric, behavioral, and genomic evidence for a new orangutan species. *Current Biology* 27: 3487–3498.e10. https://doi.org/10.1016/j.cub.2017.09.047

Online Etymological Dictionary (1993) species. *Online Etymological Dictionary.* www.etymonline.com/

Oxford University Press (1993) *The New Shorter Oxford English Dictionary.* Oxford University Press, Oxford.

Padial, J. M. and De la Riva, I. (2021) A paradigm shift in our view of species drives current trends in biological classification. *Biological Reviews* 96: 731–751. https://doi.org/10.1111/brv.12676

Pennington, R. T. and Lavin, M. (2016) The contrasting nature of woody plant species in different neotropical forest biomes reflects differences in ecological stability. *New Phytologist* 210: 25–37. https://doi.org/10.1111/nph.13724

Poelstra, J. W., Vijay, N., Bossu, C. M. et al. (2014) The genomic landscape underlying phenotypic integrity in the face of gene flow in crows. *Science* 344: 1410–1414. https://doi.org/10.1126/science.1253226

Powers, A. K., Berning, D. J., and Gross, J. B. (2020) Parallel evolution of regressive and constructive craniofacial traits across distinct populations of Astyanax mexicanus cavefish. *Journal of Experimental Zoology Part B: Molecular and Developmental Evolution* 334: 450–462. https://doi.org/10.1002/jez.b.22932

De Queiroz, K. (1989) The general lineage concept of species, species criteria, and the process of speciation: A conceptual unification and terminological recommendations. In: S. H. B. D. J. Howard (ed.) *Endless Forms: Species and Speciation.* Oxford University Press, New York, pp. 57–75.

(1999) The general lineage concept of species and the defining properties of the species category. In: R. A. Wilson (ed.) *New Interdisciplinary Essays.* MIT Press, Cambridge, MA, pp. 49–89.

(2007) Species concepts and species delimitation. *Systematic Biology* 56: 879–886. https://doi.org/10.1080/10635150701701083

Raff, R. A. (1996) *The Shape of Life: Genes, Development, and the Evolution of Animal Form.* University of Chicago Press, Chicago.

Rensch, B. (1938) Some problems of geographical variation and species-formation. *Proceedings of the Linnean Society of London* 150: 275–285. https://doi.org/10.1111/j.1095-8312.1938.tb00182k.x

Ride, W. D. L., Cogger, H. G., Dupuis, C. K. O. et al. (1999) *International Code of Zoological Nomenclature. Fourth.* The Natural History Museum, London, London.

Robinson, K. J., Twiss, S. D., Hazon, N. et al. (2015) Conspecific recognition and aggression reduction to familiars in newly weaned, socially plastic mammals. *Behavioral Ecology and Sociobiology* 69: 1383–1394. https://doi.org/10.1007/s00265-015-1952-7

Slater, M. H. (2015) Natural kindness. *The British Journal for the Philosophy of Science* 66: 375–411. https://doi.org/10.1093/bjps/axt033

Smith, J. M. (1981) Macroevolution. *Nature* 289: 13–14. https://doi.org/10.1038/289013a0

Stanley, S. M. (1979) *Macroevolution, Pattern and Process.* W. H. Freeman, San Francisco, 332 pp.

Struck, T. H., Feder, J. L., Bendiksby, M. et al. (2018a) Cryptic species – more than terminological chaos: A reply to Heethoff. *Trends in Ecology & Evolution* 33: 310–312. https://doi.org/10.1016/j.tree.2018.02.008

(2018b) Finding evolutionary processes hidden in cryptic species. *Trends in Ecology & Evolution* 33: 153–163. https://doi.org/10.1016/j.tree.2017.11.007

Templeton, A. R. (1989) The meaning of species and speciation title. In: D. Endler and J. A., Otte (eds.) *Speciation and Its Consequences.* Sinauer Associates, Sunderland, MA, pp. 3–27.

Turland, N., Wiersema, J., Barrie, F. et al. eds. (2018) *International Code of Nomenclature for Algae, Fungi, and Plants.* Koeltz Botanical Books, Oberreifenberg, Germany.

Williams, P. H., Berezin, M. V., Cannings, S. G. et al. (2019) The arctic and alpine bumblebees of the subgenus Alpinobombus revised from integrative assessment of species' gene coalescents and morphology (Hymenoptera, Apidae, Bombus). *Zootaxa* 4625: 1–68. https://doi.org/10.11646/zootaxa.4625.1.1

Williams, P. H., Cannings, S. G., and Sheffield, C. S. (2016) Cryptic subarctic diversity: A new bumblebee species from the Yukon and Alaska (Hymenoptera: Apidae). *Journal of Natural History* 50: 2881–2893. https://doi.org/10.1080/00222933.2016.1214294

Xavier, J. R., Rachello-Dolmen, P. G., Parra-Velandia, F. et al. (2010) Molecular evidence of cryptic speciation in the "cosmopolitan" excavating sponge Cliona celata (Porifera, Clionaidae). *Molecular Phylogenetics and Evolution* 56: 13–20. https://doi.org/10.1016/j.ympev.2010.03.030

Yuan, J., Zhang, X., Gao, Y. et al. (2020) Adaptation and molecular evidence for convergence in decapod crustaceans from deep-sea hydrothermal vent environments. *Molecular Ecology* 29: 3954–3969. https://doi.org/10.1111/mec.15610

2

Cryptic Species

A Product of the Paradigm Difference between Taxonomic and Evolutionary Species

Simon J. Mayo

Introduction

Taxonomic and 'real' evolutionary species (T-species and E-species respectively) can be viewed as distinct ontological entities with different epistemological roles. Conventional (Candollean) taxonomic species are cognitive concepts, refined by prevailing biological theory, which tesselate the living world into comprehensible but imperfectly delimited units in order to provide a general overview of biodiversity – the public domain *status quo* species system. A single named taxonomic species may consist of multiple species taxon concepts, each of which is objectified by the specimen sets used, that is, G. G. Simpson's *hypodigms*. The resulting compound species taxon may be vague, since there is usually no up-to-date overall taxon concept nor an agreed reference hypodigm. Evolutionary species are concepts predicted from evolutionary theory that serve as models for investigating evolutionary processes rather than the construction of an overall reference system of biodiversity. They are tools that enable biologists to probe deeper into biological reality and generate new hypotheses of noumenal patterns of biodiversity that lie beyond current confines of human cognition. Cryptic species of various kinds result from the feedback to the *status quo* taxonomic species system of discoveries in evolutionary biology *sensu lato*. They may differ from conventional species taxa in data type, methods of analysis, or different criteria of what a species should be. Their recognition as taxonomic species requires social mechanisms legitimated by convention among biologists, especially systematists. When included in the *status quo* system they should be objectified as published taxon concepts founded on the data that generated them, using conventional taxonomic protocols that require a description, diagnosis, hypodigm, and correctly formulated name. For the public domain *status quo* system to function effectively it is probably indispensable that species taxon concepts (including cryptic species taxon concepts) should eventually be based on an interacting system of online databases, created and managed collectively by systematic and evolutionary biologists themselves.

2.1 Species as Taxa and Species as Evolutionary Groups

In his thought-provoking book on the species problem, Hey (2001) argued for an ontological distinction between species as taxa and species as evolutionary groups. This was later taken up by Zachos (2016) in his book on the same topic, who called them T-species and E-species, respectively. However, much earlier, in a discussion of type concepts in taxonomy, G. G. Simpson also expounded a similar view (Simpson 1940) during the second phase of the Modern Synthesis (Junker 2004) when speciation came to the fore as the primary topic of theoretical discussion (e.g. Dobzhansky 1937; Huxley 1940, 1942; Mayr 1942; Simpson 1944; Rensch 1947; Stebbins 1950).

Taxonomic species (T-species) are derived from cognitive concepts (Atran 1990, 1999; Berlin 1992) of kinds of organism (oak, tiger, date palm). They are subjective mental constructs with certain characteristics that derive from the intuitive apparatus humans use to form group concepts, e.g. fuzzy boundaries, typicality, and degrees of membership of included individuals (Smith and Medin 1981; Murphy 2002; Gärdenfors 2014; Hannan et al. 2019). T-species are subjective both in the sense that they are mental concepts and in the sense that each of us has a personal and intuitive idea of what they represent. Simpson (1940: 414) expressed this view as follows:

> A species, as it is actually defined or diagnosed and used in the literature is a subjective concept. The usual theory, often questioned but believed by me and by most taxonomists, is that this concept corresponds more or less with a real thing in nature, a group of individual animals that are truly related in a way that makes them a natural unit of a certain approximate scope. This thing in nature may be considered the real species, but it is not and cannot be the species of taxonomy. The mental concept that is the species of taxonomy cannot be shown to be coextensive with the real species and even if by chance the two were coextensive, it would be an error to suppose them identical. One, the taxonomic species, is an estimate of the other, the real species.

Species-as-evolutionary groups (E-species), however, are real; they exist, as ontological individuals (Ghiselin 1974, 1997), beyond the phenomenal world of human cognition in the noumenal world of things-in-themselves (Kant in Stang 2018). We accept this because it is what evolutionary theory predicts, producing the paradox that what is real is also beyond our reach to comprehend directly (bringing to mind Plato's theory of forms, Ross 1951; Wikipedia 2020). The characteristics of real evolutionary groups are not determined by the limitations of human perception but are the result of their being lineages (De Queiroz 2007), or more precisely, tokogenies (Hennig 1966). They are the object of research by evolutionary biology in the wide sense, including ecology, genetics, speciation, population biology, cytology, phylogenetic systematics, anatomy, and morphology.

These two kinds of species serve different but equally important purposes. Systematic biologists often work with both. The taxonomist focusses on the presentation of formally published species taxon concepts (see Section 2.2), that is, species-as-taxa. These provide a framework of identifiable organismal kinds. When acting as an evolutionary biologist, however, the systematist focusses on discovery of evolving lineages, working out how they are maintained in nature, finding their complex phenomenal contours, and developing hypotheses of their origin and their phylogenetic and ecological relationships. The priority is to understand, in as much detail as possible, how the observed grouping patterns that evolutionary processes engender in living organisms come about.

The study of E-species often reveals that the framework of species boundaries of T-species proposed by taxonomists does not fit the results of evolutionary research studies very well. There may be little or no corroborating evidence for morphologically distinct T-species (e.g. Chapter 3), or a single T-species may turn out to encompass what are generally called cryptic species, that is, lineages or interbreeding groups that are very distinct according to the new data but morphologically indistinguishable or barely so (e.g. *Drosophila pseudoobscura* and *D. persimilis*, Mayr 1963: 34–35; Hind et al. 2019; Quattrini et al. 2019). Hey (2001) argued there were grounds for supposing that evolutionary ('real') species would prove to be too vague and complex to form the basis of a workable reference system. Evolutionary biologists, however, tend not to be so pessimistic. Their research generates group concepts based on a range of data types from various contexts, particularly genetic, ecological, phylogenetic, and behavioural. When these data do not correspond to the T-species, the groups so formed are termed cryptic species (or clades (Chapter 3)).

T-species and E-species play interdependent roles in biology. They can be seen as the end points of a spectrum, or better, a production line. Because they are primarily interested in constructing a system of taxa, taxonomists' focus is above all on establishing basic units that will last as intellectual concepts, that is, robust hypotheses of groups. The purpose of these T-species is to act as 'approved' units of biodiversity for general societal use, and their formalisation is taken as a judgement of their adequacy. Evolutionary biologists are also interested in the latter, but their primary focus is on hypotheses of the processes that engender the groups or patterns of diversification. E-species research constantly revises and expands the criteria needed for T-species to be regarded as scientifically acceptable. All investigations of evolutionary biology subject T-species to testing and very often new patterns arise that demand recognition. These cryptic species can be seen as taxonomic species 'in the waiting room', pending formalisation. Taxonomists, conscious of the huge scale of their task, chafe at the ever-increasing technical demands put on them, while evolutionary biologists, impatient with the slow recognition of their cryptic species, in the excitement of new discoveries, often in taxonomically well-studied groups such as birds and Lepidoptera, tend to ignore the vast landscape of formalised species that are based only on fragmentary observations. For good or ill, these poorly delimited entities are also part of the taxonomists' brief and must be dealt with somehow, when upgrading the *status quo*.

Despite such often-expressed frustrations, what the history of systematics clearly shows is a feedback from advances in biological theory to taxonomy (Williams and Ebach 2020). This may not be very obvious: while geneticists, ecologists, and phylogeneticists probe biodiversity differentiation from cryptic behavioural levels (Chapter 7) to global scales (Baker et al. 2019), taxonomists continue to generate descriptive treatments of species-as-taxa in monographs, floras, and faunas at many scales, and in the case of higher plants they use, for the most part, standard methods of description and documentation that have changed surprisingly little in the two hundred years since A. P. de Candolle laid the foundations of modern formal phytography (Candolle 1818). Nevertheless, ever since Darwin (1859), taxonomists have continually re-interpreted their taxa in the light of findings from evolutionary biology, and this continues in the present day.

If T-species and E-species express a certain separation of scientific roles, cryptic species represent the feedback mechanism that binds them together. Controversies over cryptic species can be seen as the smoke that indicates the fire of changing paradigms, the best-known example being the birth and nurturing of the 'biological species' (Dobzhansky 1935, Mayr 1940), which spawned biosystematics – they are in fact the means by which taxonomic species are refined as scientific concepts. In pondering cryptic species today, it is clear that greater cooperation between taxonomists and evolutionary biologists is needed in the building and rebuilding of the formal *status quo* system of species. But before looking at cryptic species further, it is necessary to discuss in more detail just what T-species are, and this turns out to be somewhat more complex than might be expected.

2.2 Species Taxon Concepts and Taxonomic Species

Despite the previously mentioned subjectivity of taxonomic species, there is a sense in which they are objective, and this is when they are conceived as *taxon concepts*, an idea introduced by Berendsohn (1995) and Geoffroy and Berendsohn (2003) to tackle problems in combining different sources into single taxonomic database systems. The taxon concept of a species is the product of the classical taxonomic process in which an intuitive idea of a species-as-a-group is objectified by a published description and diagnosis and documented by a selected number of individuals represented as specimens in public reference collections. This is a scientific process to the extent that the mental concept of the author is presented in the public domain using a standard set of protocols and is verifiable from objects labelled as standard exemplars. Where it falls short is in explaining how the taxonomist arrived at the group so described, that is, the methodological details of the analysis and the decision algorithms that enabled the author to discern and describe the species. These are subjective cognitive processes that are still poorly understood (Hannan et al. 2019).

A species taxon concept (henceforth STC) thus consists of the standardised taxonomic treatment of a species published with a scientific name by a particular authorship at a particular time in a particular journal or book. When several such accounts exist with the same species name, they are to be regarded as different taxon concepts because they almost always differ in some significant aspect such as sample size, characters, or geographical scope.

By referring to these different treatments as distinct taxon concepts, Geoffroy and Berendsohn (2003) drew attention to the fact that in each act of formal description the taxonomist presents an account of a distinct sample of what is supposed to be a single entity in nature, that is, an E-species. The different STCs of a T-species can thus be seen as objectivised 'freeze-frames' that together form a broader and less well-organised concept (the T-species), which the taxonomist supposes to be an estimate of a real, objectively existing E-species. Simpson (1961: 114) expressed a somewhat similar view of this oscillation between subjectivity and objectivity:

> Any definition of a taxon . . . distinguishes a set of real, objective organisms. On the other hand, the concept of a taxon, the thing really present in the classifier's mind and named and referred to by him, is invariably subjective . . . the definitions proposed and the names applied . . . have no real relationship to any objects in nature except through the subjective processes of a taxonomist. Thus any taxon whatever is in some respects or in some application of the words real or objective and in others simultaneously unreal or subjective.

2.2.1 The Early Subjective Stage

The subjectivity of a species taxon derives from the initial sorting process that creates a set of individuals forming the basis for its delimitation. Various authors have described this taxonomic sorting in some detail, for example, Diels (1921), Hitchcock (1925), Sprague (1940: 448–449), Mayr (1942: 12–13), Mayr et al. (1953: 72–73; 1969: 5, 9, 112, 144–147, 182), Simpson (1961: 108–109), Cullen (1968: 176–177), Leenhouts (1968: 26–27) and Jeffrey (1982: 13–17). Davis and Heywood (1963: 11) described the subjective mental structure that corresponds to each such sorted group as a *neotypological concept*, formed in the taxonomist's mind at that time and later elaborated and refined into a published taxonomic account of a species. Mayr (1969) called these phenotypic entities *phena* (*phenon*).

This phase of the taxonomic process (dubbed 'tâtonnement' [= trial and error] by A. P. de Candolle 1813: 67) is poorly understood and would merit further study. For example, the cognitive psychologist P. Gärdenfors (2014: 42–47) has modelled the learning of mental concepts within a conceptual space as a Voronoi tesselation in which the exemplar individuals identified as typical of each concept are used to mentally compute a prototype for each. Understanding the formation and testing of STCs as the result of cognitive processes is important, both in reaching a more balanced view of their scientific value and for training taxonomists themselves. Williams and Ebach (2020: 13–19) point out that the systematic literature largely fails to answer the question 'How is taxonomy done?', and I would further add, in regard to species, 'How is species taxonomy learned?'

2.2.2 Objectification into a Species Taxon Concept

The construction of the STC involves both further tatônnment as well as objectification. The taxonomist uses the exemplar individuals for description, and during this phase investigation of characters is required and comparison with a wider range of individuals. This may often necessitate juggling individuals between groups and adjustment of boundaries between putative species. But in essence, what the taxonomist does is refine an initial, internal mental concept into an explicit, documented description and delimitation that has as its objective the communication of this taxon concept to other people.

In its usual standard format (first developed by A. P. de Candolle 1818, see also Drouin 2008), a species taxon concept consists of nomenclature, description, diagnosis (including key entries), cited specimens, images, and geographical and ecological characterisation, but its most important components are the cited specimens, the description, and the diagnostic information, in that order of priority.

Hypodigm: The foundation of the STC: The set of individuals – usually but not always preserved – on which the description and diagnosis are based is conceptually prior because it results from the taxonomist's initial intuitive sorting of individual organisms. G. G. Simpson (1940: 417–420; 1961: 183–186) called this specimen set the *hypodigm* (from Greek ὑπόδειγμα, meaning 'token, example') of the species taxon: it consists of 'All the specimens used by the author of a species as his basis for inference, and this should mean all the specimens that he referred to the species, . . . ' (Simpson 1940: 418). The hypodigm is itself the result of the taxonomist's mental concept of the species, which may also be based on wider experience of living organisms in the field or other preserved material. Hypodigms of better-known species are usually selections.

In Simpson's theory of a species taxon, the hypodigm is in principle a sample of a real species (Simpson 1940: 418) and becomes more tractable statistically as it grows in size. Simpson was an early advocate of the use of biometry in taxonomy and with Anne Roe (a research psychologist with a statistical background) wrote one of the few texts on the subject that deals with biometry from a primarily taxonomic viewpoint (Simpson and Roe 1939; Simpson et al. 1960). Because he regarded the hypodigm as a population sample, he argued (Simpson 1940) that all included specimens were equally important for the taxon concept. This makes sense when the taxon concept is viewed dynamically, that is, as a concept that is continually being reassessed as new material becomes available – the process of taxonomic revision. He also emphasised that specimens can only be regarded as part of a hypodigm when cited in a published species taxon treatment, and this requirement constitutes the most important claim that an STC has for objectivity. The act of publication reflects a conscious choice of the most representative specimens of the species taxon as conceived by the taxonomist, who may have also studied and determined as such many other individuals.

Simpson thus arrived at the same key conclusion as Berendsohn's later taxon concept, that each taxonomist has a unique view of a given taxonomic species at a particular time and this is subject to constant change:

> According to this theory of taxonomy, any identification or reference of specimens to a species makes possible and implicitly or explicitly involves redefinition of that species. Two workers seldom base their [taxon] concepts of a species on exactly the same suites of specimens. Their concepts therefore will seldom be exactly the same and since these concepts, and not the natural or real groups, are the species of taxonomy, it follows that two original workers seldom use the same species. (Simpson 1940: 416)

Every other aspect of the STC is contingent on the hypodigm; it defines the group of individuals from which are derived the information presented in the description and diagnostics and the data points on which the geographical distribution and ecological profile are based. It is also the most important initial sampling framework for evolutionary biologists – molecular, ecological, genetic, morphometric etc. In the parlance of data-mining (Ripley 1996; James et al. 2013), it is the basic training set for the STC of which it is a part, and provides the means for the clearest test of the taxonomist's prior classification of individuals into species.

Description: An account of central tendency: The objective of the taxonomic species description is three-fold; (1) to present the most typical values and states of the characters used, based on direct observations of the hypodigm, (2) to summarise the range of variation of these characters, and (3) to provide a word painting that awakes in the reader's imagination a picture of the most typical members of the species.

The description emphasises the modal ranges of the characters and treats them individually and independently. It represents an analysis of the taxonomist's earlier subjective grouping concept into those individual characters that the taxonomist found to be the most objectively tractable; this may result in the exclusion of some information that the taxonomist used subjectively in recognising the STC. However, the description is objective to the extent that it is grounded on the hypodigm and can be checked against it.

Unlike the hypodigm, in which all individuals have equal status, by focussing on a set of central character values the description conforms better to the prototypical effects described by cognitive psychologists for mental concepts (e.g. Lakoff 1987; Murphy 2002; Hannan et al. 2019). This facilitates the recreation in the reader's mind of the multivariate taxon concept of the author, characterised by a core of typical members and fuzzy boundaries. It appears that this is the brain's way of performing multidimensional computation. Anyone who has written a taxonomic species description knows that this is a mental act that clarifies and fixes the group concept in the taxonomist's mind.

Diagnosis: Delimitation sensu stricto: The diagnostic element of the species taxon concept consists of a set of properties that an individual must have in order to be assigned to the species in question rather than some other (it is similar to an intensional class definition, Kripke 1972; Putnam 1975; Richards 2010: 185–186), and thus emphasises the boundary (delimitation) of the STC rather than its core (typicality). It is downstream from the recognition and description of the taxon and focuses on point-for-point comparison with other species. The taxonomist works to ensure that the diagnosis provides the most discrete delimitations possible with the material examined, but like the description, it is contingent. It can be no more than a 'best fit' to the concept represented by the hypodigm and description.

Names and other components of the species taxon concept: The indispensable nomenclatural element of the STC consists of its correct binomial name, and this is determined by the inclusion in the hypodigm of the nomenclatural type specimen that bears the earliest name that is legitimate according to the nomenclatural code (Turland et al. 2018 for plants). When sampling a taxonomic species for evolutionary biological research, it should always be borne in mind that the holotype specimen of the correct name of a species taxon is unlikely to be typical of the taxon (see e.g. Simpson 1940). It is the hypodigm rather than the nomenclatural type that determines the choice of any images included in the STC; images are usually intended to express typicality rather than variation. The specimens of the hypodigm determine the data points on which the geographical and ecological profiling of the species are based; this is a crucially important information source for other scientists.

2.2.3 Taxonomic Species: A Return to Subjectivity

The previous section on STCs describes a rather precise concept for a species taxon, and when a taxonomic species is published for the first time, it does indeed begin as a single STC with a hypodigm that may consist of just a single nomenclatural type specimen. As time passes, however, other collections are determined by taxonomists as this species and those that are preserved become part of a growing specimen base in herbaria and museums. As they have not yet been formally cited in an STC, these newer collections are not part of a hypodigm, but with time new STCs with the same name will be published. These may be from different geographical regions, as in Flora treatments, or they may be monographic, but they almost always differ in their hypodigms. The link between them is the binomial species name, based on the nomenclatural type specimen. The use of the same name for a variety of STCs means that their different authors all agree (in principle) that the type specimen of that name falls somewhere within the range of variation (i.e. within the hypodigms) of each of these taxon concepts. But until these various hypodigms have been combined into a new overall monographic treatment, the taxon as a whole remains vaguer than any of the component STCs because none of the three key elements (hypodigm, description, diagnostics) are well defined at this pre-monograph stage. The function of taxonomic revision is precisely this – to redefine these elements as a new general STC for the species.

A taxonomic species is thus an aggregate of STCs with the same currently accepted binomial name. Unless it has been very recently monographed, there will be no overall hypodigm nor description or diagnosis. It is a more subjective concept than its component STCs since different users will take their idea of the species taxon from different sources.

In this situation, the wider specimen base in biological collections assumes the role of an informal, general 'hypodigm'. Specimens named by taxonomists in reference collections are taken to represent a given taxonomic species. Systematists are aware, however, that great caution is needed, as not all such determinations are equally reliable. People who misunderstand taxonomic methods often think that one way out of this problem is to focus on the type specimen of the species name. However, the type specimen is just a single data point of a taxon concept and cannot even be said to be typical without looking at an adequate sample of the combined hypodigms of that species as currently understood.

Simpson had an essentially statistical idea of a species taxon (T-species) based on a gradually changing hypodigm, but he also formulated a reasonably clear idea of the E-species he was trying to estimate – his evolutionary species concept (Simpson 1951, 1961). He thus seems to have been alone amongst the major figures of the Modern Synthesis and later, in attempting a theory of taxonomic species rather than simply casting them to one side. He insisted that taxonomic species are not sets of characters (Simpson 1961: 65) but samples drawn from populations. Focussing on a character set freezes the taxon concept on just the diagnostic element of the STC. This procedure reinforces our cognitive bias towards prototypicality, and this in turn has led in the past to accusations of typology and essentialism, although these have recently been shown to be misguided (Winsor 2003, 2006; Mueller-Wille 2011; Witteveen 2015, 2016). In fact, a taxonomic species is constantly changing with advances in field exploration. By focussing on the specimens rather than the diagnostics it becomes clear that taxonomic species are dynamic not static concepts and are highly dependent on their hypodigms.

2.3 The *Status Quo* Species System (SQSS): The Global Reference Framework

The constantly increasing representation of species by the expansion of reference collections has posed what appears to be an intractable problem for taxonomists. The greater the pace of field exploration the quicker the hypodigms go out of date, and the process of taxonomic revision cannot keep pace overall. Taxonomic species thus remain *de facto* poorly delimited at any given time, except for a few years after they have been monographed (Candolle 1880: 55).

This in turn greatly hampers the second of the two general societal goals of taxonomy. The first is to explore the biological universe and understand its structure in the shape of organismal groups. But the second, just as pressing, is to present a scientific classification of species taxa, a species system, for scientists and the wider society. At any given time there exists in the public domain a *status quo* of taxonomic knowledge in some form or another, as a global reference system of species taxa – here dubbed the SQSS, the *status quo* species system.

In the days before the Internet, taxonomists attempted from time to time to create a single such reference system in book form. In the case of plants, this began with Linnaeus (1753) and ended with A. Engler's *Pflanzenreich* (Engler 1900–1953), which although never completed had reached 107 volumes by 1953 (Stafleu and Cowan 1976). Since then, no such project has managed to progress, despite valiant attempts such as the *Species Plantarum* project (Brummitt et al. 2001). The result has been that the STCs of species are to be found scattered throughout taxonomically oriented publications in journals, Floras, and monograph series, comprising a formidably complex literature, the navigation of which requires training and experience. The vital importance of this literature is that it is the only route to discovering the STCs – and hypodigms – of the species taxa. Without it, not even the reference collections can be relied on, and understanding described species taxa is impossible.

The advent of the Internet offered the potential for radical improvements to this scenario. One of the most important for taxonomists globally has been the creation of the Biodiversity Heritage Library (BHL 2020), which made virtually all pre-1930s taxonomic literature freely available and enormously eased the problems confronted by the great majority of taxonomists around the world who have no access to a first-class taxonomic library.

Another key development has been the growth of online catalogues of accepted species names. This is not merely a technological question of mining existing literature but requires a process of filtration through interaction with taxonomic specialists around the world to determine the correct names and synonyms of taxonomic species, usually in the absence of up-to-date revisions or monographs. These projects, such as Catalogue of Life (2020, Roskov et al. 2020), Species 2000 (2020), ITIS (2020), WCSP (2020), The Plant List (2013), and Tropicos (2020), provide a framework of names that are considered to be accepted by the taxonomic community (often using informal and *ad hoc* procedures). It will be recognised by any taxonomist who worked in the pre-internet era what a tremendous step forward this internet infrastructure has been in providing an organised jumping-off point for taxonomists to generate monograph-standard STCs. It is a key infrastructure for linking together taxonomists and their institutions, but it should be emphasised that this network of databases also falls short of the ultimate taxonomic goal of a species system. I only stress this point because many non-taxonomist biologists are under the impression that once its name is known, the taxonomic identity of a species can easily be found. In reality the name can lead not to a single, easily assimilated taxon concept but to several, which may be of different ages, sizes, and scope; citing the original author may point to a very early and now inadequate taxon concept, for example, Linnaeus (1753). Taxonomy is only a restricted niche in biology, even within systematics itself, and so perhaps lack of a deeper understanding of its practice is not so surprising, given the withering away of the discipline in universities, as noted by Williams and Ebach (2020).

An important point about the online network of taxonomic resources now available is that they have enormously improved the tools available to taxonomists themselves for finding information, and the speed with which this can be done. However, the next phase in building a fully functioning SQSS is to bolt onto this network the activity and products of the taxonomists themselves, that is, online construction, revision, and publication of monograph-standard STCs, and the creation of a global workflow that will keep them up to date with advances in field exploration and evolutionary research.

The SQSS is a vital infrastructure for biology and the source of species taxa for investigation by other scientists. Evolutionary biologists sometimes express frustration with taxonomy because of the difficulty of reliably sourcing the full scientific detail of taxonomic species. But taxonomists themselves suffer from this problem and have always struggled with the immensity of the task. What seems plain is that to achieve an adequate SQSS – one that will serve the needs of evolutionary biology as well as those of taxonomists – some significant changes are needed to the methodology and publication of STCs: (1) the most important is to publish the metadata and images of the hypodigm specimens online as downloadable data. These are the objective basis of taxonomic species and should be the primary sampling frames for all kinds of systematic studies; (2) The traditional descriptive information of STCs should be supplemented by data sets in matrix form based on the hypodigm and published online so that computational analysis by other workers is possible based on the source material. This will also require further development of analytical methods because of the special problems of this kind of data; (3) Taxonomists of each major group should be responsible for uploading and maintaining the online STCs, using agreed protocols for revision and versioning to present them as a consensus among experts subject to regular modification. However, this should not be taken to mean the imposition of doctrine. The SQSS should ideally represent the optimal view of the systematics community at any one time in order to provide for the general need of the public – including other scientists outside systematics. Conventional species taxa as presented here are dynamic concepts. Instead of a pyramid of stone blocks, a better metaphor for the SQSS is a honeycomb in a continual state of reconstruction. STCs are not monumental end points but contingent starting points for scientific exploration. Formal publication of STCs should not be seen as a kind of sacral anointing but rather as an acclamation by a sufficiently legitimate group of the relevant scientists that a provisional consensus has been reached – the STCs therein can be accepted as 'facts' (Fleck 1980) for the present. This is no different in principle to any other scientific publication.

Radical changes to working practices in Candollean taxonomy are unlikely without some change to the conventions of career rewards, but it is a reasonable assertion to make that the output of taxonomists that has the greatest impact on science and society is through STCs, and this would be best delivered through online databases. Some system of agreed convention is needed to achieve an optimal view of species taxa at any given time, but this is of little value unless the data underlying these consensus taxa is also available online for continual scientific reappraisal.

Many taxonomists have already pioneered these objectives in part. In the earliest pre-internet phase of taxonomic databasing and computer-aided taxonomy, for example, Pankhurst (1970, 1991), ILDIS (2020, Roskov et al. 2006), DELTA (2020, Dallwitz 1974, 1984, 2018), Allkin and Bisby (1984), and PRECIS (Russell and Arnold 1989), there was a focus on descriptive and diagnostic data. Later on, in the early 2000s, a range of e-taxonomy websites (Scoble et al. 2007; Clark et al. 2009) began to emerge, in which STCs and related information of individual plant and animal families and genera is published and maintained by taxonomists themselves: for example, Solanaceae Source (PBI *Solanum* Project 2020), CATE (Haigh et al. 2008; Smith et al. 2008; CATE-Araceae 2020; Kitching 2020), Palmweb (2020), Cichorieae Portal (Kilian et al. 2020), and a great many others, for example, at the Scratchpads website (Smith et al. 2012). The largest monographic e-taxonomy yet to appear is probably GrassBase, created and developed by Derek Clayton and collaborators at Kew and in Australia (Clayton et al. 2006) and containing full descriptive STCs in DELTA for the over 12,000 species that are widely recognised in the Poaceae. However, only in a very few cases have taxonomists generated their STCs through an explicit computational procedure from specimen character vectors and released their matrices online, the best examples being the monographs of Andrew Henderson on genera of the palm family (Henderson 2005, 2006, 2011, 2012).

Modern online identification key systems have promoted the habit of matrix-building for morphological characters in many plant and animal groups, for example, Lucid (Lucidcentral 2020) and Xper (Ung et al. 2010a, b), building on the earlier foundations laid by Pankhurst (1970, 1991), Dallwitz (1974), and ILDIS (Allkin and Bisby 1984).

Much-needed modern tools designed to help taxonomists to produce monographs have begun to appear in recent years, after a long gap since Skov (1989) pioneered the

Hypertaxonomy software. Borsch et al. (2015) and Kilian et al. (2015) have designed detailed workflows aimed at collective online construction of large-scale monographs, and an R package (R Core Team 2020) has been published by Reginato (2016) to facilitate production of taxonomic species treatments.

The importance of morphological features for ecological and biogeographical research and environmental monitoring has given impetus to the appearance of online trait databases such as TRY (Kattge et al. 2011) and PalmTraits (Kissling et al. 2019), with the latter being based on selected morphological data from all palm species. These online international collaborations between ecologists and taxonomists represent a filtering and repackaging of data from the SQSS and point the way towards a more ambitious and coordinated online framework for the STCs themselves.

Major global scale projects now exist for online presentation of plant species descriptions for all plants: Encyclopedia of Life EoL (2020), Plants of the World online (POWO 2020), and World Flora Online (2020). These have significantly advanced the international institutional frameworks needed for a global SQSS online. However, the coordinated involvement of the specialist taxonomists themselves in creating and maintaining the STCs presented has not yet been achieved at a global scale, and this remains a complex but important problem for the future.

Many herbaria and museums are building online databases with images of their reference collections, thus making possible the remote presentation of species hypodigms, for example, Paris (mnhn.fr/en/collections/collection-groups/botany/vascular-plants), Berlin (ww2.bgbm.org/herbarium/default.cfm), and Kew (apps.kew.org/herbcat). Herbarium databases have been linked through the global-scale GBIF project (GBIF 2020), which presents the geographic ranges of species using specimens as data points - a project which highlights the importance for general scientific goals of constant attention to improving the quality of specimen identifications in public reference collections. Brazilian taxonomists have been especially innovative in creating an online STC system for the flora of Brazil (Flora do Brasil 2020 under construction) with supporting online herbarium databases (INCT Herbário Virtual 2020; SpeciesLink 2020) linking specimen images and metadata from national and international herbaria. These Brazilian initiatives have shown that a large-scale e-taxonomy of STCs is a feasible proposition: the flora of 49,255 currently recognised species is being built by the collaboration of a team of over 900 taxonomists individually responsible for content creation and maintenance of their specialist groups and administered through the national herbarium at Rio de Janeiro.

There is thus progress towards a more effective global SQSS. There are still many technological and indeed administrative issues involved in remotely linking databases, some of which are being solved by the global databases previously mentioned. But there are other more important obstacles. One is the current climate of biodiversity science, which has weakened the focus on taxonomy, exacerbated by the prevailing paradigms of scientific publication, which tend to maximise competition amongst individuals, institutions, and the publication vehicles themselves. These forces can confound attempts to promote international collective actions by taxonomic institutions even though most global taxonomic infrastructures that currently exist are in fact the result of collective action undertaken by these same institutes and their (often governmental) funding bodies. Here again, Brazil has shown the way by securing the commitment of an entire national community of taxonomists for database content-building online, despite a prevailing professional climate in which their individual careers depend, as elsewhere, on maximising individual publication in high-impact journals. Recent papers show an increasing concern to promote global taxonomy as a collective enterprise (e.g. Garnett et al. 2020).

2.4 Cryptic Species: Taxonomic Species in Disguise

Cryptic species result from different approaches to the investigation of E-species than that described here for conventional taxonomic species. These studies are based on predictions from evolutionary theory that give the resulting groups a better claim to represent the lineages of real E-species than just morphological similarity and geographical and ecological coherence. This is, for example, the basis of the argument for the superiority of the biological species concept made by Dobzhansky (1935, 1937) and Mayr (1942), in which Mayr's 'sibling species' played a critical role (Mayr 1963).

However, ever since evolutionary biologists began to show that other kinds of data can contradict or alter the circumscription of conventional STCs, finding ways of integrating this new knowledge into the formal structure of the SQSS has proven problematic. In the 1920s and 1930s, before the term biosystematics had been coined (Camp and Gilly 1943; Camp 1951), various authors proposed more or less informal frameworks for recognising such cryptic species in parallel to standard taxonomy, for example, Turesson (1922a; b; 1925), Du Reitz (1930), Turrill (1938, 1940), Gilmour and Gregor (1939), Gilmour and Heslop-Harrison (1954), and Heslop-Harrison (1953). The results of cytogenetics had a major influence on assessing species limits (e.g. Grant 1981), and this eventually resulted in the inclusion of chromosome numbers as part of the standard description of many taxonomic treatments. Pollen structure visualised by scanning electron microscopy also became an additional character field but has rarely been incorporated into STCs. More recently, species circumscription based on molecular data has produced a much greater preoccupation with combining different data types into single taxon concepts (e.g. Sites and Marshall 2003; Dayrat 2005; Derkarabetian and Hedin 2014; Pante et al. 2015; Yang et al. 2019) and the burgeoning field of integrative species delimitation has resulted.

While these are important advances that promise to gradually engineer species taxon concepts towards a better fit to real natural lineages, the requirements of a fully effective SQSS should also not be ignored. Cryptic STCs arise from patterns in various kinds of usually non-morphological data and are also often the result of computational analysis. They nevertheless need to be based on the same objectified components as a conventional taxonomic species: a hypodigm, a description, and a diagnosis constructed from the data of the study itself (Hassemer et al. 2020). An integrative STC should provide online access to all the data sources from which it has been built, and this can only be done in the form of databases. As with classical STCs, conventional publication paradigms provide only a superficial view of the result, when what is needed for the SQSS is continual access to the underlying data. Probably one of the basic reasons for difficulties in integrating cryptic species into formal taxonomic frameworks has been to do with the persistence of an old publication paradigm as the only means of presenting the SQSS. At the same time, it is evident that unless the morphological descriptions are also presented as matrices of individual data vectors, they will not be included in integrative STCs, risking the future disengagement of conventional taxonomic species concepts. Integration of data types also requires further development of computational methods. Often only very small samples are represented by the hypodigms of STCs, and this poses computational problems that can probably only be overcome by adopting some kind of modelling approach.

Using DNA data has greatly accelerated the discovery of cryptic species, but the problem of deciding whether a particular clade of individuals should be recognised as a species rather than just a component population is in principle no different from that confronted by taxonomists in deciding if two morphologically recognisable populations are different species. To make this decision, the existence of corresponding data suggesting isolation or autonomy in other ways has always been most influential, that is, the consilience of different data sets (Wilson 1998; Bookstein 2014). Interlinked publication of the data sets underlying integrative STCs could make this easier. Computational methods as well as the hypodigms will change continually, and re-analysis of the data is needed regularly to provide coherent update and progression in the SQSS.

Unlike most previous methods in systematics, molecular methods have the potential to reach or exceed sampling levels of species that exist in reference collections, and it is now becoming possible to sample directly the DNA of specimen hypodigms of taxonomic species (i.e. herbarium or museum specimens, Brewer et al. 2019). This offers the prospect of an unprecedented level of integration between data sets that underlie formal species taxon concepts provided the morphological information is also available in matrix form.

Much evolutionary biological research requires living individuals, but the inclusion of these samples into integrative species taxa is problematic, for example, cytogenetics, reproductive biology, etc. Occasionally, a correlation between a character preserved in specimens and one observable only in living individuals has been used, for example, stomatal cell size and polyploidy (Babcock and Stebbins 1938 for *Crepis*) but usually in these kinds of study, the experimental material is a separate sample from the hypodigm of the taxonomic species. The link between them is usually made by preserving a representative specimen as a herbarium voucher and relying on taxonomic determination of the rest of the sample using the conventional STC, for example, a key or by comparison between herbarium specimens and the experimental material.

In this situation, it is not possible to compare directly patterns of variation in, say, DNA and overall morphology because the latter may be represented typologically by one or few voucher specimens that will not usually be part of the hypodigm of the species. There may be ways to mitigate this problem, for example, by documenting each sampled individual using images that can be compared statistically with the hypodigm. Given that natural populations are almost always distinct to some, at least statistically describable, extent, there is a compelling reason to try to link more closely the hypodigms and sample sets involved. Decisions on species limits - necessary for the SQSS - will be made whenever possible based on the consilience of different data sets, and the closer the links between them, the more robust will be the argument for the resulting formal circumscription.

To sum up, cryptic species in the sense understood here are cryptic only in respect to currently formatted taxonomic species. If the latter were in computable form and if taxon concepts were published as data bases rather than only as texts - virtual or otherwise - there would be no need for the term cryptic species. The most objective representation of a species would then be an online taxon concept, founded when possible on a combination of data types, available as matrices for recomputation, based on an overall hypodigm managed by a dispersed international team of taxonomists and accessible to all interested users.

2.5 E-species: Species-as-Evolutionary-Groups

Hey (2001: 67–87) argued that biological diversity and its evolution can be studied without recourse to species. While evolutionary theory predicts that 'evolutionary groups exist - as real dynamic entities, consisting of multiple interacting, but unconnected parts', it also promotes the view that 'What we can see readily with hypothetical evolutionary groups are the ways these groups may not be distinct. Because of hierarchical structure within them, and partial boundaries between them, we would expect evolutionary groups to sometimes be uncountable', that is, vague entities.

Evolutionary biology has been able to show that evolution can produce distinct detectable groups but also that these are not necessarily as distinct or as delimitable as we would like, that is, our human cognitive apparatus tends to exaggerate distinctions and sometimes forces group structure onto data. This applies as much to interpreting a multivariate ordination as it does to intuitive sorting of herbarium specimens.

E-species can be viewed as hypotheses of evolutionary group-forming processes The debate over species concepts as formulated by De Queiroz (2007) illustrates this: a fundamental idea - the lineage - is the causal agency of groups that can be detected and delimited to a greater or lesser extent by using the data and methods of the different disciplines that make up evolutionary biology in the wide sense.

The relationship of taxonomy to evolutionary biology thus seems to revolve around the cognitive necessity to model the data of biodiversity into basic packages. In the early years of scientific systematics in the eighteenth century this data consisted almost entirely of the visible features of macrostructure. To a great extent the groups recognised were equally obvious to scientist and lay person. What was known of the patterns of biodiversity at that time consisted of little else than these easily seen morphological characters. Since then, evolutionary biology has revealed a much richer diversity of structures and processes and shown that the species most easily discerned are only one of many kinds of pattern that have evolved, from genes to megaclades.

Insisting on the existence of species as a distinct and definable category – a stage in the evolutionary process through which all organisms must pass – seems today almost a hangover from pre-scientific folk taxonomies, in which one level of diversity is cognitively prioritised as basic (Berlin 1992; Rosch 1978). The results of evolutionary biological research surely indicate otherwise. But the usefulness and indeed the necessity of recognising basic scientific 'kinds' as if they were units of biodiversity remains. This is shown not only by the debate on species concepts but also by that on 'population thinking' promulgated by such authors as Mayr (Witteveen 2015, 2016) and still a crucial paradigm for systematic and evolutionary biology today. The concept of the population seems to involve the same kind of *a priori* expectation of a group-in-nature as does the initial taxonomic sorting process described earlier. It is analogous to the Linnean species in the sense that the idea of a population acquires its ontological veracity through our scientific understanding of reproduction, dispersal, population genetics, and evolutionary processes – given all this, the existence of a population as a meaningful group seems a reasonable *a priori* assumption. In Linnaeus's time the premises were the prevailing religious and philosophical doctrines, and the available information was largely restricted to macromorphological comparison – and these made a species likewise a reasonable *a priori* assumption.

This is why separating E-species and T-species makes sense. It is clear, as Simpson stated in 1940, that they are distinct. But it is also necessary that they are given equal weight as each depends on the other. Taxonomy is the discipline that exists to provide a scientific (= formal, orderly, documented, reproducible) formulation of units of biodiversity as species (= kinds). This is a process that is self-evidently artificial, in the sense that in view of what we know about evolution the units will never match reality beyond a certain limited extent. On the other hand, these units – taxonomic species – are (or should be) objective in the sense that they are documented and reproducible, while E-species are more like a sandbox of hypotheses that serves primarily to stimulate evolutionary research.

Cryptic species is a term that can be applied not only to groups that, it may be argued, should be recognised formally as taxonomic species but also to groups that are discerned by evolutionary biologists without this intention. If a phylogenetic study shows the presence of three subclades within a clade that corresponds to a previously recognised taxonomic species, the question whether they should be recognised as species is a matter of convention, and indeed of the motivation of the researcher. The best forum for debate and decision-making on such issues is the community of systematists who work on the taxonomic group concerned.

Formally recognised species taxa should not be divided into cryptic and non-cryptic species. Any species whatever that is formalised as a species taxon concept is a taxonomic species, and since it becomes thereby objectified it can no longer be termed cryptic. Equally, any species that has not yet been so formalised, no matter how conspicuous, may be considered cryptic in that it is not yet formally visible in the public domain (i.e. absent from the SQSS). However, this also means that STCs must be more than text presentations – they must become online databases in which all types of data that compose the STC of a species are available in computable form.

References

Allkin, R. and Bisby, F. A. (1984). *Databases in Systematics*. Academic Press, London.

Atran, S. (1990). *The Cognitive Foundations of Natural History*. Cambridge University Press, New York.

(1999). The universal primacy of generic species in folkbiological taxonomy: Implications for human biological, cultural and scientific evolution. In: R. A. Wilson (ed.) *Species: New Interdisciplinary Essays*. Bradford/MIT Press, Cambridge, MA, pp. 231–261.

Babcock, E. B. and Stebbins, G. L. (1938). *The American Species of Crepis, Their Interrelationships and Distribution as Affected by Polyploidy and Apomixis*. Carnegie Institution of Washington Publication No. 504, Washington, DC.

Baker, W., Barber, V., Forest, F. et al. (2019). *PAFTOL – Plant and Fungal Trees of Life*. Third Annual Report. Royal Botanic Gardens Kew, Richmond. Available at: https://kew.iro.bl.uk/work/8cfad0f0-9a5f-489c-9c99-d1c92782f939 (accessed 10 February 2020).

Berendsohn, W. G. (1995). The concept of "potential taxa" in databases. *Taxon* 44: 207–212.

Berlin, B. (1992). *Ethnobiological Classification: Principles of Categorization of Plants and Animals in Traditional Societies*. Princeton University Press, Princeton, NJ.

BHL (2020). Biodiversity Heritage Library. Available at www.biodiversitylibrary.org/ (accessed 29 January 2020).

Bookstein, F. L. (2014). *Measuring and Reasoning: Numerical Inference in the Sciences*. Cambridge University Press, Cambridge.

Borsch, T., Hernández-Ledesma, P., Berendsohn, W. G. et al. (2015). An integrative and dynamic approach for monographing species-rich plant groups: Building the global synthesis of the angiosperm order Caryophyllales. *Perspectives in Plant Ecology, Evolution and Systematics* 17(4): 284–300.

Brewer, G. E., Clarkson, J. J., Maurin, O. et al. (2019). Factors affecting targeted sequencing of 353 nuclear genes from herbarium specimens spanning the diversity of Angiosperms. *Frontiers in Plant Science* 10: 1102. https://doi.org/10.3389/fpls.2019.01102.

Brummitt, R. K., Castrovieja, S., Chikuni, A. C. et al. (2001). The Species Plantarum Project: An international collaborative initiative for higher plant taxonomy. *Taxon* 50, 1217–1230. www.jstor.org/stable/1224752

Camp, W. H. (1951). Biosystematy. *Brittonia* 7 (3): 113–127.

Camp, W. H. and Gilly, C. L. (1943). The structure and origin of species. *Brittonia* 4 (3): 323–385.

Candolle, A. L. P. P. de (1880). *La phytographie*. G. Masson, Paris.

Candolle, A. P. de (1813). *Théorie élémentaire de la botanique*. Déterville, Paris.

(1818). *Regni vegetabilis systema naturale*. Vol. I. Treuttel and Würz, Paris.

Catalogue of Life (2020). www.catalogueoflife.org

CATE-Araceae (2020). CATE Araceae. http://cate-araceae.myspecies.info

Clark, B. R., Godfray, H. C. J., Kitching, I. J., Mayo, S. J., and Scoble, M. (2009). Taxonomy as an eScience. *Philosophical Transactions of the Royal Society A* 367: 953–966.

Clayton, W. D., Vorontsova, M. S., Harman, K. T., and Williamson, H. (2006 onwards). GrassBase – the online world grass flora. www.kew.org/data/grasses-db.html

Cullen, J. (1968). Botanical problems of numerical taxonomy. In: V. H. Heywood, (ed.) *Modern Methods in Plant Taxonomy*. Academic Press, London, pp. 175–183.

Dallwitz, M. J. (1974). A flexible computer program for generating diagnostic keys. *Systematic Zoology* 23: 50–57.

(1984). Automatic typesetting of computer-generated keys and descriptions. In: R. Allkin and F. A. Bisby (eds.) *Databases in Systematics*. Academic Press, London, pp. 279–290.

(2018). Overview of the DELTA system. DELTA – DEscription Language for TAxonomy. Available at www.delta-intkey.com/www/overview.htm (accessed 10 February 2020).

Darwin, C. (1859). *On the Origin of Species by Means of Natural Selection, or the Preservation of Favoured Races in the Struggle for Life*. John Murray, London.

Davis, P. H. and Heywood, V. H. (1963). *Principles of Angiosperm Taxonomy*. Oliver & Boyd, Edinburgh.

Dayrat, B. (2005). Towards integrative taxonomy. *Biological Journal of the Linnean Society* 85: 407–415.

DELTA (2020). DELTA – DEscription Language for TAxonomy. www.delta-inkey.com

De Queiroz, K. (2007). Species concepts and species delimitation. *Systematic Biology* 56 (6): 879–886.

Derkarabetian, S. and Hedin, M. (2014). Integrative taxonomy and species delimitation in harvestmen: A revision of the western North American genus *Sclerobunus* (Opiliones: Laniatores: Travunioidea). *PLoS ONE* 9(8): e104982, 25 p.

Diels, L. (1921 separate, 1924 book chapter). Die Methoden der Phytographie und der Systematik der Pflanzen. In: E. Abderhalden (ed.) *Handbuch der biologischen Arbeitsmethoden, Abteilung XI: Methoden zur Erforschung der Leistungen des Pflanzenorganismus*. Teil 1: Allgemeine Methoden zur Untersuchung des Pflanzenorganismus. Urban & Schwarzenberg, Berlin and Vienna, pp. 67–190.

Dobzhansky, T. (1935). A critique of the species concept in biology. *Philosophy of Science* 2: 344–355.

(1937). *Genetics and the Origin of Species*. Columbia University Press, New York.

Drouin, J.-M. (2008). *L'herbier des philosophes*. Éditions du Seuil, Paris.

Du Rietz, G. E. (1930). The fundamental units of biological taxonomy. *Svensk Botanisk Tidskrift* 24(3): 333–428.

Encyclopedia of Life (2020). http://eol.org

Engler, A. (1900–1953). *Das Pflanzenreich: Regni vegetabilis conspectus*. W. Engelmann, Berlin.

Fleck, L. (1980 [1935]). *Entstehung und Entwicklung einer wissenschaftlichen Tatsache*. Suhrkamp, Frankfurt am Main.

Flora do Brasil (2020) under construction. Jardim Botânico do Rio de Janeiro. Available at: http://floradobrasil.jbrj.gov.br/ (accessed 1 February 2020).

Gärdenfors, P. (2014). *The Geometry of Meaning: Semantics Based on Conceptual Spaces*. The MIT Press, Cambridge, MA.

Garnett, S. T., Christidis, L., Conix, S. et al. (2020). Principles for creating a single authoritative list of the world's species. *PLoS Biology* 18(7): e3000736. https://doi.org/10.1371/journal.pbio.3000736.

GBIF (2020). Global Biodiversity Information Facility. Available at www.gbif.org/ (accessed 20 January 2020).

Geoffroy, M. and Berendsohn, W. G. (2003). The concept problem in taxonomy: importance, components, approaches. Pp. 5-14 In: W. G. Berendsohn, W.G. (ed.) *MoReTax: Handling Factual Information Linked to Taxonomic Concepts in Biology*. Schriftenreihe für Vegetationskunde vol. 39. Bundesamt für Naturschutz, Landwirtschaftsverlag, Münster.

Ghiselin, M. T. (1974). A radical solution to the species problem. *Systematic Zoology* 23: 536–544.

(1997). *Metaphysics and the Origin of Species*. State University of New York Press, Albany.

Gilmour, J. S. L. and Gregor, J. W. (1939). Demes: A suggested new terminology. *Nature* 144: 333–334.

Gilmour, J. S. L. and Heslop-Harrison, J. (1954). The deme terminology and the units of micro-evolutionary change. *Genetica* 27: 147–161.

Grant, V. (1981). *Plant Speciation.* Second edition. Columbia University Press, New York.

Haigh, A., Bogner, J., Boyce, P. C. et al. (2008). A new website for Araceae taxonomy on www.cate-araceae.org. *Aroideana* 31: 148–154.

Hannan, M. T., Le Mens, G., Hsu, G. et al. (2019). *Concepts and Categories: Foundations for Sociological and Cultural Analysis.* Columbia University Press, New York.

Hassemer, G., Prado, J., and Baldini, R. M. (2020). Diagnoses and descriptions in plant taxonomy: Are we making proper use of them? *Taxon* 69(1): 1–4.

Henderson, A. (2005). The methods of herbarium taxonomy. *Systematic Botany* 30 (2): 456–469.

(2006). Traditional morphometrics in plant systematics and its role in palm systematics. *Botanical Journal of the Linnean Society* 151: 103–111.

(2011). A revision of *Geonoma* (Arecaceae). *Phytotaxa* 17: 1–271.

(2012). A revision of *Pholidostachys* (Arecaceae). *Phytotaxa* 43: 1–48.

Hennig, W. (1966). *Phylogenetic Systematics.* University of Illinois, Urbana.

Heslop-Harrison, J. (1953). *New Concepts in Flowering Plant Taxonomy.* Heinemann, London.

Hey, J. (2001). *Genes, Categories, and Species: The Evolutionary and Cognitive Causes of the Species Problem.* Oxford University Press, Oxford.

Hind, K. R., Starko, S., Burt, J. M. et al. (2019). Trophic control of cryptic coralline algal diversity. *Proceedings of the National Academy of Sciences* 116(30): 15080–15085.

Hitchcock, A. S. (1925). *Methods of Descriptive Systematic Botany.* John Wiley & Sons Inc., New York.

Huxley, J. S. (1940). *The New Systematics.* Systematics Association special volume no. 1. Oxford University Press, Oxford.

(1942). *Evolution: The Modern Synthesis.* Allen & Unwin, London.

ILDIS (2020). ILDIS (International Legume Database and Information Service). www.ildis.org

INCT Herbário Virtual (2020). INCT Herbário Virtual da Flora e dos Fungos. http://inct.florabrasil.net

ITIS (2020). ITIS: Integrated Taxonomic Information System. http://itis.gov

James, G., Witten, D., Hastie, T., and Tibshirani, R. (2013). *Introduction to Statistical Learning with Applications in R.* Springer, New York.

Jeffrey, C. (1982). *An Introduction to Plant Taxonomy.* Second edition. Cambridge University Press, Cambridge.

Junker, T. (2004). *Die zweite darwinsche Revolution: Geschichte des synthetischen Darwinismus in Deutschland* 1924 bis 1950. Basilisken-Presse, Marburg.

Kattge, J., Diaz, S., Lavorel, S. et al. (2011). TRY A global database of plant traits. *Global Change Biology* 17: 2905–2935. https://doi.org/10.1111/j.1365-2486.2011.02451.x

Kilian, N., Hand, R., and Raab-Straube, E. von (2020). Cichorieae Portal. http://cichorieae.e-taxonomy.net/portal/

Kilian, N., Henning, T., Plitzner, P. et al. (2015). Sample data processing in an additive and reproducible taxonomic workflow by using character data persistently linked to preserved individual specimens. *Database* 2015: bav094, https://doi.org/10.1093/database/bav094

Kissling, W. D., Balslev, H., Baker, W. J. et al. (2019). PalmTraits 1.0, a species-level functional trait database of palms worldwide. *Scientific Data* 6: 178: 1–12. https://doi.org/10.1038/s41597-019-0189-0

Kitching, I. J. (2020). Sphingidae Taxonomic Inventory. http://sphingidae.myspecies.info/ Accessed on 11 September 2020.

Kripke, S. A. (1972). *Naming and Necessity.* Harvard University Press, Cambridge, MA.

Lakoff, G. (1987). *Women, Fire, and Dangerous Things: What Categories Reveal about the Mind.* University of Chicago Press, Chicago.

Leenhouts, P. W. (1968). A guide to the practice of herbarium taxonomy. *Regnum Vegetabile* 58. International Bureau for Plant Taxonomy and Nomenclature of the

International Association for Plant Taxonomy, Utrecht. 60 p.

Linnaeus, C. (1753). *Species plantarum.* Impensis Laurentii Salvii, Stockholm.

Lucidcentral (2020). Lucidcentral identification and diagnostic tools. Available at www.lucidcentral.org/ (accessed 4 February 2020).

Mayr, E. (1940). Speciation phenomena in birds. *American Naturalist* 74: 249–278.

(1942). *Systematics and the Origin of Species: From the Viewpoint of a Zoologist.* Columbia University Press, New York.

(1963). *Animal Species and Evolution.* Belknap Press, Harvard University Press, Cambridge, MA.

(1969). *Principles of Systematic Zoology.* McGraw-Hill Book Company, New York.

Mayr, E., Linsley, E. G., and Usinger, R. L. (1953). *Methods and Principles of Systematic Zoology.* McGraw-Hill Book Company, New York.

Mueller-Wille, S. (2011). Making sense of essentialism. *Critical Quarterly* 53(4): 61–67.

Murphy, G. L. (2002). *The Big Book of Concepts.* MIT Press, Cambridge, MA.

Palmweb (2020). Palmweb: Palms of the World Online. www.palmweb.org/

Pankhurst, R. J. (1970). A computer program for generating diagnostic keys. *Computer Journal* 12: 145–151.

(1991). *Practical Taxonomic Computing.* Cambridge University Press, Cambridge.

Pante, E., Schoelinck, C., and Puillandre, N. (2015). From integrative taxonomy to species description: one step beyond. *Systematic Biology* 64(1): 152–160.

PBI *Solanum* Project (2020). Solanaceae Source. http://solanaceaesource.org Accessed on 11 September 2020.

POWO (2020). Plants of the World Online. Royal Botanic Gardens Kew, Richmond. Available at www.plantsoftheworldonline.org/ (accessed 29 January 2020).

Putnam, H. (1975). The meaning of "meaning". In: K. Gunderson, Language, mind, and knowledge. Minnesota Studies in the Philosophy of Science. Vol. 7. Reprinted in Putnam, H., *Mind, Language and Reality* (1975), pp. 215–271. Cambridge University Press, Cambridge., pp. 131–193.

Quattrini, A. M., Wu, T., Soong, M.-S., Benayahu, Y., and McFadden, C. S. (2019). A next generation approach to species delimitation reveals the role of hybridization in a cryptic species complex of corals. *BMC Evolutionary Biology* 19(116): 19.

R Core Team (2020). *R: A Language and Environment for Statistical Computing.* R Foundation for Statistical Computing, Vienna. www.R-project.org/.

Reginato, M. (2016). monographaR: An R package for the production of plant taxonomic monographs. *Brittonia* 68(2): 212–216.

Rensch, B. (1947). *Neuere Probleme der Abstammungslehre: Die transspezifische Evolution.* F. Enke, Stuttgart.

Richards, R. A. (2010). *The Species Problem: A Philosophical Analysis.* Cambridge University Press, Cambridge.

Ripley, B. D. (1996). *Pattern Recognition and Neural Networks.* Cambridge University Press, Cambridge.

Rosch, E. (1978). Principles of categorization. In: E. Rosch and B. B. Lloyd (eds.) *Cognition and Categorization.* Lawrence Erlbaum Associates, Hillsdale, NJ, pp. 28–48.

Roskov, Y. R., Bisby, F. A., Zarucchi, J. L., Schrire, B. D., and White, R. J. (2006). ILDIS world database of Legumes: draft checklist, version 10. Reading, UK (CD-ROM).

Roskov, Y., Ower, G., Orrell, T. et al. eds. (2020). Species 2000 & ITIS Catalogue of Life, 25th March 2019. Digital resource at www.catalogueoflife.org/col. Species 2000: Naturalis, Leiden, the Netherlands. ISSN 2405-8858.

Ross, W. D. (1951). *Plato's Theory of Ideas.* Clarendon Press, Oxford.

Russell, G. E. G. and Arnold, T. H. (1989). Fifteen years with the computer: Assessment of the

PRECIS taxonomic system. *Taxon* 38: 178–195.
Scoble, M. J., Clark, B. R., Godfray, C. J., Kitching, I. J., and Mayo, S. J. (2007). Revisionary taxonomy in a changing e-landscape. *Tijdschrift voor Entomologie* 150: 305–317.
Simpson, G. G. (1940). Types in modern taxonomy. *American Journal of Science* 238: 413–431.
(1944). *Tempo and Mode in Evolution.* Columbia University Press, New York.
(1951). The species concept. *Evolution* 5: 285–298.
(1961). *Principles of Animal Taxonomy.* Columbia University Press, New York.
Simpson, G. G. and Roe, A. (1939). *Quantitative Zoology: Numerical Concepts and Methods in the Study of Recent and Fossil Animals.* McGraw-Hill Book Company, New York.
Simpson, G. G., Roe, A., and Lewontin, R. C. (1960). *Quantitative Zoology.* Revised Edition. Harcourt, Brace, New York. Dover Publications Edition, 2003.
Sites, J. W. and Marshall, J. C. (2003). Delimiting species: A renaissance issue in systematic biology. *Trends in Ecology and Evolution* 18 (9): 462–470.
Skov, F. (1989). HyperTaxonomy: A new computer tool for revisional work. *Taxon* 38: 582–590.
Smith, C. R., Godfray, H. C. J., Scoble, M. J. et al. (2008). Introducing CATE: A model for moving taxonomy to the web. In: O. Yata (ed.) *The 2nd Report on Insect Inventory Project in Tropical Asia (TAIIV) "The development of insect inventory project in Tropical Asia (TAIIV)"*, pp. 137–144. Kyushu University, Fukuoka, Japan.
Smith, E. E. and Medin, D. L. (1981). *Categories and Concepts.* Harvard University Press, Cambridge, MA.
Smith, V. S., Rycroft, S., Scott, B. et al. (2012). Scratchpads 2.0: a virtual research environment infrastructure for biodiversity data. Available at http://scratchpads.eu (accessed 11 September 2020).
Species 2000 (2020). www.sp2000.org
SpeciesLink (2020). O projeto *species* Link. http://splink.cria.org.br
Sprague, T. A. (1940). Taxonomic botany, with special reference to the angiosperms. In: J. Huxley (ed.) *The New Systematics.* The Systematics Association/Oxford University Press, Oxford, pp. 435–454.
Stafleu, F. A. and Cowan, R. S. (1976). *Taxonomic Literature.* Second Edition. Volume I: A – G. Bohn, Scheltema & Holkema, Utrecht.
Stang, Nicholas F. (2018). "Kant's Transcendental Idealism", *The Stanford Encyclopedia of Philosophy* (Winter 2018 Edition), Edward N. Zalta (ed.) https://plato.stanford.edu/archives/win2018/entries/kant-transcendental-idealism/.
Stebbins, G. L. (1950). *Variation and Evolution in Plants.* Columbia University Press, New York.
The Plant List (2013). Version 1.1. Published on the Internet; www.theplantlist.org (accessed 29 January 2020).
Tropicos (2020). Tropicos.org. Missouri Botanical Garden. Available at: www.tropicos.org/ (accessed 29 January 2020)
Turesson, G. (1922a). The species and variety as ecological units. *Hereditas* 3: 100–113.
(1922b). The genotypical response of the plant species to the habitat. *Hereditas* 3: 211–350.
(1925). The plant species in relation to habitat and climate. *Hereditas* 6: 147–236.
Turland, N. J., Wiersema, J. H., Barrie, F. R. et al. (eds.) (2018) *International Code of Nomenclature for Algae, Fungi, and Plants (Shenzhen Code) Adopted by the Nineteenth International Botanical Congress Shenzhen, China, July 2017.* Regnum Vegetabile 159. Koeltz Botanical Books, Glashütten.
Turrill, W. B. (1938). The expansion of taxonomy with special reference to Spermatophyta. *Biological Reviews* 13: 342–373.
(1940). Experimental and synthetic plant taxonomy. In: J. S. Huxley (ed.) *The New Systematics.* Clarendon Press, Oxford, pp. 47–71.
Ung, V., Dubus, G., Zaragüeta-Bagils, R., and Vignes-Lebbe, R. (2010a). Xper^2: introducing

e-taxonomy. Bioinformatics 26: 703–704. Available at http://dx.doi.org/10.1093/bioinformatics/btp715

Ung, V., Causse, F., and Vignes-Lebbe, R. (2010b). Xper2: Managing descriptive data from their collection to e-monographs. In: P. L. Nimis and R. Vignes-Lebbe (eds.) *Tools for Identifying Biodiversity: Progress and Problems*. EUT Edizioni Università di Trieste, Trieste, pp. 113–120. Available at http://dbiodbs1.units.it/bioidentify/files/volumebioidentifylow.pdf

WCSP (2020). World Checklist of Selected Plant Families. Facilitated by the Royal Botanic Gardens, Kew. Published on the Internet; http://wcsp.science.kew.org/ Retrieved 29 January 2020.

Wikipedia (2020). Theory of forms. http://en.wikipedia.org/wiki/Theory-of-forms

Williams, D. M. and Ebach, M. C. (2020). *Cladistics: A Guide to Biological Classification*. Third edition. Cambridge University Press, Cambridge.

Wilson, E. O. (1998). *Consilience: The Unity of Knowledge*. Little, Brown and Company, Boston.

Winsor, M. P. (2003). Non-essentialist methods in pre-Darwinian taxonomy. *Biology and Philosophy* 18: 387–400.

(2006). The creation of the essentialism story: An exercise in metahistory. *History and Philosophy of the Life Sciences* 28(2): 149–174.

Witteveen, J. (2015) "A temporary simplification": Mayr, Simpson, Dobzhansky, and the origins of the typology/population dichotomy (Part 1 of 2). *Studies in History and Philosophy of Biological and Biomedical Sciences* 54: 20–33. https://doi.org/10.1016/j.shpsc.2015.09.007

(2016) "A temporary simplification": Mayr, Simpson, Dobzhansky, and the origins of the typology/population dichotomy (Part 2 of 2). *Studies in History and Philosophy of Biological and Biomedical Sciences* 57: 96–105. https://doi.org/10.1016/j.shpsc.2015.09.006

World Flora Online (2020). www.worldfloraonline.org

Yang, L., Kong, H., Huang, J.-P., and Kang, M. (2019). Different species or genetically divergent populations? Integrative species delimitation of the *Primulina hochiensis* complex from isolated karst habitats. *Molecular Phylogenetics and Evolution* 132: 219–231.

Zachos. F. E. (2016). *Species Concepts in Biology: Historical Development, Theoretical Foundations and Practical Relevance*. Springer, Switzerland.

3

Species Circumscription in Cryptic Clades

A Nihilist's View

Richard M. Bateman

Cryptic: expressing something in a mysterious or indirect way so that it is difficult to understand (Macmillan Open Dictionary, January 2020)

3.1 What Is a Species?

There seems little point in attempting to define conceptually a cryptic species until we have first defined species per se, given that the species is the fundamental currency of all biological research. Reviews seeking the optimal species concept were a common feature of systematic biology literature throughout the second half of the twentieth century but their frequency has diminished in the twenty-first century, implying general agreement that either the optimal species concept had been found or an optimal species concept does not exist and so cannot be found. Ongoing debates regarding the relative merits of contrasting species concepts tend to overlook the fact that almost all formal taxonomic literature describing species (or redescribing species, or undescribing species) continues to be published in the absence of any explicit statement regarding whether any particular species concept has been employed; bizarrely, an explicit statement of the species concept used is not a requirement for legitimate nomenclatural publication (Bateman 2011).

Should a biologist have the uncommon wisdom to actively seek an optimal species concept, a wide range of concepts are readily available in the scientific literature. Mayden (1997) not only identified and summarised but even classified approximately 24 apparently distinct species concepts, and these have since been supplemented with several additional proposals or, alternatively, attempts at reductionism (e.g. De Queiroz 2007) that are easily

Table 3.1 The approximately 30 species concepts thus far described in the scientific literature can arguably be reduced to various combinations of just three fundamental concepts – similarity/diagnosability, monophyly, and reproductive isolation – each of which should reflect a perceived discontinuity. Modified after Bateman (2012).

Criterion	Herbarium/laboratory data	Field data
Similarity/ diagnosability	(a) Specimen comparison (b) DNA bar-coding – candidate gene[2] – 'whole genome'[1]	(a) Key-based identification (b) Image-recognition identification[1] (c) Morphometrics[3] (d) Field sequencer[1]
Monophyly	(a) DNA sequence tree (b) Morphology tree[2,3]	[Not applicable]
Reproductive isolation	Indirect historical signal (e.g. genetics) Direct observation through experimental crossing[3]	Direct autecological observation (e.g. pollination) Indirect observation through natural hybrids

[1] Approaches under development
[2] Approaches in decline
[3] Approaches currently under-utilised

refuted (Bateman et al. 2021a). My attempt to boil down these species concepts to their bare essentials (Bateman 2012) condensed about 30 species concepts into what I viewed as three fundamental criteria: similarity/diagnosability, monophyly, and reproductive isolation. Each of these three concepts can usefully draw on multiple lines of evidence derived from studies based in the field, laboratory, and/or preserved collections (Table 3.1).

Traditional plant taxonomy relies primarily on perceived similarity, which is pursued by comparing dried or, less often, pickled specimens in the herbarium, using morphological differences observed within sets of specimens to circumscribe species, then diagnosing those species using formal technical descriptions of those morphological characters that can still be assessed in preserved materials (e.g. Mayo 2022). Users of such taxonomies then subsequently identify those species, typically employing various written and pictorial aids such as keys and images. Both keys and images are increasingly likely to be digital and potentially interactive. But the main analytical tool enacting species delimitation remains the taxonomist's brain, often operating under a complex suite of biases imposed implicitly by earlier taxonomic studies.

It is seldom expressly stated that the more recently developed DNA 'barcoding' methods are also rooted in the concept of similarity and diagnosability. The approach most frequently advocated identifies an 'unknown' plant by obtaining a DNA sequence from a commonly analysed gene and comparing it with a master data set – typically those held in

GenBank and/or EMBL – to list in decreasing order the most similar 'known' species (admittedly, specimens sequenced for GenBank are likely to have been identified by traditional morphological methods; unfortunately, a significant percentage of sequences have been assigned the wrong name). Sadly, molecular research is still largely confined to the laboratory, as attempts to develop an effective portable field-sequencer have progressed more slowly than I, for one, expected (Bateman 2016).

A more recent concept than similarity, monophyly (an inclusive group of organisms that consists of all the descendants of a single hypothetical ancestor) can only be assessed after a rooted phylogenetic tree has been constructed from an explicit numerical data matrix. In early phylogenetic studies of the late 1970s and 1980s, the matrix was likely to consist of morphological characters. However, DNA data overtook morphology as the most commonly produced category of matrix by the late 1990s, and today morphological cladistic studies are disappointingly rare. Although the principle of monophyly now dominates supraspecific classification, its general applicability at and below the species level is more complex and ambiguous (e.g. Wheeler and Meier 2000). One key aspect of these ongoing debates that is central to this chapter concerns whether monophyly can be expected in cases where the putative species under scrutiny have not yet acquired the last of our three primary criteria – reproductive isolation. Initially, reproductive isolation was the preserve of population biologists studying evolution and ecology, but in the mid-twentieth century the 'New Synthesis' brought isolation to the forefront of systematic biology, most famously in Mayr's (1942, *et seq.*) concept of the biological species: 'a group of interbreeding natural populations that is reproductively isolated from other such groups'. Even today, this remains the explicit species concept most likely to be invoked by a taxonomist, though often acting in ignorance of the fact that *the* 'biological species concept' has itself speciated into several competing versions (Mayden 1997; Hausdorf 2011).

Indeed, two highly contrasting approaches that view the world from very different perspectives are available to characterise reproductive isolation. Genetic analyses of plant populations – no longer the exclusive preserve of the specialist geneticist – can inform us about the degree of gene flow that these lineages have experienced in the past, albeit averaged over long periods of time. Mathematical analyses of whole-genome data are rapidly increasing in apparent sophistication and today lay claim to a much broader panoply of potential inferences. In contrast, ecological observation (for example, transfer of pollen between plants by pollinating animals) provides direct evidence of presumed gene flow, but practical constraints mean that such 'ethological' evidence is worryingly localised in both space and time. Both approaches can usefully be informed by studies of hybrids, either those that are found in nature or those of known parentage generated through artificial captive breeding. But no plant species is wholly immune to gene-flow, given growing evidence of gene exchange even occurring between contrasting kingdoms (Bergthorsson et al. 2004; Morjan and Rieseberg 2004). It is increasingly obvious that the key question is no longer whether gene-flow occurs between lineages but rather how infrequent gene flow must become before a lineage can be viewed as *effectively/statistically* isolated and thus qualifies as a *bona fide* 'biological' species.

3.2 What Is a Cryptic Species?

This question has been rigorously explored by Struck et al. (2018), who rightly noted massive inconsistences of definition since the term cryptic species was first coined (reputedly by Darlington 1940 – a decade before the structure of DNA was elucidated by Watson and Crick). The majority of published definitions approximate that popularised by Bickford et al. (2006: 148): 'two or more distinct species that are erroneously classified (and hidden) under one species name'. However, in my opinion, such definitions merely describe perplexities that have perennially dogged traditional taxonomy. Who is to say whether a particular classification is erroneous when comparing competing classifications? Which (if any) species concept has been chosen? Which (if any) sampling strategy was employed? Which (if any) categories of data have been brought to bear on the problem? Has there been any kind of attempt at quantitative analysis of the available data categories that might permit a legitimate claim to have scientifically circumscribed the range of species in question rather than merely awarding each of them a formal name consistent with the international laws of nomenclature? And, perhaps most importantly, are the results obtained from two or more contrasting data categories convincingly congruent?[11]

The relevance of each of these questions is itself influenced strongly by the chosen definition of cryptic species. Some definitions make clear that the strong similarity responsible for 'hiding' the two or more species under one taxonomic roof is morphology (e.g. Pfenninger and Schwenk 2007: 1; Tronteij and Fišer 2009: 1; Fišer et al. 2018: 627). Other authors have included the ability to interbreed, presumably inspired by Mayr's (1942, 1982) (in)famous 'biological species concept'. Bernardo (2011: 389) later expanded the reproductive element in the definition, presumably aiming to achieve compatibility with one or more of the many variants of the phylogenetic species concept (reviewed by Wheeler and Meier 2000). Thus, Bernardo defined cryptic species as 'species that are morphologically similar to such a degree that their distinct evolutionary trajectories cannot be readily recognised based on quantitative analysis of morphological features alone'. This definition presents us with three additional challenges: (1) performing a quantitative analysis of the morphology of the study organisms, (2) deciding how best to recognise an 'evolutionary

[1] Note that within this text I frequently use two terms in an undesirably broad manner: congruence and typology. In the case of 'typology', a strict definition would be 'non-hierarchical system of groupings, the members of which are identified by postulating specific attributes that are mutually exclusive and collectively exhaustive'. My use is less philosophical and more biological, being rooted in the systematic concept of a single type specimen as being considered sufficient for the formal establishment of a taxon and for encapsulating its essential features (thus representing a single point in space and time). The concept contrasts radically with the demographic approach based on organised sampling that seeks to determine through circumscription the boundaries of a *demonstrable* taxon, rather than to define the essence of a *putative* taxon. When discussing levels of 'congruence' observed between patterns obtained from contrasting categories of data, it is arguable that it would have been more accurate for me to employ the term 'consilience'. However, even brief consideration of the contrasting treatments of this term by Wilson (1998), Gould (2003), and Bookstein (2014) suggested to me that its meaning is no less ambiguous than that of 'congruence'.

trajectory', and (3) deciding what evidence is needed before that trajectory can legitimately be described as sufficiently 'distinct' from all others to qualify as a genuine species.

Today, it is congruence between data categories that lies at the heart of most debates surrounding cryptic species. In theory, there are many categories of data that could usefully be compared with morphology, but at present, it is in most cases nucleic acid sequence data that are gathered as the primary comparator. And the DNA data are increasingly generated in vast quantities through next-generation sequencing (NGS: e.g. Olson et al. 2016) rather than through older candidate gene approaches.

In practice, most current authors treat the molecular data as providing the 'correct' species circumscriptions, though Jorger and Schroedl (2013: 1) made a potentially important conceptual distinction between species that are genuinely cryptic and those that are pseudo-cryptic. If a single species circumscribed using traditional morphological approaches is later re-circumscribed into two or more species on the basis of newly gathered population genetic data, it behoves the researchers to perform a quantitative (preferably in situ morphometric) analysis of morphology at the population level, which will often reveal phenotypic distinctions whose existence (or, at least, whose significance) passed unrecognised during earlier, more typological studies. Using the terminology of Jorger and Schroedl (2013), the two or more newly recognised species would then be regarded as pseudo-cryptic rather than truly cryptic. Before evaluating this potentially important distinction, I will outline three case studies that have influenced my own opinions on these matters. Note that all of my examples are botanical case studies, whereas the bulk of the published case studies that purport to explore cryptic species (and the bulk of the underlying conceptual discussions) are zoological or microbial in nature.

3.3 (Pseudo)Cryptic Species within the European Orchid Flora

I have spent much of the last four decades studying problematic complexes of closely similar putative species in several genera of north-temperate terrestrial orchids – genera that collectively reflect several different drivers of speciation. Our early studies (e.g. Bateman and Denholm 1983) were based only on in situ morphometric surveys conducted at the population level, but molecular data were later brought to bear, first Sanger sequencing (Bateman et al. 1997, *et seq.*) and later next-generation sequencing (Bateman et al. 2018, *et seq.*). Here, I briefly review insights gained from recent studies of three such genera, before attempting to draw more general conclusions regarding the nature of supposedly cryptic species and critiquing the value of the underlying concept(s).

3.3.1 Epipactis Section Epipactis

Epipactis section *Epipactis* (Epidendroideae: Neottieae) is a monophyletic group of between three (Sundermann 1980) and 65 (Delforge 2016) putative species of helleborine orchids that appears derived relative to the remainder of the genus. A recent study (Sramkó et al. 2019; Bateman 2020b) used restriction site-associated sequencing (RAD-seq) to analyse 108 plants representing 29 named taxa of *Epipactis*. By applying the principles of monophyly supported by substantial underpinning branch lengths, Sramkó et al. (2019) resolved the 27 named ingroup taxa into what they considered to be 11 genuine species plus three subspecies (Figure 3.1), suggesting that there are no more than 20 species of section *Epipactis* in Europe and Asia Minor. The excess of species recognised by most taxonomic authorities represented two main trends. First, species initially named in western Europe were later awarded novel species epithets by eastern European taxonomists that, in the light of recent RAD-seq analyses, are now seen to be junior synonyms. Second, none of the rare local endemics newly described or raised to species level during the last few decades appeared sufficiently distinct; all of the species recognised by Sramkó et al. (2019) and Bateman (2020b) are both geographically widespread and widely accepted by the taxonomic community.

In terms of evolutionary interpretation, Sramkó et al. (2019) were able to both support and elaborate a scenario advanced by previous authors, notably Hollingsworth et al. (2006). A single geographically widespread and comparatively ecologically tolerant species, *E. helleborine*, proved to be ancestral to the majority of the remaining species and thus was resolved as paraphyletic rather than monophyletic. Anecdotal evidence suggests that *E. helleborine* is the most ecologically tolerant of the *bona fide* species, and pseudo-F statistics showed that it experiences significantly greater allogamy than any of the other species (Bateman 2020b). Among the genuine species, most of those that are the more strongly autogamous originated independently of each other. Nonetheless, most have acquired heritable floral morphologies and/or occupy categories of habitat that increase the probability of self-pollination.

Unfortunately, Sramkó et al. (2019) did not gather any category of data from *Epipactis* other than RAD-seq, perpetuating a trend that has consistently hampered research into this perennially troublesome genus. Data collected by Hollingsworth et al. (2006) were also entirely genetic, though they were able to compare allozyme profiles with candidate gene sequences for substantial samples of most of the species occurring in western Europe. The largest scale morphometric study performed on the group (Tyteca and Dufrene 1994) was largely confined to population-level studies of a few species in Iberia and similarly involved no other data category. Brys and Jacquemyn (2016) performed artificial crossing experiments between a total of eight populations encompassing five species; the resulting fruit set and seed viability data showed that the species tested lacked significant post-zygotic isolating mechanisms, irrespective of the degree of autogamy inferred. In developing arguably the most integrated study conducted within the genus, Jacquemyn et al. (2018) compared only four Belgian populations of two infraspecific ecological variants within *E. helleborine* that differ substantially in frequency of autogamy. Fairly detailed morphometrics were combined with Genotyping-By-Sequencing, reciprocal in situ seed germination experiments and ex situ crossing followed by seed viability testing. Genetic differentiation between the ecotypes proved greater than morphological differentiation; modest decreases in estimated fitness of both hybrids and migrants were ascribed tentatively to mycorrhizal differentiation.

In this context, Schiebold et al. (2017) sampled six *Epipactis* species at nine sites in Germany and the Netherlands, using nrITS sequences to identify peleton-forming

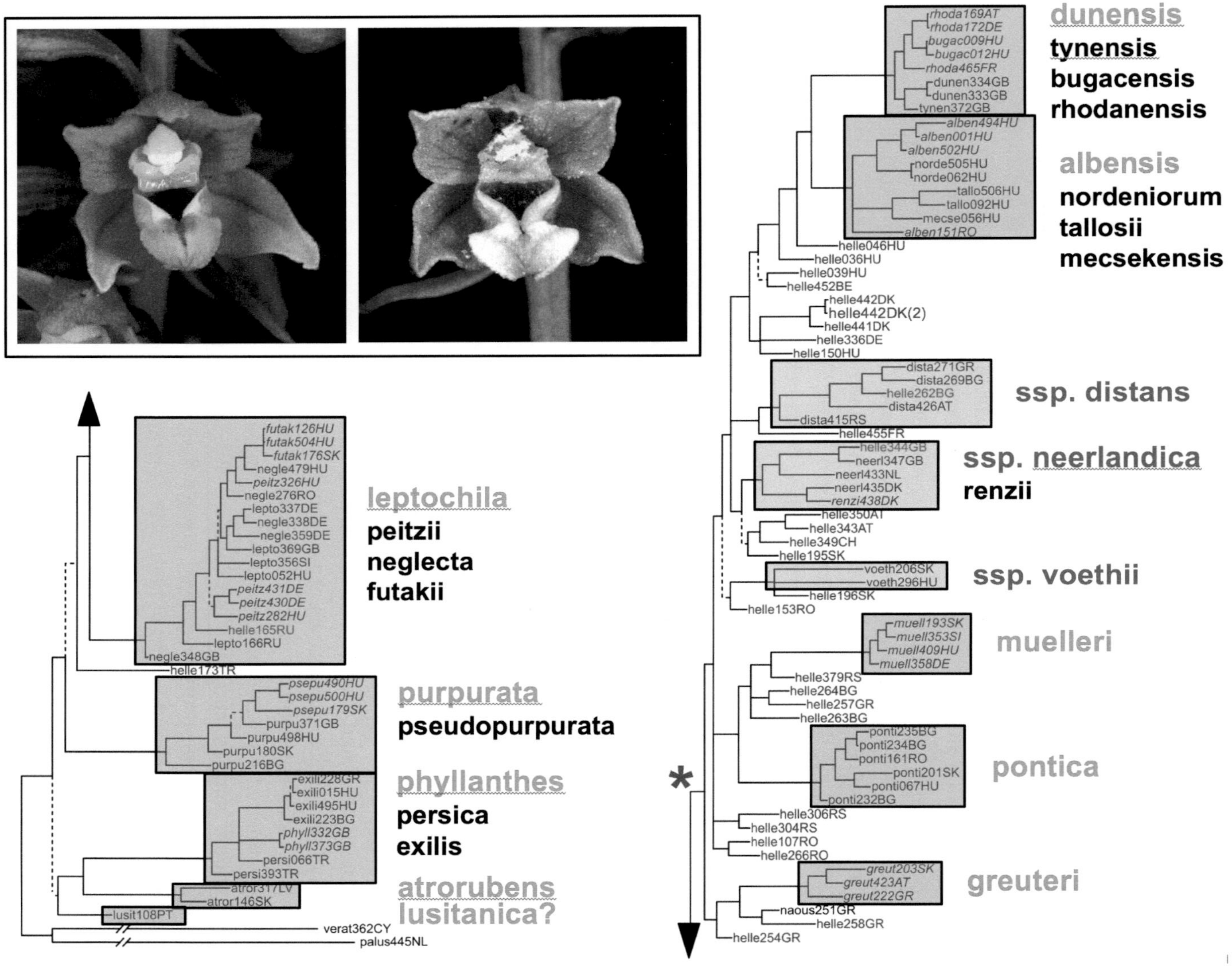

Figure 3.1 Maximum Likelihood phylogram depicting the evolutionary relationships of 108 *Epipactis* plants based on RAD-seq data analysed using RAxML. Samples of the geographically widespread, locally common ancestral species *E. helleborine* subsp. *helleborine* lie outside the boxes. Species re-circumscribed according to a combination of monophyly and branch length are placed in shaded boxes and subspecies in open boxes. Collapsed branches attracted support values of less than 80 per cent and dashed branches achieved only 80–90 per cent support (approximate likelihood ratio test). The asterisk marks the only branch to have received meaningful support from the previous candidate gene study of nrITS by Bateman et al. (2005). Inset: putative ancestral species *E. helleborine* (left) and descendant species *E. dunensis* (right) in the UK. Source country codes: AT, Austria; BE, Belgium; BG, Bulgaria; CH, Switzerland; CY, Cyprus; DE, Germany; DK, Denmark; FR, France; GB, Britain; GR, Greece; HU, Hungary; LV, Latvia; NL, Netherlands; PT, Portugal; RO, Romania; RS, Russia; SK, Slovakia; TR, Turkey.

endomycorrhizae in the roots and employing carbon and nitrogen isotopes to assess levels of transfer of nutrients from surrounding trees via the fungi to the roots of these rhizomatous orchids. They detected what can broadly be described as a cline of enrichment in both carbon and, especially, nitrogen that extended from preferred habitats of marshes through open woodland (including *E. helleborine*) to closed woodland, concomitant with gradations from endomycorrhizae to ectomycorrhizae and from basidiomycetes to ascomycetes, and congruent with increasing frequencies of self-fertilisation. Bateman (2020b) later speculated that mycorrhizal switching may have contributed to habitat diversification among the multiple derivatives of *E. helleborine*, though molecular data remain too patchy to test this hypothesis with sufficient rigour.

After adjustment for latitude and altitude, all species flower within a summer window of approximately seven weeks. A wide range of pollinating insects have been observed to visit the flowers of the *E. helleborine* complex (e.g. Claessens and Kleynen 2011), presumably attracted by the secretions of both osmophores and nectaries that generate substantial quantities of nectar into the cup-shaped proximal portion of the labellum (hypochile). Social wasps are regarded as the primary pollinators, though understanding of pollination in the group would benefit greatly from large-scale, long-term, quantified field observations.

3.3.2 Dactylorhiza Section Dactylorhiza

Dactylorhiza section *Dactylorhiza* is undoubtedly a monophyletic group, provided that the under-researched Asiatic species *D. aristata* (formerly treated as a separate Section *Aristae*) is included. The clade encompasses between four (Sundermann 1980) and about 65 (Averyanov 1990) putative species, and is undoubtedly derived relative to the remainder of the genus (Brandrud et al. 2020). As with *Epipactis*, *Dactylorhiza* has recently been subjected to detailed genetic analysis using RAD-seq (Brandrud et al. 2020), analysing 207 *Dactylorhiza* plants plus 18 outgroup samples representing the four most closely related genera. The 28 named entities of section *Dactylorhiza* analysed resolved into approximately 21 species: seven diploid and 14 tetraploid. Extrapolating these results across the genus, Bateman (2019) estimated that the entire genus encompasses about 30 *bona fide* species and section *Dactylorhiza* about 24 – a figure midway along the scale of previous estimates. Most local endemics and ecotypes failed to provide convincing cases for recognition as *bona fide* species.

However, the main ongoing area of contention once again surrounds the primary mechanism of speciation operating within the clade, in this case whole-genome duplication; the genus has become a model system for the study of polyploidy. Chromosome counts meant that polyploidy was suspected in the genus by the mid-twentieth century, early hypotheses (e.g. Heslop-Harrison 1954) focusing on the likelihood of dominant allopolyploidy (hybridisation combined with chromosome doubling). Subsequent genetic studies, initially using allozymes (e.g. Hedrén 1996) and later using candidate-gene sequencing (e.g. Pillon et al. 2007), confirmed these hypotheses, showing that most of the western European allotetraploids originated through hybridisation between the same pair of parental species, *D. fuchsii* (or, less often, its autopolyploid derivative, *D. maculata*) and *D. incarnata*. Moreover, *D. fuchsii* was reliably the 'mother' and *D. incarnata* the 'father' of the allopolyploids. Subsequent studies in eastern Europe and Asia Minor revealed a similar pattern, but with *D. saccifera* and *D. euxina* taking over the roles previously performed by their close relatives *D. fuchsii* and *D. incarnata*, respectively (Hedrén et al. 2007; Bateman 2019; Brandrud et al. 2020).

Increasingly deep understanding of this evolutionary mechanism has caused two contrasting concepts of species circumscription to develop within the same broader research group, both concepts being logical and rooted in evolutionary biology. Some team members argue that all allopolyploid lineages derived from the same pair of parental species should by definition be assigned to a single species, which in western Europe is *D. majalis* s.l. (Hedrén et al. 2008; Kühn et al. 2019). Other team members believe that each independent successful allopolyploidy event constitutes a separate speciation event – effectively a monophyletic bottleneck – and that the allotetraploid lineages that result can, with care, be differentiated through subtle differences in morphology and ecological preference. They further argue that these differences among allopolyploids reflect their independent origins from at least subtly different ecotypes that are documented within each of the two parental species (Pillon et al. 2007; Bateman and Denholm 2012; Brandrud et al. 2020). Interestingly, contrasting spectra of allopolyploid taxa are concentrated on either side of the maximum extent of the last glaciation, the most recently evolved species occurring to the north of the boundary (Bateman 2011, 2020; Brandrud et al. 2020).

Unlike *Epipactis*, *Dactylorhiza* has been subject to well-sampled and detailed morphometric studies, albeit confined to either the British Isles (Bateman and Denholm 1983, *et seq.*) or Belgium (Gathoye and Tyteca 1987). These studies demonstrated the absence of phenotypic discontinuities between closely related species, particularly allotetraploid derivatives of similar parents. They also illustrated how, among closely related lineages, subtle differences in phenotype that directly reflect genotype can largely be masked by non-genetic factors: differences in epigenetic properties such as methylation (Paun et al. 2010), developmental status/plant maturity, and local environmental influences that are often mediated through the aversion of most species to dry soils (Bateman and Denholm 1989).

All species flower within a window of approximately six weeks. In terms of symbiotic partners, species of *Dactylorhiza* section *Dactylorhiza* are food-deceptive (non-rewarding) generalists, attracting a wide range of putative pollinators but apparently relying most heavily on bumble bees (Claessens and Kleynen 2011). They also lack significant post-zygotic reproductive barriers, readily generating hybrid swarms even between contrasting ploidy levels (De hert et al. 2012). Mycorrhizal partners are also comparatively diverse taxonomically (Bailarote et al. 2012).

Interestingly, the sister genus to *Dactylorhiza*, *Gymnadenia* sensu lato, shows a similar enthusiasm for polyploid speciation (Travnicek et al. 2012), in this case aided by a presumed radical evolutionary-developmental shift to generate a clade that is far more divergent phenotypically than genotypically and is still regarded by many non-cladists as a separate genus, '*Nigritella*' (cf. Bateman et al. 2018; Brandrud et al. 2019). And their shared sister-genus, *Platanthera* sensu lato, also includes – in the form of Section *Platanthera* (e.g. Bateman et al. 2014) and Section *Hyperborea* (e.g. Wettewa et al. 2020) – allopolyploid complexes of putative species that are difficult to distinguish both phenotypically and genotypically. Both *Gymnadenia* and *Platanthera* have fragrant flowers that offer substantial nectar rewards in elongate labellar spurs (Bell et al. 2009); these features best fit the majority of *Gymnadenia* and *Platanthera* species for lepidopteran pollinators (Claessens and Kleynen 2011). Both genera have also been the subjects of fairly large-scale molecular and morphometric studies (Bateman et al. 2013; Valuyskikh et al. 2019; Bateman et al. 2021b). The aggregate clade formed by these three genera plus *Galearis*, collectively termed the taper-tubered clade, has a derived chromosome number ($2n = 40$, derived from $2n = 42$) and exhibits a strong tendency towards ploidy change.

3.3.3 Ophrys (Especially the *O. sphegodes* Complex)

The genus *Ophrys* (bee orchids) has the dubious honour of being the most taxonomically controversial among the three troublesome genera reviewed here (cf. Bateman et al. 2011; Vereecken et al. 2011), the number of species that it encompasses residing anywhere between nine (Bateman et al. 2018) and upwards of 353 (Delforge 2016). Clearly, such a great taxonomic discrepancy could not have emerged unless radically different species concepts had been applied. In order to discuss the products of contrasting species concepts, Bateman (2018; also Bateman et al. 2018) felt obliged to establish a pragmatic working terminology based on three hierarchical levels of putative 'species': numerous 'microspecies ' recognised by ethologically oriented taxonomists such as Delforge (2016) on the basis of intuited morphological differences, 23 'mesospecies ' representing the 23 categories into which Delforge (2016) grouped these microspecies on supposedly reliable (if typological) morphological distinctions, and just nine macrospecies that could be distinguished on the basis of one or two fairly reliable (though imperfectly differentiated) base-pair differences within the (rapidly mutating) nrITS region (first established as ten potential macrospecies by Devey et al. 2008). New 'microspecies' continue to be described within the genus at an average rate of approximately ten per annum, typically in the absence of any accompanying laboratory-based analyses and creating such profound challenges to field identification that anyone routinely claiming success with confidence is, in my carefully considered opinion, delusional.

As with the previous genera, the genus *Ophrys* has long been subject to molecular surveys, initially using ITS (e.g. Bateman et al. 2003; Devey et al. 2008), later using multiple low-copy nuclear genes (e.g. Breitkopf et al. 2015), and latterly using next-generation sequencing. Bateman et al. (2018) applied RAD-seq to plants of 32 microspecies representing all nine macrospecies. The resulting molecular phylogeny, based on 4,159 heavily filtered SNPs, yielded only nine groups that were both monophyletic and acceptably statistically robust. Phylogenetic relationships among the nine macrospecies were well-supported, though the topology contrasted in places with those resulting from previous molecular phylogenetic and morphological phylogenetic studies (Bateman et al. 2018). Karyotypes have been characterised for representatives of seven of the nine macrospecies (D'Emerico et al. 2005). Polyploid lineages have been reported, but on present evidence they appear uncommon. Artificial crosses have revealed negligible post-zygotic reproductive barriers (Scopece et al. 2007), while self-pollination experiments identified sufficient phenotypic differences among genetically identical progeny to match levels of phenotypic difference observed among some of the more closely similar microspecies (Malmgren 2008; Bateman et al. 2011). Natural hybrids are frequently recorded, though levels of natural hybridisation within the genus are difficult to estimate due to the exceptional morphological similarity of the microspecies; this issue remains highly contentious (cf. Bateman 2018; Paulus 2018). An extensive morphometric survey conducted across the genus suggests that the microspecies represent attempts to arbitrarily divide what is actually a phenotypic continuum (Bateman and Rudall unpublished; see Section 3.8.2).

The genus as a whole tolerates a wide range of soil moisture contents and pH values, though dry, disturbed limestone soils are favoured by most microspecies. Current evidence of their mycorrhizal associates is limited but suggests that *Ophrys* species are generalists (Liebel et al. 2010). In contrast, conventional wisdom states that the bee orchids are an archetypal example of specialist pollination, in this case by means of pseudo-copulation. Through a complex series of olfactory, visual, and tactile cues, the orchid deceives naïve male insects into attempting to mate with its flowers rather than with conspecific females, thereby occasionally transferring pollinaria to the visiting male insect. The pollinaria are then (only very occasionally) deposited on the stigma of another conspecific flower if the insect is foolish enough to repeat its misconceived sexual advances on a nearby inflorescence (e.g. Paulus and Gack 1990; Vereecken and Francisco 2014). It has even been argued that most *Ophrys* microspecies are so well adapted that they attract only a single species of bee or wasp, though other observers have questioned both the evidence that such perfect symmetry of relationship has been achieved and the likelihood that any such exclusive plant–pollinator relationships that do exist would permit the pattern of explosive speciation envisaged by proponents of the microspecies concept (e.g. Bateman et al. 2011; Bateman 2018; Kuhn et al. 2019).

The most informative of recent studies of *Ophrys* have integrated multiple lines of evidence and focused on just one of the nine molecularly circumscribed macrospecies, *O. sphegodes* – a macrospecies that contains at least 113 microspecies species according to Delforge (2016). Breitkopf et al. (2013) investigated both pure and mixed populations of *O. sphegodes* s.s. and *O. exaltata* in central Italy, studying their population genetics through 322 AFLP markers, labellar scents through gas chromatography, and pollinator choice through field observation of translocated plants. These two *Ophrys* microspecies proved barely differentiable genetically but more readily distinguished using the detailed composition of their fragrances. Pollinator data maintained the possibility of there being a single dominant pollinator for at least some *Ophrys* microspecies, but only in any one geographic area at any one time of the day and day of the year, and always with other insect species also visiting the flowers. Sedeek et al. (2014) extended the earlier study of Breitkopf et al. (2013) to include two further microspecies, *O. garganica* and *O. incubacea* in five populations, each consisting of two or more microspecies. They analysed plants for three-dimensional labellum shape, fragrance composition, and genetic composition through AFLP markers, again showing that clear differentiation was confined to fragrance composition. Later reanalysis of the NGS genetic data of Sedeek et al. (2014) by Cozzolino et al. (2020) arguably extracted more precise indications of relationship. However, genome skimming data obtained from 41 microspecies by Bateman et al. (2021a) demonstrated that even whole plastomes cannot distinguish microspecies within *O. sphegodes* s.l. and showed poor correlation between plastid genotype and phenotype of individual plants.

Flowering in the genus as a whole spans a period of about five months, though within each macrospecies the period is considerably shorter. Having long been viewed as a textbook case of highly specialised pseudo-copulatory pollination, *Ophrys* has now become established by researchers categorised as 'ethologists' by Bateman et al. (2011) as a classic example of 'ecological speciation' through sexual deception. Minute genetic changes subtly alter pseudo-pheromone cocktails that in turn modify pollinator preference. This phenomenon, sometimes assisted by a degree of phenological divergence, is hypothesised to permit co-existence of multiple 'distinct' lineages (e.g. Sedeek et al. 2014; Baguette et al. 2020). But are the microspecies truly 'distinct', and if not, can an incohesive aggregate of populations that are still regularly experiencing gene flow legitimately be viewed as a *bona fide* species that, along with hundreds of other brethren, apparently forms part of an ongoing evolutionary radiation in this remarkable genus?

3.4 Seeking Common Patterns in (Pseudo)Cryptic Clades

Each of the three orchid clades reviewed in Section 3.3 continues to generate multiple, radically contrasting taxonomies. Choosing a preferred taxonomy from among those available perennially perplexes all other biologists – at least, all other biologists who care whether the species that they are studying are real or imaginary. The issues are sufficiently complex to encourage many observers towards simply taking 'intuitive' (i.e. arbitrary and authoritarian) decisions, thereby making a mockery of the intensive efforts currently made by several research groups to better understand the systematic biology of these reputedly cryptic plants. If genuine progress is to be made, we should now ask whether there are any features shared by these three clades that could helpfully explain their apparently cryptic nature?

Epipactis plants are rhizomatous and bear leaves of a texture resembling parchment, whereas the remaining genera possess tuberoids and produce fleshier leaves. Flowers are numerous and small in *Dactylorhiza*, moderate in number and size in *Epipactis*, and moderate to large but few in number in most *Ophrys* species. *Epipactis* has a bipartite, hinged, concave labellum, *Dactylorhiza* has a unipartite flat labellum expanded into a spur, and *Ophrys* features a unipartite convex labellum of great micromorphological complexity. *Dactylorhiza* and *Ophrys* have pollinaria with long caudicles and viscidia enclosed in a bursicle, whereas *Epipactis* has caudicles that are barely visible, both pollinaria being largely enclosed within an operculum. In summary, unifying morphological features are few, excepting the ubiquitous family-wide constraint of a characteristic orchid floral architecture that consists of six-part perianth of two closely spaced whorls surrounding the characteristic congenitally fused gynostemium (e.g. Rudall and Bateman 2002).

The dominant speciation mechanism might be predicted to be a more fruitful area for comparison. However, inferred dominant speciation mechanisms similarly differ greatly among the three case studies: iterative transitions towards greater autogamy in *Epipactis* (possibly assisted by mycorrhizal switching), ploidy change in *Dactylorhiza* (possibly assisted by contrasting soil pH preferences), and dominant pollinator switching through modified pseudo-pheromone cocktails in *Ophrys* (possibly assisted by phenological shifts). There is evidently more than one way to generate a cryptic clade.

Could the explanation for cryptic speciation reside in the much-discussed symbiotic relationships of orchids? The limited evidence currently available suggests a cline in specificity of mycorrhizal partners from generalised in *Ophrys* through to highly specialised in at least some *Epipactis* species. Considered together, the three case studies span almost the full range of soil types, shade conditions, and climatic conditions represented in Europe. There also exists a cline of increasing specificity in pollinators, but in contrast, it runs from Section *Dactylorhiza* through Section *Epipactis* to *Ophrys*. All three of the three main pollination syndromes evident among European terrestrial orchids are represented in these three case studies (especially if consideration of *Dactylorhiza* is expanded to encompass the most closely related genera, *Gymnadenia* and *Platanthera*): in order of increasing frequency of successful pollination these syndromes are sexual deception, food deception, and nectar-rewarding (cf. Cozzolino and Widmer 2005). All three clades share the possession of at best only weak post-zygotic isolation, which is imperfect even in the case of inter-ploidal crosses. However, this statement appears to apply to the majority of clades within the orchid family, and so cannot legitimately be cited as a likely driver of cryptic speciation in these particular clades. Similarly, all three case studies have moderately sized genomes (1C = 3–13 pg, depending largely on ploidy level) that are typical of the orchid family in general (Pellicer and Leitch 2020).

The main feature distinguishing each of these three cryptic clades from its close relatives is actually comparatively strong genetic similarity between putative species. This fact is most readily illustrated by considering levels of divergence in nrITS sequences evident within the three case studies. Distances from the crown-group node to the tips of terminal branches are 4–9 bp within *Dactylorhiza* section *Dactylorhiza* (Pillon et al. 2007), 1–5 bp within the *Ophrys sphegodes* complex (Devey et al. 2008), and 0–3 bp within *Epipactis* section *Epipactis* (Bateman et al. 2005; Tranchida-Lombardo et al. 2012). In these cases, at least, the problematic clades that are most open to accusations of being cryptic all combine broadly similar morphologies with molecular trees in which terminal branches are short relative to those subtending other species included in the analysis. Moreover, the relationships between the cryptic species are rendered ambiguous by even shorter interconnecting internal branches that, from a statistical viewpoint at least, form a near-polytomy. In other words, the most commonly held view of cryptic species as being reliably distinguishable using DNA but not distinguishable using morphology is contradicted here by three case studies in which the putative species are difficult to distinguish using either category of data.

Before we consider in greater detail the implications of such near-polytomous topologies, it is necessary to reappraise the interpretation of molecular trees in the light of recent advances in genomics.

3.5 Reappraising Constraints on Molecular Phylogenetics

3.5.1 Substantial Incongruence Has Become Predictable between Nuclear and Organellar Genomes

Recent studies of closely related species pursued using next-generation sequencing techniques such as RAD-seq (and its close relative Genotyping-By-Sequencing), Hyb-seq, and Genome Skimming have, in my opinion, provided the systematics community with a profound challenge to our confidence in the likely topological accuracy of molecular phylogenies (admittedly a property that is, of course, ultimately untestable in nature). Most importantly, it has increasingly become clear to me that, among the closely related taxa that constitute most supposedly cryptic clades, phylogenetic signals from organellar and nuclear genomes consistently show substantial incongruence when compared in sufficient detail. At this point, it is necessary to distinguish between two different kinds of incongruence that usually become apparent when comparing two trees constructed from the same spectrum of taxa (preferably also the same spectrum of analysed individuals). The most commonly discussed of the two forms is topological incongruence, when two trees differ in the relationships inferred between the analysed individuals. Less often discussed is branch-length incongruence, when radical differences become apparent between trees in the relative lengths of comparable branches within the tree. Such contrasts in disparity measures suggest substantial rate differences between the categories of data under comparison.

Moreover, when the phylogenetic signal from each category of genome – nuclear and plastid – is compared with morphometric similarities, nuclear data correlate more closely with morphological similarity, whereas plastid data correlate more closely with geographic location. Although I have drawn these far-reaching conclusions primarily from studies of orchids, including the *Dactylorhiza* and *Ophrys* examples outlined in Section 3.4, similar patterns are evident in other taxonomic groups of plants that have been explored with sufficient rigour using both next-generation sequencing and morphometric data (e.g. in the herbaceous legume *Astragalus*: Zaveska et al. 2019).

An excellent illustration supporting these conclusions was provided by Cozzolino et al. (2019), who conducted a more sophisticated reanalysis of the Genotyping-By-Sequencing data first published by Sedeek et al. (2014). Studying data from five geographically close populations of *Ophrys sphegodes* s.l., they found strong correlation between similarities in the nuclear genome and assignment to four morphologically determined co-occurring microspecies, which were revealed to be potentially monophyletic within the constraint of their geographically localised sampling. However, this tentative congruence between nuclear genome and traditional morphology was achieved at the expense of revealing strong incongruence between nuclear sequences and both plastid and mitochondrial patterns. Frequent failure in next-generation sequencing studies to obtain 100% bootstrap support from individual branches representing hundreds of SNPs suggests to me the presence of remarkably extensive homoplasy – or, to put it another way, remarkably little phylogenetic signal within the concatenated matrix being analysed. Cozzolino et al. (2019)

surprisingly found that bootstrap support peaked when their data were only lightly filtered for gap reduction in the matrix by employing a comparatively low minimum threshold of 30 per cent shared loci per plant. This phenomenon is difficult to explain, as the plants involved appear far too genetically similar to be prone to locus dropout (mutations accumulating in many of the restriction sites that permit RAD-seq analyses: Lee et al. 2018).

The implications of these observations for the discipline of molecular phylogenetics in general are immense. Most importantly, the first quarter-century of molecular phylogenetics relied heavily on candidate gene sequencing of particular plastid regions that have gained considerable fame in botanical circles, notably *rbcL*, *matK*, and *trnL*. Consequently, these genic regions constitute the bulk of phylogenetically oriented data deposited in GenBank and have come to dominate supraspecific plant classification, primarily via syntheses published by the Angiosperm Phylogeny Group (e.g. APG 2016). As recently as 2009, the combination of *matK* and *rbcL* was designated as the internationally agreed standard for DNA barcoding of green plants (Hollingsworth et al. 2009, 2016), thereby excluding from the molecular top-table competing nuclear regions such as high-copy ribosomal ITS and low-copy MADS-box genes such as *LEAFY* (admittedly, the decision to omit nuclear regions prompted some dissenting voices: e.g. Fazekas et al. 2012; Li et al. 2015).

So much for organelles – we should now move on to consider the nuclear genome in the (apparently justified) hope that it is more reliable. Piniero-Fernández et al. (2019) sequenced entire transcriptomes of four members of the *Gymnadenia* clade (sister-genus to *Dactylorhiza*, as discussed in Section 3.3.2). Previous studies had shown this clade to exhibit considerable topological incongruence between all six of the data sets accrued by a raft of earlier studies of the genus: RAD-seq, candidate-gene nuclear, candidate-gene plastid, morphological cladistic, morphometric, and genome size (cf. Travnicek et al. 2012; Bateman et al. 2018; Hedrén et al. 2018; Brandrud et al. 2019). The four *Gymnadenia* species chosen by Pineiro-Fernández et al. (2019) together spanned much of the morphological and molecular diversity present in the clade, such that three (arguably all four) of the species selected are easily distinguished morphologically. These authors constructed individual four-taxon trees from each of several hundred families of single-copy nuclear genes and found support from at least one gene family for six of the theoretically possible 15 topologies. Even the most frequent topology was recovered in only 33 per cent of the single-gene-based trees generated, emphasising the absence of a strong phylogenetic signal in this seemingly straightforward matrix – a simple four-taxon statement consequently featuring an exceptionally high ratio of informative SNPs to scored taxa and thus in theory promising a clear, statistically unequivocal topology. But in practice, a minimum of 67 per cent of gene families yielded incorrect topologies, implying that the considerable majority of candidate-gene studies that could be performed using single-copy nuclear genes would actually be seriously misleading. When appraised statistically in a study of cryptic clades, the majority of genes are seen to provide little if any phylogenetic signal, topological instability being caused largely by a minority of genes that offer stronger but contradictory signals (e.g. Pérez-Escobar et al. 2020).

Recent papers applying next-generation sequencing to other angiosperm families suggest that statistically hard polytomies are not wholly confined to recent speciation events. To cite one example, Koenen et al. (2020) sampled numerous species across the legume family, sequencing 72 plastid genes through genome skimming and 1,103 nuclear orthologues through transcription sequencing. But even at the level of the six widely accepted subfamilies of legumes, the plastid data yielded very short branches and thus failed to provide reliable resolution. Nuclear regions offered greater sequence variation but when analysed individually, only a small percentage of genes yielded trees that featured bootstrap values exceeding 80 per cent for critical nodes directly linking subfamilies. Moreover, a large proportion of the few regions that did yield bootstrap values exceeding 80 per cent contradicted each other topologically, signalling an overall level of incongruence that led Koenen et al. (2020) to infer that subfamily relationships within legumes effectively constitute a hard polytomy. They further argued that the nuclear genome still most likely preserves substantial levels of incomplete lineage-sorting (ILS) events, despite the fact that those events are estimated to have occurred at least 50 million years ago. If so, ILS has in theory the ability to undermine the entire discipline of molecular phylogenetics.

So many studies in comparative biology that employ DNA sequencing, when unable to reach conclusions as concrete as they initially anticipated, complain of having foundered between the Scylla of hybridisation *sensu lato* and the Charybdis of ILS. Hybridisation and introgression are undeniably widespread and perhaps near-ubiquitous among higher plants. In contrast, ILS – reflecting the persistence of genetic polymorphisms through multiple speciation events – is a phenomenon so often invoked by phylogeneticists yet so rarely convincingly demonstrated in particular case studies. These phenomena need to be explored and explained rather than simply invoked when operating under duress.

In the case of the orchid studies reported in Section 3.3, it is difficult to explain such results through any phenomenon other than significant levels of ongoing gene flow among putative species whose levels of sequence divergence suggest recency of origin (assuming that these putative species can be said to have a recognisable origin at all, rather than simply participating in ongoing infraspecific gene exchange – tokogeny). First principles predict stronger positive correlation between morphology and the nuclear genotype that ultimately encodes it, compared with the maternally inherited organelles whose functions are primarily physiological. I conclude that periodic introgression will inevitably reduce the clarity of evolutionary relationships inferred from the nuclear genome, but that frequent capture of 'foreign' haploid plastids has an even more catastrophic impact on attempts at phylogeny reconstruction, rapidly erasing the contribution made to the offspring by other recent maternal parents in its genealogy.

In short, plastid sequences – for so long providing the core skeleton of plant molecular phylogenetics – have been found wanting. Among closely similar molecular lineages, the strongest remaining potential phylogenetic value of plastid sequences is to identify, through strong incongruence with nuclear sequences, recent cases of wide hybridisation – hybridisation with a maternal species that lies outside the cryptic clade under scrutiny and whose plastome origin can therefore be identified with confidence.

3.5.2 The Central Role of Extinction in Enhancing Perceived Phylogenetic Signal

Why then do higher level phylogenies based on (typically typological) comparison of plastid haplotypes yield *any* credible relationships and thus permit formal classifications that bear considerable similarity to their traditional predecessors? I can conceive of no realistic fundamental explanation other than the extinction of intermediate lineages.

Consider the hypothetical scenario summarised in Figure 3.2. Single representative individuals of ten cryptic species have been sequenced and their phylogeny reconstructed from SNPs to yield Figure 3.2a. A long branch subtends the clade of interest but much

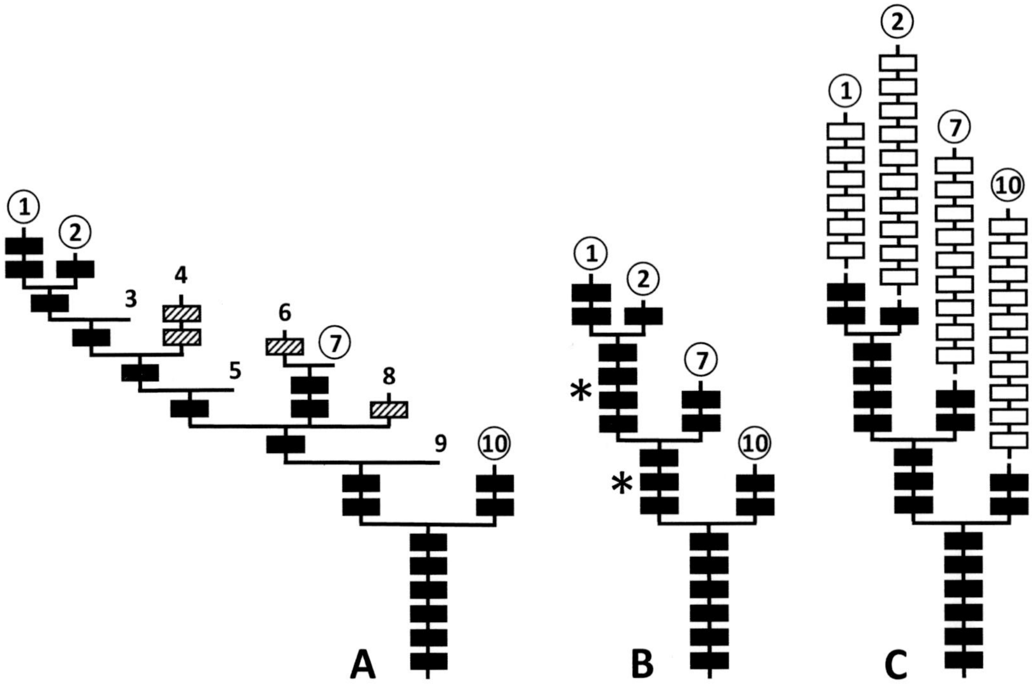

Figure 3.2 Hypothetical scenario illustrating how extinction of intermediate lineages strengthens perceptions of phylogenetic resolution. (a) Single representative individuals of ten putative species of a typical cryptic clade have been sequenced and their phylogeny reconstructed from the detected SNPs. A long, well-supported branch subtends the clade, but much shorter, poorly supported branches are evident within the clade. From a statistical viewpoint the samples form a de facto polytomy. (b) Six of the ten genotypes then become extinct, either through extinction of their 'host' species or through being eliminated within their 'host' populations through selection or drift. The extinctions increase statistical support for the asterisked branches, despite the fact that the information content supporting those relationships has not increased. The more typological the approach to a particular study, and the greater the number of putative species belonging to the clade that are omitted from the analysis, the greater is the likelihood that it will yield a statistically robust topology – but it is no more likely to be accurate. (c) Further evolution adds numerous autapomorphies to the terminal branches; this increases the amount of data in the matrix but should have no impact on the relationships inferred.

shorter branches are evident within, both internal and terminal (thus mirroring the actual situation in our orchid case studies). The few perceived synapomorphies mean that the internal branches will not attract strong statistical support (and in addition are especially likely to suffer from considerable incomplete lineage sorting). Although some topological resolution could in theory be claimed for the tree, from a statistical viewpoint the samples form a de facto polytomy.

Six of the ten genotypes (3–6, 8, 9) then become extinct, either through extinction of their 'host' species or through being driven out of their 'host' populations through selection or drift. This results in the reduced tree shown as Figure 3.2b. Most of the character-state changes from Figure 3.2a are retained, losing only any autamorphies detected in the extinct genotypes (cross-hatched in Figure 3.2a), and no new mutations occur within the four surviving 'species'. Nonetheless, aggregation of formerly dispersed synapomorphies into far fewer internal branches means that those asterisked branches attract much stronger statistical support, and the resulting topology is likely to be interpreted with much greater confidence. Whereas in fact, Figure 3.2b is no more likely to represent an accurate topology for the surviving lineages than is Figure 3.2a. Of course, the same effect of increased statistical confidence can be achieved simply by reducing the range of genotypes sampled during a particular study; the more typological the approach to a particular study, and the greater the number of putative species in the clade that are absent from the analysis, the greater is the likelihood that it will yield a statistically robust (but erroneous) topology among those samples that have been selected for inclusion.

Returning to our hypothetical example summarised in Figure 3.2, in Figure 3.2c, those four lineages surviving from Figure 3.2a have passed through sufficient additional evolutionary time since Figure 3.2b was generated for each to gain many more mutations (open boxes). Although these additional autapomorphies would not offer any further assistance in resolving the initial 'statistical polytomy', they could justifiably increase our confidence that each of the four lineages represents a *bona fide* species.

All three of the genuine case studies summarised in Section 3.3 most closely resemble the fictitious example in Figure 3.2a, because they represent recent evolutionary events, are suspected to have experienced comparatively little extinction, and have been thoroughly sampled on a monographic basis to minimise the artificial increases in terminal branches that reflect failures to analyse intervening taxa. Consider the rooted RAD-seq tree for Section *Epipactis* (Figure 3.1), which strongly suggests that, perhaps within the last million years, at least four species and three subspecies of *Epipactis* have originated independently in Eurasia from different local populations of a single ancestral species – the geographically widespread and frequently occurring *E. helleborine*. 'Backbone' branches of the tree are on average considerably shorter, and thus attract less statistical support, than the remaining branches, such that the derived species form a topology that, from a statistical viewpoint, qualifies as a polytomy. And if the analyst regards the polytomy as 'hard' (effectively intractable) rather than 'soft' (metastable, simply reflecting inadequate supporting data), conventional wisdom states that the polytomy can potentially translate into an evolutionary radiation.

3.5.3 Evolutionary Radiation or Evolutionary Experimentation?

Larridon et al. (2020) recently used two universal targeted sequencing kits to characterise large numbers of low-copy nuclear genes in the taxonomically challenging sedge genus *Cyperus*, which is species rich and includes cryptic C4 clades of species that are closely similar in both morphology and molecules and are viewed by the authors as representing 'recent rapid species radiations'. The resulting molecular trees were both well-resolved and statistically well-supported compared with polytomies generated by earlier barcoding-style studies of the clade, leading to the claim of having used the gene-rich matrices to resolve successfully a recent radiation.

However, Larridon et al. (2020) offered no definition of either a species or an evolutionary radiation. My preferred definition is 'a large surplus in the rate of natality over the rate of mortality for species and/or character states within a specified clade through a specified time interval' (Bateman 1999). The criterion of subtracting deaths from births of species is required if the extensive diversification required for a radiation is to be distinguished with confidence from mere species turnover that entails no overall increase in species number through time. In any case, recent mathematical reappraisal (Louca and Pennell 2020) has cast considerable doubt on the validity of the currently popular pastime of using molecularly dated nodes in trees to estimate speciation and extinction rates. And of course, the second essential criterion for estimating speciation rate is that the species that constitute the 'cryptic' clades must also be genuine. High turnover of lineages that are only subtly distinct, both morphologically and molecularly, is exactly what should be expected in a case where evolution is generating multiple potential species, but such evolutionary experimentation undoubtedly fails far more often than it succeeds in generating *bona fide* species; most such 'prospecies' soon either become extinct or are reintegrated into their parental lineage(s) through excessive gene-flow (e.g. Bateman 2012).

This leads to a crucial conundrum of whether statistical polytomies indicate the presence of multiple 'cryptic' species or a single species. This ambiguity places any ensuing evolutionary interpretation in a classic 'Catch 22' scenario, because the use of terms such as hybridisation to describe gene-flow between such lineages is legitimate only if they represent genuinely distinct species. Similarly, suggestions of morphological convergence between lineages – offered as one possible explanation of at least some cryptic species by Fišer et al. (2018) – imply species-level divergence, as well as requiring an assumption of sister-species relationship that cannot be supported in a polytomy.

To summarise, unless the species involved in a putative radiation have already been convincingly circumscribed, and their levels of ongoing gene flow have already been shown to fall below a specified minimum level of acceptability, it is not logical to refer to multiple plants found to be closely similar (i.e. near-polytomous) in DNA-based trees as species, and therefore not logical to regard that clade as constituting an evolutionary radiation.

3.5.4 Countering Concatenation

I have never made any secret of my disdain for the trend of concatenation long favoured in molecular phylogenetics (e.g. Bateman 1999, 2020a). Combining contrasting categories of data is an approach driven by the painfully simplistic logic that the larger is the assembled data set the more likely one is to obtain an accurate result (or, more precisely, is conducted in the hope of obtaining greater resolution and greater statistical support). *In extremis*, this so-called total evidence approach combines organellar genes, nuclear genes, and morphological observations to generate what I have described as 'brown phylogenies' (the inevitable outcome when several colours are mixed in an *ad hoc* manner). Amalgamating contrasting categories of data that evolve under fundamentally different constraints prioritises superficial statistical rigour while simultaneously masking various forms of incongruence, thereby squandering the opportunity to understand the clade under scrutiny at a deeper, process-based level by seeking explanations for incongruence among genes, genomes, or between genetic and other categories of data.

If (as is usually the case) there is actually substantial incongruence between morphology and molecules, or nuclear and plastid genes, or molecular dates and stratigraphic ranges, those incongruences require recognition and subsequent process-based explanation. Indeed, without such comparison, it is difficult to argue for the presence of cryptic species, given that the term and the underlying concept are firmly rooted in perceived incongruence in general and branch-length incongruence in particular. Many of the more insightful observations on plant evolution made in recent years owe their existence to incongruence, and the recent development of data analysis techniques that explore genomic incongruence gene by gene are, in my opinion, a genuine advance that could in time relegate concatenation to the backwaters of molecular data-analysis.

3.5.5 Revisiting the 'Genetic Divergence Lag'

Viewed from a phylogenetic perspective, I see the crux of cryptic clades as being the mode and recency of ancestry. Multiple origins of lineages from within the same ancestral species, as in Section *Epipactis*, multiple origins of lineages through allopolyploidy, as in Subgenus *Dactylorhiza*, and phenotypic convergence encouraging ongoing gene-flow, as in the *Ophrys sphegodes* aggregate, should all from first principles be expected to generate clades that will appear cryptic for a significant period of evolutionary time.

I have repeatedly argued in print that the raw materials of speciation are first a (generally small scale) heritable genetic or epigenetic innovation that causes a phenotypic change. The scale of that change differs from case to case but must at least be sufficient to alter the way that the organism interacts with its biotic and/or abiotic environment. In contrast, the accumulation of the much larger-scale genetic changes that are needed to allow us to detect differences between an ancestral and a derived lineage requires a much longer time period. I termed this period separating genetic innovation from eventual genetic isolation the 'genetic divergence lag' (fig. 6 of Bateman 2009). The reliance of higher plants on error-prone abiotic or biotic intermediaries to effect sexual reproduction commonly greatly prolongs this lag phase compared with typical macro-evolutionary patterns documented in higher animals. The lag phase for any particular lineage can be shortened through the evolution of strong post-zygotic sterility barriers, or through large-scale physical separation over geological timescale, or through the extinction of the closely related lineages that are competent to exchange genes with the lineage under scrutiny. But even if such processes eventually confer full reproductive isolation, that event cannot be equated with complete genetic isolation; there are plenty of mechanisms other than sex that are capable of splicing 'foreign' genes into a host plant's genome (e.g. Bergthorsson et al. 2004; Bock 2010). A further problem posed by the lag phase is its negative impact on any attempt to use molecular clocks to determine the age of the putative species (e.g. Fišer et al. 2018) – estimates that will already be statistically invalidated by the comparatively large size of the error bars inevitably incurred by recently originated lineages.

Ultimately, for most lineages that by one means or another eventually achieve a considerable degree of evolutionary independence, the key question becomes determining precisely what level of incoming gene flow from other lineages ('semi-permeability' *sensu* Hausdorf 2011) we consider sufficiently low for us to regard that particular lineage as sufficiently evolutionarily independent to be regarded as a novel species. This crucial issue is surprisingly rarely discussed, perhaps because it is so difficult to provide a convincing answer.

3.6 Integrated Monography Allows Genuine Species Circumscription

3.6.1 What Is Integrated Monography?

While drafting a 1998 policy document I coined the term 'integrated monography' to describe approaches to comparative biology that explore a substantial clade by extending traditional herbarium-based morphological taxonomy to encompass data sets gathered in the microscopy laboratory (e.g. anatomy, palynology) and the molecular laboratory (e.g. nucleic acid sequences, biochemistry). There is no necessity for the resulting work to be published as a single cohesive 'classic' monograph, provided that by the end of the project an objective, balanced taxonomic overview of the clade has been achieved. But why emphasise a monographic approach, and why should it be 'integrated'?

My life-long advocacy of monography reflects my belief, born of harsh experience, that piecemeal taxonomy does more harm than good. Certainly, a fondness for the *ad hoc* approach to taxonomy within the orchidological community has generated an exceptional number of unproven species and an associated exceptional level of synonymy that, through ambiguity, seriously impedes the work of more rigorous researchers. Short papers entitled 'A new species of taxon X from geographical location Y' often prove to represent a strange specimen encountered in a preserved collection (or, less often, a strange population encountered in the field) that looked a little different from material previously seen by the taxonomist or descriptions encountered by her/him in the literature. Or it may be prompted by a single plant or population encountered in the field that appears somewhat different from a perceived, previously assumed 'norm'. Such papers almost always lack mention of an explicit species concept, present a description that is not only wholly morphological but also based on a very limited amount of (often preserved) material, and rarely include any kind of comprehensive comparison with many other broadly similar species that is a pre-requisite to convince sceptical readers such as myself that a new species had indeed been found, rather than merely hypothesised. As noted by Williams et al. (2014), only unusually brave (or foolhardy) monographers are willing to follow their admirable example by tackling genera supposed to contain several hundred species. Yet it is these supposedly species-rich genera that often contain large cryptic clades in greatest need of in-depth study.

Equally unscientific is the emphasis placed by biologists (and not just those in the taxonomic community) on the 'discovery' of new species at the expense of their 'undiscovery'. For example, Royal Botanic Gardens Kew issues an annual summary of the plant 'species' discovered during each calendar year, but Kew stays resolutely silent on the topic of how many putative species it has reduced to synonymy during that same year. When discussing species discovery, Bebber et al. (2014: 701) argued that 'changes of taxonomy where species, subspecies and varieties are reassigned to a different rank ... are an important aspect of taxonomic activity but represent opinion and changes of status rather than new discoveries'. Yet taxon description and synonymisation are the yin and yang of biodiversity assessment; removing a non-existent species from further consideration as a species is a contribution to biodiversity studies every bit as important as 'discovering' a genuine new species, as in theory it allows future research and resources to be focused on challenges that are real rather than imaginary. Unfortunately, in practice, it is far easier to coin a new species name than it is to drive a stake through the heart of a demonstrable non-species. It is important to note that both activities are equally open to accusations of constituting mere authoritarian 'opinion' – I am at a loss to understand why Bebber et al. (2014) viewed species 'discovery' as somehow more objective than a scientifically justified demotion in rank. The only real difference is that species 'discovery' is seen as a more charismatic pursuit.

As for my determination to embed laboratory-generated data sets within monographic treatments, this reflects my belief that, given twenty-first-century technology and accepting the importance of relating phenotype to genotype, such integrated approaches should be obligatory for any future taxonomic publication. Congruence between contrasting categories of data is the only meaningful test of earlier attempts at classification. Now that we have reached the 2020s, none of the relevant categories of data remains prohibitively expensive to gather, and any local research deficiencies can in theory be remedied through collaboration.

3.6.2 What Is Foundation Monography?

A broadly similar concept developed by Robert Scotland and colleagues was termed the 'foundation monograph' by Scotland and Wood (2012). Aiming to make faster progress in monographic revisions of species-rich plant genera, they argued for convening small teams of researchers to pursue well-organised herbarium-based taxonomy to be supplemented with molecular data derived from the herbarium specimens as a 'single character' test of what they regarded as prior species delimitation through a classical herbarium approach. Initially, the molecular data were envisaged to be DNA barcodes (Scotland and Wood 2012); the first foundation monograph, of *Ipomoea*, used nrITS as a test of herbarium-based species (Williams et al. 2014; Wood et al. 2017). In the subsequent, more ambitious foundation monograph of *Convolvulus*, the NGS approach Hyb-seq was used to supplement barcodes from nrITS and the plastid regions *matK* and *trnH* (Munoz-Rodriguez et al. 2019). But despite the considerable progress made through these projects, I believe that the foundation monograph – although undeniably falling within the broad remit of integrated monography – suffers from three serious drawbacks that certainly render it unsuitable for addressing especially challenging cryptic clades.

Traditional herbarium approaches to formally establishing (or, less frequently, revising or rejecting) species rarely begin serious study until after a living – and thus potentially infinitely testable – plant has been reduced to a desiccated fossil. Even in an age when technological advances increasingly permit the extraction of at least some genomic information from at least some dried plant specimens (e.g. Munoz-Rodriguez et al. 2019), it has always struck me as extraordinarily counter-productive to voluntarily kill your intended study organism. Even at the basic level of morphology, a significant proportion of the characters available for study in a living plant are either lost or seriously degraded. And even when relevant features are retained on a specimen, they are no longer directly comparable with homologous features of a living plant; for example, shrinkage of organs such as petals not only reduces their overall size but does so anisotropically, thereby also distorting their shape (e.g. Bateman et al. 2013; Parnell et al. 2013).

In other words, on the rare occasions when herbarium specimens are used to form the basis of a morphological matrix that also includes thoroughly described living plants, the matrix is inevitably riddled with gaps and distorted measurements. A further weakness of a typical herbarium-based study is that comparative matrices are not produced, or if they are, they are not subjected to any form of objective algorithmic synthesis summarising the morphology of the entire plant. In essence, under such circumstances, the perceived species boundary lies somewhere between the human eye and the human brain. Existing classifications are therefore likely to be perpetuated unless the contradictory evidence is stark, and morphological characters previously regarded as offering comparatively strong taxonomic discrimination ('conservative' *sensu* Wood et al. 2017) within the chosen taxonomic group receive preferential attention. In short, relying on herbarium-based morphological discrimination is a worryingly subjective approach to species delimitation and offers no prospect of achieving results that are anywhere close to optimal.

The third, and perhaps most serious, constraint on herbarium-based taxonomy is typology. The difficulties of locating and collating relevant herbarium specimens were well described in the laudable foundation monographs of Williams et al. (2014) and Munoz-Rodriguez et al. (2019). But despite their heroic efforts to overcome those difficulties, the mean number of specimens examined for morphology per resulting putative species was 1.9 in Williams et al. (2014). The ratio in Munoz-Rodriguez et al. (2019) was 1.8 for NGS data, 7.4 for at least one DNA barcode, and presumably exceeded that figure for actual, or digitised images of, herbarium specimens. In other words, practical constraints of minimal numbers of preserved specimens render both the morphological and molecular aspects of foundation monography perilously close to typological. Under such circumstances, is it realistic to write in terms of having performed species delimitation ('to establish limits or boundaries') or species circumscription ('to draw a line around, encircle') when in truth only one or two data points are available per putative species? Isn't typology actually by definition restricted to simply offering a *hypothesis* of the existence of a species rather than any kind of test of that hypothesis?

The cornerstone of traditional taxonomy, a type specimen, is, in effect, a single point in both space and time. Many additional data points are needed, ideally reflecting several different kinds of data, before that putative species can be viewed as having been adequately circumscribed. Consider the imaginary case shown in Figure 3.3 (from fig. 5 of Bateman 2018). Here, six type specimens (a–f) have been deposited in various herbaria and awarded different names but, by definition, they provide us with no information on the nature of the variation that is found within these putative species. At that stage, we know little about any of the properties that might be useful in circumscribing genuine species – morphology, genetics, or ecology. When (or, more accurately, if) that variation has been adequately documented in populations across the range of the putative species (Figure 3.3b), and assuming that we use discontinuities in those data as the primary basis of our species circumscription, we can then see that in fact only three species are present rather than the originally hypothesised six; holotypes c–f must be unified under a single formal name because they actually constitute a continuum, lacking the discontinuities that should by definition separate genuine species. To emphasise the significance of conducting analyses that extend beyond typology, I have placed beneath the cut-out shapes shown in Figure 3.3a and b a well-known painting to create Figure 3.3c and d. Viewed only through the formalised typology of the type specimens (Figure 3.3c), the portrait is utterly unrecognisable. In contrast, when viewed through the much larger portals provided by well-founded knowledge of variation across the group (Figure 3.3d), the portrait is readily identified.

The high risk that a set of taxonomic decisions will artificially divide into supposed species what is actually a single morphological and molecular continuum lies at the crux of the problems posed by species viewed as cryptic. Diligent searches for discontinuities in one or more of multiple data categories are the only valid approach to achieving genuine circumscription of species.

3.7 Beyond Trees: Species Circumscription through Multivariate Discontinuity

3.7.1 What Is Demographic Monography?

The solution to circumscribing species in cryptic clades (and in many ways the antithesis to herbarium taxonomy) is what I have termed 'demographic monography' (Bateman 2001, 2011). This more strongly field-based approach to integrated monography replaces herbarium-based morphological taxonomy with detailed morphometric studies performed in situ in the study populations, typically using about 50 scored macromorphological and micromorphological characters to summarise the morphology of all above-ground organs of each plant. Substantial sampling of leaves or flowers from a representative number of

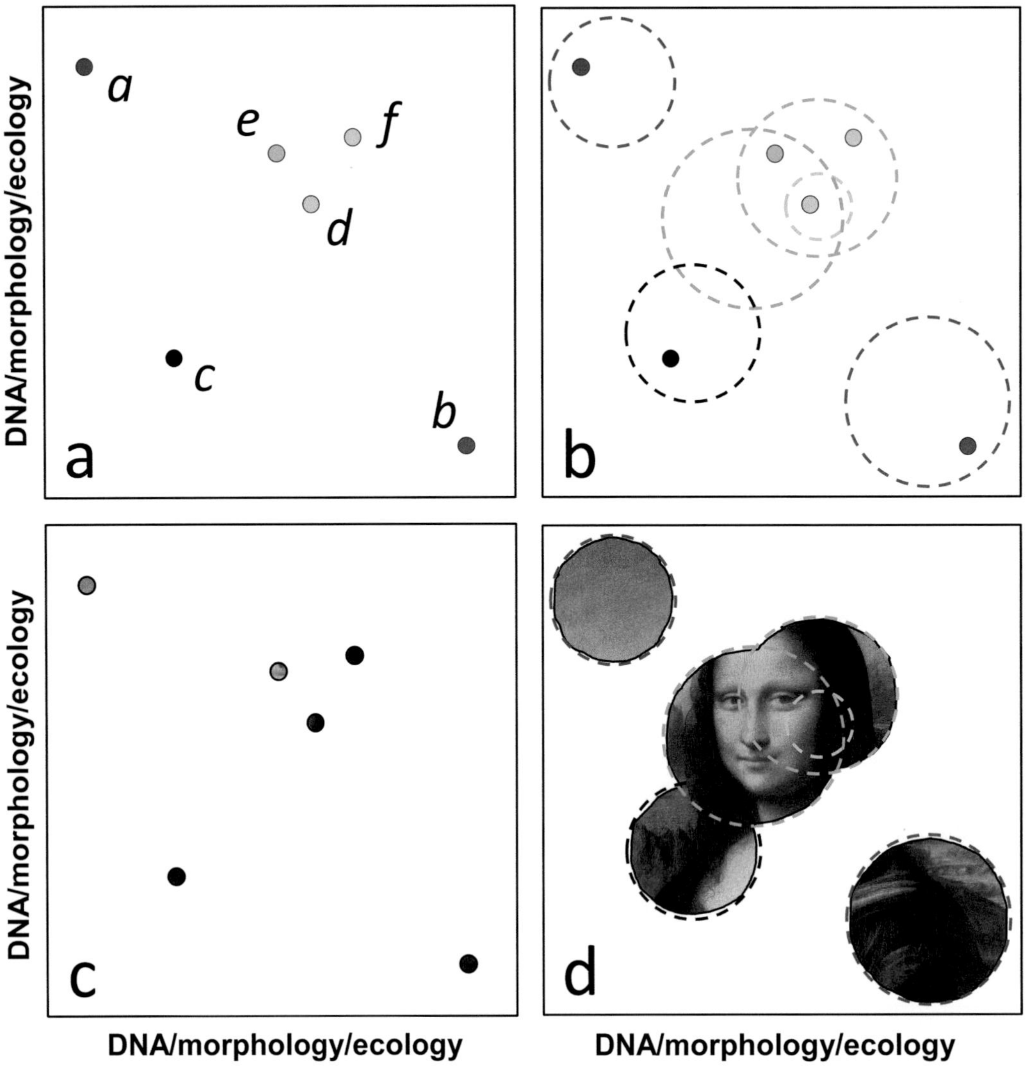

Figure 3.3 Hypothetical study aiming to circumscribe species. (a) Six holotypes (a–f) establish the typological basis for six formal nomenclatural epithets. (b) Subsequent detailed studies of variation within those putative species allow species circumscription on the basis of discontinuities separating spheres of variation; the absence of separating discontinuities requires types c–f to be unified into a single species. (c) A famous painting cannot be recognised if viewed only through the lens of the single holotype specimens. (d) Only once variation within those species has been adequately documented can the underlying pattern of the painting be recognised; the initial hypothesis of a species established by the holotype has thus been succeeded by circumscription of a convincing species.

plants in each of numerous study populations across the geographic distribution of the target clade permits later molecular analysis (possibly also microscopic examination) of the same individual plants that have already yielded morphometric data sets. Ideally, ecological data are also gathered on the study populations, covering symbionts such as pollinators and mycorrhizae, soil conditions, habitat and vegetation types, and local climate. Given that the sampled plants remain viable, they can if desired also be used in crossing experiments designed to assess genetic compatibility.

The net result of a carefully planned demographic approach is several contrasting data sets available for (a) analysis at multiple demographic levels (individual, population, infra-specific taxa, species) and (b) integration to assess levels of congruence between disparate data sets. Such an approach allows patterns to be discerned and so greatly transcends the interpretative straightjacket that is typology. Most importantly, it allows reciprocal illumination between both contrasting demographic levels and contrasting categories of data, both of which are essential for genuine species circumscription.

Traditional taxonomy is also often cited as offering reciprocal illumination, through field and/or herbarium identification of further plants purporting to belong to the same species as the holotype. But this is at best a far weaker form of reciprocal illumination, because it operates within a single category of data (classical morphology) and usually ignores demographic levels. Moreover, when faced with a plant (or specimen of a plant) to identify, it is rare that a non-taxonomist subsequently critiques in print an imperfect taxonomic description; plants that deviate modestly from the prior formal description are likely to be shoehorned into the closest existing morphological fit, utilising those few characters that were specified as most diagnostic in the original description (or in floristic accounts derived from that original description). In contrast, if the specimens under scrutiny are found by a taxonomist rather than say a field ecologist, the converse is more likely; the desire to find 'new' species is a strong motivation, and the hypothesis of a new species is likely to be published without the lengthy delay required by a rigorous monographic approach that first examines all previously known specimens and descriptions of broadly similar species before reaching taxonomic conclusions.

In short, obtaining multiple categories of data from the same range of carefully chosen plants from a known demographic context is a far stronger form of reciprocal illumination than utilising a single category of data accrued for an *ad hoc* assemblage of specimens. This statement applies not only to traditional morphological studies but also to alternative approaches to (eukaryotic) species circumscription that rely heavily or wholly on molecular data (e.g. Sites and Marshall 2003). Tracing categories of comprehensive quantitative data describing phenotype, genotype, and ecology through the demographic approach means that all three of the basic principles of taxonomy outlined in Section 3.1 – similarity, reproductive isolation, and (albeit more equivocally) monophyly – can each be deployed effectively and on multiple data sets, before some form of conceptually explicit synthesis is attempted to optimise species boundaries. Also, in addition to circumscribing *bona fide* species with confidence, the underlying speciation mechanism can often be inferred – an impossible goal given only traditional data.

Admittedly, there is a price to be paid for such a rigorous approach. It requires the scientists to be transported to the plants rather than vice versa, increasing expenditure (and carbon footprints). Unless such a study is pursued with vigour by multiple researchers, it will require a longer period of time than that spanned by a typical research grant, particularly in tropical latitudes where biodiversity is greater and poorer transport infrastructures often restrict access to sufficient numbers of representative populations. It will often require international travel, a luxury now seriously challenged by the SARS CoV-2 virus that is causing a global pandemic at the time of writing. And unless overall resourcing to biodiversity studies is increased considerably, the overall pace of species description will be reduced by adopting a demographic approach, contradicting an alternative school of thought (one advocated by proponents of the foundation monograph, among many other interest groups) that aims to accelerate monograph production by explicitly accepting a strong typology constraint and also precluding categories of data other than herbarium-based morphology and barcode-level DNA sequences (Williams et al. 2014; Wood et al. 2015; Muñoz et al. 2019).

On the other hand, the demographic approach is far more effective in detecting taxa that fail the species circumscription test and thus can be erased into synonymy; it is the only technique to effectively balance the yin and yang of erecting novel species while simultaneously eliminating older pseudo-species. And unless further unforeseen technological breakthroughs are achieved that identify new sources of biosystematic data within plants, demographic monographs will stand the test of time, rather than merely fulfilling the stated aim of foundation monographs to provide 'a solid foundation for future in-depth studies within the genus' (Williams et al. 2014: 1301). My fear is that, given the limited number of active taxonomists and the accelerating degradation of the global environment, such revisionary studies may never be initiated in practice.

3.7.2 Trees, Networks, or Ordinations?

So far, this discussion has focused almost entirely on a combination of (a) the traditional approach to morphological taxonomy that is pursued without statistical synthesis and (b) DNA-based analyses of specimens that are summarised statistically as dichotomous (and usually rooted) trees. But it is questionable whether the tree motif, and the algorithms that underlie tree generation, are the most appropriate way to explore and depict lineages that are subject to considerable ongoing gene flow and thus suffer reticulation, irrespective of the nature of the underlying data (e.g. DeSalle and Vogler 1994). Even the increasingly popular coalescence methods of constructing genetic trees by amalgamating alleles progressively backwards through evolutionary time ultimately incur the same over-simplification of imposing dichotomy on processes that are not fundamentally dichotomous.

With regard to demographic monography, estimates of overall similarity for character-rich morphometric matrices can be used to produce classical phenetic dendrograms that, like morphological and molecular phylogenies, yield a uniformly dichotomous topology with differential branch lengths. However, ordinations lie at the core of the approach, the hyperspatial arrangement of analysed plants being projected onto planes of greatest variation as individual data-points. Interestingly, clustering of plants and populations in morphometric ordinations tends to yield results that are more congruent with corresponding genetic matrices than do obligatorily dichotomous dendrograms. One problem posed by dendrograms is the fact that a tree only minutely statistically inferior to the optimal tree can differ radically from that optimal tree; ordinations offer greater stability. Yet the main power of an ordination is the ability to directly visualise discontinuities among many individuals, relate each individual to its source population, and thus detect the all-important discontinuities separating groups of populations that allow those groups to be treated with

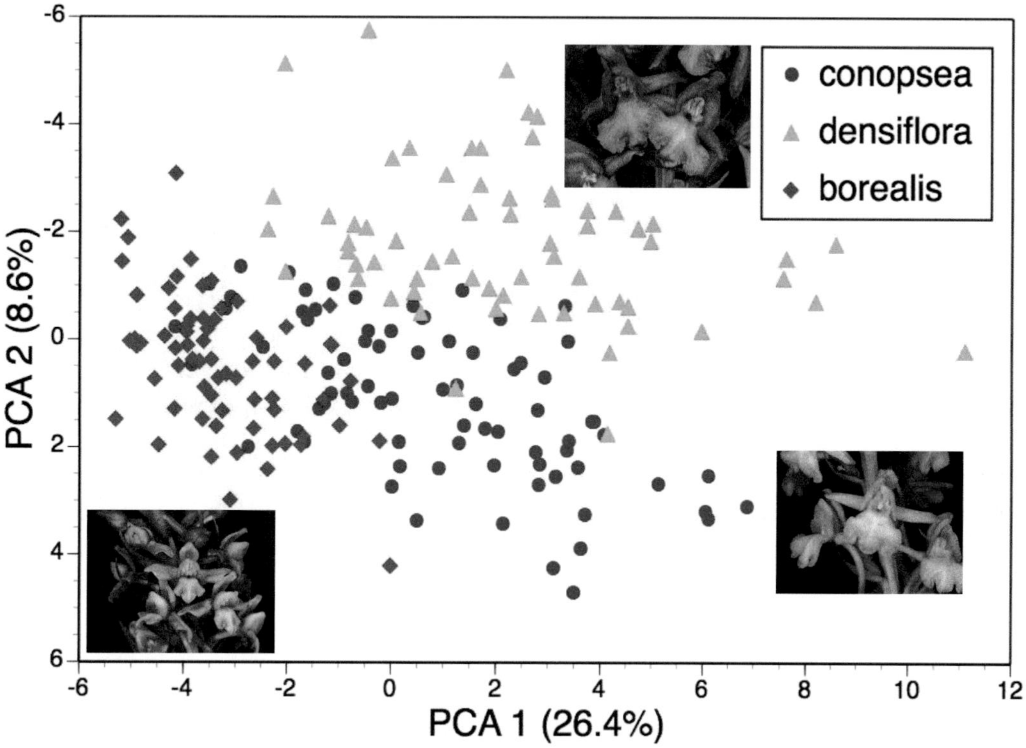

Figure 3.4 Morphometric ordination of a typical cryptic clade. The three named taxa of the Eurasian orchid genus *Gymnadenia* (inset) are genetically similar but nonetheless are reliably distinguishable using nrITS and next-generation genotypic data, and have contrasting ecological preferences. But here, the first two principal coordinates present the three species as subdivisions of a morphological continuum, despite representing morphometric data for 41 characters measured for 240 plants of eight populations per putative species. R. M. Bateman, P. J. Rudall, and I. Denholm (unpublished).

confidence as separate species. This approach literally permits species circumscription – a circle (or preferably a polygonal convex hull or 95 per cent confidence ellipse) can easily be drawn around that portion of multivariate space that is occupied by each species.

An example of such an ordination is shown in Figure 3.4. Here, three taxa within the orchid genus *Gymnadenia*, ecotypes that were regarded as subspecies throughout the twentieth century, are ordinated for 41 metric, multistate and bistate characters (Bateman, Rudall, and Denholm, unpublished). The slight overlap of circumscriptions of the three ecotypes would be consistent with division of a morphological continuum and thus subspecies status. However, simple ITS sequencing is sufficient to reliably distinguish the three taxa (Bateman et al. 2018), an observation confirmed by later RAD-seq analysis (Brandrud et al. 2019). The three taxa have distinct habitat preferences and rarely co-occur, constituting an archetypal example of a cryptic clade. Admittedly, interpretation of ordinations and networks requires careful thought because the magnitude of the discontinuity separating putative species is dependent of the kind(s) and volume of the available data, and the rigour with which those data have been analysed (Bateman 2001, 2018).

The core data describing the morphology of each plant can usefully be supplemented with anatomical data and with analytical approaches that better describe complex three-dimensional structures, such as three-dimensional scanning combined with landmark analysis of data obtained through micro-CT scanning (e.g. Sedeek et al. 2014).

Congruence between phenotype and genotype is best assessed by deriving both data sets from the same spectrum of individuals. The reliability and robustness of the discontinuities detected within a particular matrix, whether phenotypic or genotypic, are obviously influenced by the number of informative characters scored, together with the numbers of populations per putative species and of individuals per population that have been analysed. It is also important to prioritise ordination algorithms that do not distort multivariate hyperspace in favour of species status by making prior assumptions regarding group membership of plants and populations; the data should be left free to tell their story as objectively as possible.

I also note that ordination approaches are equally effective for summarising corresponding ecological data such as guilds of pollinators or mycorrhizal symbionts, associated plant species, or abiotic environmental measures such as soil composition.

3.7.3 The Problems of Scale and Comparability

When species subjected to contrasting genetic analyses are compared, it is essential to always bear in mind the scale of the observed genetic differences. I differ genetically from my mother, but only to a small degree; comparison with a bonabo is needed before it becomes clear genetically that my mother and I are conspecific. Exactly the same codicil applies to morphology and other aspects of phenotype. It is the (I hope) much greater phenotypic distance separating my mother and myself from a bonabo that identifies my mother and I as conspecific. Discontinuities – whether genotypic or phenotypic – can only be realistically assessed in a comparative framework employing strategies for sampling and data analysis that permit interpretation on a level playing field. Unfortunately, there are severe limits to how far that level playing field can be expanded.

Early proponents of phenetic trees predicted a morphometric utopia where all taxa of equivalent rank recognised in the analysed group would have approximately similar degrees of differentiation, an outcome achieved simply by cutting through the tree at a convenient percentage of similarity (e.g. Sneath and Sokal 1973). Similar approaches have since been suggested for dividing molecular trees into classifications, particularly for taxonomic groups exhibiting relatively simple morphologies (e.g. fungi, as reviewed by Matute and Sepulveda 2019). I find such approaches worryingly simplistic, and certainly not in accord with recommendations made in this chapter to explore congruence among contrasting data categories; nonetheless, I understand the appeal of the underlying concept – to establish taxa simultaneously within a single unified framework.

The problem is that both trees and ordinations present relative rather than absolute distances between analysed organisms. *Ophrys* microspecies that differ by perhaps one SNP in nrITS can differ by numerous SNPs in a RAD-seq matrix, a realistic scale of comparison only being achieved when plants from different populations of the same microspecies differ in the same tree by tens of SNPs and different from the same population differ by few if any

SNPs (Bateman et al. 2021b). Sister to *Ophrys* is the genus-level pairing of *Serapias* and *Anacamptis*. Similar levels of molecular disparity among putative species are evident in *Serapias*, but nrITS disparities within *Anacamptis* are an order of magnitude greater (Bateman et al. 2003).

Similar issues affect morphometric ordinations. For example, the ordination of three closely related species of *Gymnadenia* (Figure 3.4) suggests considerable cohesion within *G. conopsea* s.s., but if plants of that species alone are ordinated, the simplified patterns of variation within the reduced matrix allow a more nuanced assessment of relationships among the eight populations measured. The distances separating populations occupying different types of calcareous soils might be taken as evidence of potential speciation in the absence of knowledge of the broader analysis.

On current evidence it is difficult to envision how species can possibly be made in any way comparable across the vast panoply of (eukaryotic) life. But I've yet to be convinced that this pragmatic conclusion excuses us from at least giving the issue some serious thought, rather than continuing the current trend of largely ignoring the entire problem. 'Keep calm and carry on' is suboptimal advice if you've already stepped over the cliff edge.

3.7.4 Co-opting Evolutionary Biologists

I have used this review to level many criticisms at traditional descriptive approaches to plant taxonomy, preferring instead to champion the case for a more broadly based, better sampled, better integrated approach to species-level taxonomy that aims to optimally circumscribe species rather than merely establish them through formal naming and holotype-based description. The aim is in effect to pursue multidisciplinary studies that allow us to not only classify organisms but also to simultaneously infer speciation mechanisms and ecological tolerances. The result of this additional labour should be a genuinely predictive taxonomy – one that can help to foresee the future behaviour of the studied species in their attempts to respond to increasingly rapid and substantial changes in their environment (Bateman 2011).

If I am correct in my conclusion that most cryptic clades involve recent or even current evolutionary experimentation, such groups are likely to be especially attractive to evolutionary biologists studying speciation by integrating data describing genotype with those describing phenotype and behaviour.

Unfortunately, I have found most evolutionary biologists to be remarkably reluctant to draw taxonomic conclusions from their research. Attempts by some authors to sidestep the issue by arguing that they are studying speciation rather than species fall at the first logical hurdle as, by definition, only a *bona fide* species can have completed a successful speciation process. When challenged, the most frequent responses from students of evolution are 'But that's not my job – I'll leave that to the taxonomists' or 'Taxonomy's not sufficiently important to warrant my precious time'. Further questioning revealed several more explicit concerns. Negatives cited by evolutionary biologists include the low impact factors attracted by most outlets willing to publish monographic and other taxonomic treatments, the low citation rate (though long half-life) typical of taxonomic works, unwillingness to assimilate the arcane complexities of the

International Code of Nomenclature, and even what is perceived as the unusually contentious nature of a discipline that persistently generates passionately defended competing classifications.

To this list I would add one phenomenon that is probably most prevalent in especially charismatic, intensively researched taxonomic groups such as orchids: taxonomic 'gazumping'. Ideally, a biosystematics team would publish a series of papers that collectively present multiple categories of data, and would aim to publish their taxonomic conclusions only after they had completed data collection for the entire project, thereby maximising their chances of producing a robust natural classification that would not require future amendment. Unfortunately, there is a high risk of a traditional taxonomist pre-empting that team of biosystematists by appropriating their incomplete observations to produce their own formal classification of the clade in question or, worse still, issuing yet another 'New species of X from Y' paper. Under such conditions, the new taxonomy is often not only premature as a result of being data-deficient but also produced without adequate understanding of the non-traditional categories of data that have previously been published by the biosystematics team.

3.7.5 Overview

What I have described in this review is an internationally agreed system (or, more accurately, set of multiple overlapping systems) of addressing systematics issues that is disappointingly suboptimal in the way that it classifies species in general and troublesome cryptic clades in particular. Taxonomic organisations rightly pride themselves in operating as part of a genuinely global network. At the level of descriptive morphological taxonomy, that tradition continues in the form of, for example, loans of specimens, digitised images, or DNAs (despite the many bureaucratic constraints that have accumulated during the last quarter-century). But the links often remain poor between traditional taxonomists and researchers in other disciplines who are also competent to generate biosystematically useful data. Even today, monographic projects are rarely so carefully organised that the various categories of relevant data can be integrated seamlessly in the later stages of the project. Foundation monographs remain rare, and demographic monographs almost non-existent.

It is difficult to escape the conclusion that the taxonomic enterprise remains disappointingly *ad hoc*, such that opportunities for greater synergy and increased reciprocal illumination are often squandered by all potentially interested parties. Taxonomy's claim to be a fully-fledged science is not helped by the fact that taxonomists are required to follow implicitly the Codes of Nomenclature that legislate naming but, bizarrely, they are under no obligation to state the taxonomic concepts that they are employing or to make explicit the nature and volume of information that underlies the decisions made. Such statements should be made obligatory.

There is nothing inherently special about cryptic clades – they are an arbitrary and 'metastable' concept that constitutes one end of a continuous spectrum of perceived degrees of genetic differentiation of putative species from their closest relatives. The more genetically distinct the representative(s) of that putative species, the greater is the probability that the species will be natural rather than artificial (i.e. will be a cohesive biological entity), but at this severely limited level of sampling, any statement of species status nonetheless represents educated guesswork.

3.8 Synthesis

(1) The species is the basic unit of biological currency. In theory, the acceptance of the principles of organic evolution during the twentieth century transformed the species from an artificial pigeonhole, constructed by taxonomists to accommodate organisms considered acceptably similar in appearance, into a self-defining aggregate of populations, together possessing a shared and independent evolutionary trajectory. But in practice, pigeonholes predominate even today.

(2) In order to demonstrate evolutionary cohesion, it is necessary to compare and contrast patterns observed in morphology with those evident in the genomes that ultimately underpin that morphology (ideally, field data would also be gathered on habitat preferences and biotic interactions of each putative evolutionary species to provide a third category for comparison). An aggregate of populations separated from all other populations by a reliable discontinuity in both phenotype and genotype would qualify as an unequivocal species.

(3) Species delimitation becomes more challenging if either the morphological or molecular data categories fail to reveal detectable discontinuities, instead presenting as a single continuum or showing varying degrees of overlap between putatively separate species. Such cases, especially where genetics appears discriminatory but morphology does not, have increasingly been described in the literature as 'cryptic species'. However, given that the underlying question is whether the group consists of closely related, closely similar, multiple species, the taxonomically neutral term 'cryptic clade' appears more appropriate.

(4) The discontinuities that should separate genuine species can only be detected with confidence by substantial planned sampling of individuals within populations, and populations within putative species, thereby allowing reciprocal illumination between the three demographic levels. Such in-depth demographic studies, characterising both phenotype and genotype, constitute genuine species circumscription.

(5) Typical taxonomic practice deviates massively from this theoretical optimum. Most plant species are formally established on the basis of herbarium specimens, ranging from a holotype only through to typically the few specimens that can be accessed by the taxonomist within a realistic timeframe. This is an essentially typological approach that, at best, samples variation in a very limited and *ad hoc* manner, simply assuming discontinuities from all other species rather than actually testing for their presence or absence. Moreover, the all-important character differences among specimens that allow their assignment to particular species are usually determined cerebrally rather than algorithmically.

(6) In my opinion, it is not valid to describe the traditional taxonomic process as species circumscription (= delimitation). A species *hypothesis* has been made available for future scrutiny, but that hypothesis has not been meaningfully tested. Nor is it tested by subsequent attribution of further plants or specimens to that new binomial, unless quantitative

comparative analyses are conducted or other categories of data are acquired. Acceptance of this controversial conclusion would mean that most of the species thus far awarded formal Linnean binomials actually remain untested hypotheses.

(7) Molecular data must be gathered before their congruence with morphological taxonomy can be assessed. Until this is done, no species or clades can legitimately be accused of being cryptic. The majority of molecular studies published to date also resemble traditional taxonomy in being typological or near-typological; attempts to assess infraspecific variation typically range from non-existent to limited and *ad hoc*. Recent innovations in next-generation sequencing have greatly reduced sequencing costs, perhaps leaving the availability of suitable plant materials as the main limiting factor.

(8) Molecular analyses are most commonly presented as rooted trees constructed via parsimony, likelihood, or (increasingly) Bayesian or coalescence algorithms. Molecular trees are divided into supraspecific taxa by seeking groups that are monophyletic, and preferably subtended by a branch that is comparatively long and statistically robust. But in wholly typological studies, species status is typically assumed rather than tested. Doubt may arise if two or more supposedly distinct species yield DNA sequences that, within the context of the tree, are very similar and yield topologies that, from a statistical perspective, are polytomies.

(9) Statistical polytomies with comparatively short terminal branches are often interpreted as recent species-level radiations characterised by comparatively high diversification rates. Such clades are those most likely to become labelled as 'cryptic'. However, a short-toothed comb is also the pattern that one routinely detects when analysing multiple populations of the same species or multiple individuals of the same population. Genuine species circumscription is actually a pre-requisite for obtaining credible diversification rates and identifying putative radiations. And rooted trees evidently constitute a far from optimal means of circumscribing species.

(10) Both population-level genetic/genomic data and morphometric data are best analysed through multivariate ordinations and/or unrooted networks. These approaches identify planes of maximum variation most likely to be of taxonomic value, and allow genuine circumscription of individuals and populations into credible clusters representing species. Ideally, morphometric data and molecular data are derived from the same individuals. Adjacency or even modest overlap of species clusters is permissible for morphometric data, whereas on the parallel genetic plots the equivalent clusters should be separated by discontinuities; such an outcome is typical of cryptic clades. If clusters of individuals initially thought to be species show considerable overlap, by definition they do not warrant recognition as separate species.

(11) Effective morphometric circumscription requires a serious attempt to characterise all parts of a plant, thereby generating a matrix that is inevitably heterogeneous. Studies that use only metric/meristic characters at the expense of multistate (scalar) and bistate characters are overly subjective, as are those that purport to be morphometric but focus on morphological characters already believed to be most diagnostic. Such studies are not competent to adequately circumscribe species.

(12) Genome-wide data generated through next-generation sequencing have further exposed the weaknesses of previous genetic studies based on one–few candidate genes.

Even given nuclear genomic data, a majority of the constituent genes can support topologies other than the most strongly supported topology, especially in cryptic clades – a level of understanding that can be achieved only by analysing genomes gene by gene (or better still, analysing transcriptomes on a transcriptome by transcriptome basis), rather than following the popular but damagingly optimistic trend of concatenating genes. Concatenation of genes sequenced from multiple genomes is especially difficult to justify, as they operate under radically different functional and evolutionary constraints.

(13) Despite residing at the core of molecular phylogenetics for the last 30 years, organellar genomes are proving even less reliable for building species trees than are nuclear genomes. Until recently, plastid genes were the 'great green hope' of DNA barcoding, but it is now clear that the main surviving utility of plastomes is identifying recent hybridisation under uniparental inheritance of organelles. And even for this limited purpose, plastomes are unreliable within cryptic clades.

(14) Most models of speciation assume a fairly straightforward and rapid transition from a freely interbreeding to a wholly isolated status as lineages diverge to form species, but they are reticent (or, more often, silent) on the topic of what level of ongoing gene-flow is the minimum that can be accepted for a daughter lineage to be recognised as a newly born species. I regard this decision as crucial because (a) I believe that the vast majority of attempts to speciate fail, most temporarily isolated metapopulations soon being reintegrated into their parental lineages, and (b) it is increasingly clear that all putative plant species are subject to some level of acquisition of 'foreign' genes. In order to determine this crucial threshold of gene-flow it will be necessary to find more effective methods of distinguishing recent gene-flow from incomplete lineage sorting (ILS).

(15) By presenting a direct and urgent challenge to meaningful species circumscription, cryptic clades are the key to determining that crucial threshold of gene-flow. Given that cryptic clades are potentially actively diversifying, they also represent our best opportunity to not only recognise but also understand speciation, as most speciation mechanisms are capable of producing clades likely to be labelled as 'cryptic'. Older lineages, rendered more distinct by extinction of intermediates, offer poorer case studies.

(16) Among the three fundamental criteria for species circumscription summarised in Table 3.1, monophyly – its assessment so heavily dependent on the motif of repeatedly dichotomous trees – is arguably the least useful. Essential for determining supraspecific taxa, monophyly remains a desirable property of species, but it will not characterise all; for example, it cannot be expected in cases where one ancestral species has generated more than one descendant species. Quantified similarity of genotype and phenotype within and among populations is a more informative measure. Genotype also indirectly suggests levels of reproductive isolation, though accompanying ecological observations and/or breeding experiments are desirable to provide a more nuanced assessment.

(17) Substantial technological innovations are imminent in the field of genotyping, particularly if third-generation nanopore sequencing technology can soon be deployed in the field; a breakthrough that would allow all kinds of field biologist to become parataxonomists, and would greatly alleviate the main ongoing constraint – the number of plants sequenced. Unfortunately, characterising the phenotype is less amenable to technological

advance. Automated image identification and interactive keys are no substitute for a thorough morphometric description of a living plant if the primary objective is to better circumscribe species by relating well-characterised phenotype directly to well-characterised genotype.

(18) In my opinion, the entire *modus operandi* of the systematics community should be reviewed, aiming to replace piecemeal taxonomy with carefully planned global monography of prioritised clades. In as far as is practicable, the monographic approach chosen should transcend typology in order to allow genuine species circumscription through morphological, molecular and ecological data sets derived from the same individuals, and referable to particular source populations. This is the only effective way to demonstrate the valid status of a putative species, to refute that status for other supposed species (currently a greatly under-rated activity), and to circumscribe accurately the boundaries of those *bona fide* species that survive the circumscription process.

(19) I recognise that achieving this goal would require greater resourcing and much greater coordination among groups such as herbarium taxonomists, molecular systematists, evolutionary biologists, and field ecologists. Technological advances in genomics/transcriptomics make it essential that traditional taxonomists recognise the need to operate in partnership with 'harder' sciences, and that genomics specialists lose their fear of active involvement in taxonomy. Replacing the Codes of Nomenclature with Codes of Biosystematics that place obligatory constraints on taxonomic practice as well as naming (i.e. encompass genuinely scientific issues) would be a good start to the suggested rapprochement.

(20) I concede that, even if such cultural concessions are made, demographic monography will probably slow the pace of so-called species discovery. But the species thus recognised will have the advantage of standing the test of time, and they will be sufficiently well-understood as biological entities to make taxonomy predictive. Most species viewed today as cryptic would no longer be so, having been shown to be either readily differentiable from their close relatives (and thus pseudo-cryptic) or figments of a taxonomist's imagination. During my lifetime, I have 'discovered' many more of the latter than of the former.

Postscript

As I write, I have just learned that

> a global effort has been officially launched to sequence all living organisms on Earth including animal, plant, protozoan and fungal species. The project, known as the Earth BioGenome Project (EBP), will generate the largest data set of its kind from the 1.5 million species of eukaryotes worldwide. EBP will provide new insights into a range of scientific fields including; evolution, biodiversity conservation, agriculture and medicine . . . The UK arm, known as the Darwin Tree of Life Project [DToL] will form part of the EBP in its effort to sequence 66,000 species . . . This year [2019], a proof of concept pilot project has been carried out by the Sanger Institute and its collaborators to sequence 25 iconic UK species (http://genomics.ed.ac.uk/news-events/news/darwin-tree-life-project).

This highly collaborative project is in some ways impressively ambitious. But how many of the UK's eukaryotes (constituting an impoverished and phylogenetically eclectic biota, largely assembled through dominantly northward migration only during the postglacial respite of the last 11,500 calibrated years BP) are properly circumscribed species? And can their relevance be understood in the absence of global analyses of the relevant taxa? Ironically, one of the four plant species among the first 25 species to be sequenced was the blackberry, *Rubus fruticosus* agg. (Rosaceae). Beneath the veneer of the seemingly innocent 'agg.' reside, according to some authorities (e.g. Taylor 2005), at least 320 species, reflecting the unusual reproductive biology of this aggregate macrospecies. Even more ironically, the blackberry was classed by DToL as 'iconic' rather than being placed with other 'cryptic' species analysed during the initial trial. Given many such radically contrasting taxonomies, and the weak scientific framework underpinning most of them, do we really know whether a single objective global monograph of all eukaryotes would recognise anywhere close to 1.5 million species? Or could the more accurate figure be 15 million? Or perhaps just 150 thousand? (bearing in mind that the results of well-sampled next-generation sequencing studies often suggest major reductions to previous estimates of species numbers).

If cryptic clades are rare, the massive guesswork inherent in most extant species names may have some basis in biological reality. But if, as I suspect, clades that could on present evidence legitimately be classed as 'cryptic' are common, we are still operating within a taxonomic framework that is largely artificial – a salutary warning to the increasingly popular and often high-profile biodiversity meta-analyses that use species-level classifications as their raw data. A typological approach to assessing biodiversity that relies on applying the 'technological fix' of next-generation 'genotypology' to a list of formal 'accepted names' will effectively take as read both species circumscription and species-level phenotypic differentiation; it is therefore likely to yield results that are rather less profound than anticipated.

Acknowledgements

I thank Oscar Pérez-Escobar and Paula Rudall for critiquing an earlier version of this manuscript, and Simon Mayo and Alex Monro for providing constructive reviews.

References

Angiosperm Phylogeny Group (APG). (2016) An update of the Angiosperm Phylogeny Group classification for the orders and families of flowering plants: APG IV. *Botanical Journal of the Linnean Society* **181**: 1–20.

Averyanov, L. V. (1990) A review of the genus *Dactylorhiza*. In: J. Arditti (ed.) *Orchid Biology: Reviews and Perspectives, V*. Timber Press, Portland, OR, pp. 159–206.

Baguette, M., Betrand, J., Stevens, V. M. et al. (2020) Why are there so many bee-orchid species? Adaptive radiation by intraspecific competition for mnemonic pollinators. *Biological Reviews* **95**: 1630–1663.

Bailarote, B. C., Lievens, B., and Jacquemyn, H. (2012) Does mycorrhizal specificity affect orchid decline and rarity? *American Journal of Botany* **99**: 1655–1665.

Bateman, R. M. (1999) Integrating molecular and morphological evidence for evolutionary radiations. In: P. M. Hollingsworth, R. M. Bateman, and R. J. Gornall (eds.) *Molecular Systematics and Plant Evolution*. Taylor & Francis, London, pp. 432–471.

(2001) Evolution and classification of European orchids: insights from molecular and morphological characters. *Journal Europäischer Orchideen* **33**: 33–119.

(2009) Evolutionary classification of European orchids: The crucial importance of maximising explicit evidence and minimising authoritarian speculation. *Journal Europäischer Orchideen* **41**: 243–318.

(2011) The perils of addressing long-term challenges in a short-term world: making descriptive taxonomy predictive. In: T. R. Hodkinson, M. B. Jones, S. Waldren, and J. A. N. Parnell (eds.) *Climate Change, Ecology and Systematics*. Systematics Association Special Volume 78. Cambridge University Press, Cambridge, pp. 67–95.

(2012) Circumscribing species in the European orchid flora: multiple datasets interpreted in the context of speciation mechanisms. *Berichte aus den Arbeitskreisen Heimische Orchideen* **29**: 160–212.

(2016) Après le déluge: Ubiquitous field barcoding should drive 21st century taxonomy [Chapter 6]. In: P. D. Olson, J. Hughes, and J. A. Cotton (eds.) *Next Generation Systematics*. Systematics Association Special Volume 85. Cambridge University Press, Cambridge, pp. 123–153.

(2018) Two bees or not two bees? An overview of *Ophrys* systematics. *Berichte aus den Arbeitskreisen Heimische Orchideen* **35**: 5–46.

(2019) Next-generation dactylorchids. *Journal of the Hardy Orchid Society* **16**(4): 114–128.

(2020a) Hunting the Snark: The flawed search for mythical Jurassic angiosperms. *Journal of Experimental Botany* **71**: 22–35.

(2020b) Implications of next-generation sequencing for the systematics and evolution of the terrestrial orchid genus *Epipactis*, with particular reference to the British Isles. *Kew Bulletin* **75**(4): 1–22.

Bateman, R. M. and Denholm, I. (1983) A reappraisal of the British and Irish dactylorchids, 1. *The tetraploid marsh-orchids. Watsonia*, **14**, 347–376.

Bateman, R. M. and Denholm, I. (1989) A reappraisal of the British and Irish dactylorchids, 3. The spotted-orchids. *Watsonia* **17**: 319–349.

Bateman, R. M. and Denholm, I. (2012) Taxonomic reassessment of the British and Irish tetraploid marsh-orchids. *New Journal of Botany* **2**: 37–55.

Bateman, R. M., Pridgeon, A. M., Chase, M. W. et al. (1997) Phylogenetics of subtribe Orchidinae (Orchidoideae, Orchidaceae) based on nuclear ITS sequences. 2. Infrageneric relationships and taxonomic revision to achieve monophyly of Orchis sensu stricto. *Lindleyana* **12**: 113–141.

et al. (2003) Molecular phylogenetics and evolution of Orchidinae and selected Habenariinae (Orchidaceae). *Botanical Journal of the Linnean Society* **142**: 1–40.

et al. (2005) Tribe Neottieae: Phylogenetics. In: A. M. Pridgeon, P. J. Cribb, M. W. Chase, and F. N. Rasmussen (eds.) *Genera Orchidacearum: Volume 4. Epidendroideae (Part One)*. Oxford University Press, Oxford, pp. 487–495.

et al. (2011) Species arguments: Clarifying concepts of species delimitation in the pseudo-copulatory orchid genus *Ophrys*. *Botanical Journal of the Linnean Society* **165**: 336–347.

Bateman, R. M., Rudall, P. J., Moura, M. et al. (2013) Systematic revision of *Platanthera* in the Azorean archipelago: Not one but three species, including arguably Europe's rarest orchid. *PeerJ* **1**: doi 10.7717/peerj.218 [86 pp.]

et al. (2014) Speciation via floral heterochrony and presumed mycorrhizal host-switching of endemic butterfly orchids on the Azorean archipelago. *American Journal of Botany* **101**: 979–1001.

et al. (2018) Integrating restriction site-associated DNA sequencing (RAD-seq) with morphological cladistic analysis clarifies evolutionary relationships among major species groups of bee orchids. *Annals of Botany* **121**: 85–105.

Bateman, R. M., Rudall, P. J., Murphy, A. R. M. et al. (2021a) Whole plastomes are not enough: Phylogenomic and morphometric exploration at multiple demographic levels of the bee orchid clade *Ophrys* sect. *Sphegodes*. *Journal of Experimental Botany* 72: 654–681.

et al. (2021b) In situ morphometric survey elucidates the evolutionary systematics of the orchid genus *Gymnadenia* in the British Isles. *Systematics and Biodiversity* **19**: 571–600.

Bebber, D. P., Wood, J. R. I., Barker, C. et al. (2014) Author inflation masks global capacity for species discovery. *New Phytologist* **201**: 700–706.

Bell, A. K., Roberts, D. L., Hawkins, J. A. et al. (2009) Comparative morphology of nectariferous and nectarless labellar spurs in selected clades of subtribe Orchidinae (Orchidaceae). *Botanical Journal of the Linnean Society* **160**: 369–387.

Bergthorssen, U., Richardson, A. O., Young, G. J. et al. (2004) Massive horizontal transfer of mitochondrial genes from diverse land plant donors to the basal angiosperm Amborella. *Proceedings of the National Academy of Sciences of the USA* **101**: 17747–17752.

Bernardo, J. (2011) A critical appraisal of the meaning and diagnosability of cryptic evolutionary diversity, and its implications for conservation in the face of climate change. In: T. R. Hodkinson, M. B. Jones, S. Waldren, and J. A. N. Parnell (eds.) *Climate Change, Ecology and Systematics*. Systematics Association Special Volume 78. Cambridge University Press, Cambridge, pp. 380–438.

Bickford, D., Lohman, D. J., Sodhi, N. S. et al. (2006) Cryptic species as a window on diversity and conservation. *Trends in Ecology and Evolution* **22**: 148–155.

Bock, R. (2010) The give-and-take of DNA: Horizontal gene transfer in plants. *Trends in Plant Science* **15**: 11–22.

Bookstein, F. L. (2014) *Measuring and Reasoning: Numerical Inference in the Sciences*. Cambridge University Press, Cambridge.

Brandrud, M. K., Baar, J., Lorenzo, M. T. et al. (2020) Phylogenomic relationships of diploids and the origins of allotetraploids in *Dactylorhiza* (Orchidaceae): RADseq data track reticulate evolution. *Systematic Biology* **61**: 91–109.

Brandrud, M. K., Paun, O., Lorenz, R. et al. (2019) Restriction-site associated DNA sequencing supports a sister group relationship of *Nigritella* and *Gymnadenia*. *Molecular Phylogenetics and Evolution* **136**: 21–28.

Breitkopf, H., Onstein, R. E., Cafasso, D. et al. (2015) Multiple shifts to different pollinators fuelled rapid diversification in sexually deceptive *Ophrys* orchids. *New Phytologist* **207**: 377–389.

Breitkopf, H., Schlüter, P. M., Xu, S. et al. (2013) Pollinator shifts between *Ophrys sphegodes* populations: Might adaptation to different pollinators drive population divergence? *Journal of Evolutionary Biology* **26**: 2197–2208.

Brys, R. and Jacquemyn, H. (2016) Severe outbreeding and inbreeding depression maintain mating system differentiation in *Epipactis* (Orchidaceae). *Journal of Evolutionary Biology* **29**: 352–359.

Claessens, J. and Kleynen, J. (2011) *The Flower of the European Orchid: Form and Function*. Published by the authors, Voerendaal, Netherlands.

Cozzolino, S., Scopece, G., Roma, L et al. (2020) Different filtering strategies of genotyping-by-sequencing data provide complementary resolutions of species boundaries and relationships in a clade of sexually deceptive orchids. *Journal of Systematics and Evolution* **58**: 133–144.

Cozzolino, S. and Widmer, A. (2005) Orchid diversity: An evolutionary consequence of deception? *Trends in Ecology and Evolution* **20**: 487–494.

Darlington, C. D. (1940) Taxonomic species and genetic systems. In: C. D. Darlington and J. Huxley (eds.) *The New Systematics*. Clarendon Press, Oxford, pp. 137–160.

De hert, K., Jacquemyn, H., Van Glabeke, S. et al. (2012) Reproductive isolation and hybridization in sympatric populations of three *Dactylorhiza* species (Orchidaceae) with different ploidy levels. *Annals of Botany* **109**: 709–720.

De Queiroz, K. (2007) Species concepts and species delimitation. *Systematic Biology* **56**: 879–886.

De Salle, R. and Vogler, A. (1994) Phylogenetic analysis on the edge: The application of cladistics techniques at the population level. In: Bo Golding (ed.) *Non-Neutral Evolution*. Springer, Boston, MA, pp. 154–174.

Delforge, P. (2016) *Orchidées d'Europe, d'Afrique du Nord et do Proche-Orient* (4th ed.). Delachaux et Niestle, Paris.

D'Emerico, S., Pignone, D., Bartolo, G. et al. (2005) Karyomorphology, heterochromatin patterns and evolution in the genus *Ophrys* (Orchidaceae). *Botanical Journal of the Linnean Society* **148**: 87–99.

Devey, D. S., Bateman, R. M., Fay, M. F. et al. (2008) Friends or relatives? Phylogenetics and species delimitation in the controversial European orchid genus *Ophrys*. *Annals of Botany* **101**: 385–402.

Fazekas, A. J., Kesanakurti, P. R., Burgess, K. S. et al. (2009) Are plant species inherently harder to discriminate than animal species using DNA barcoding markers? *Molecular Ecology Resources* **9**: 130–139.

Fazekas, A. J., Kuzmina, M. L., Newmaster, S. G. et al. (2012) DNA barcoding methods for land plants. In: *DNA Barcodes: Methods and Protocols* W. J. Kress and D. L. Erikson (eds.) Humana Press, Totawa, NJ, pp. 223–252.

Fišer, C., Robinson, C. T., and Malard, F. (2018) Cryptic species as window into the paradigm shift of the species concept. *Molecular Ecology* **27**: 613–635.

Gathoye, J.-L. and Tyteca, D. (1987) Étude biostatistique des *Dactylorhiza* (Orchidaceae) de Belgique et des territoires voisins. *Bulletin du Jardin Botanique National de Belgique* **57**: 389–424.

Gould, S. J. (2003) *The Hedgehog, the Fox and the Magister's Pox*. Harmony Books, New York.

Hausdorf, B. (2011) Progress toward a general species concept. *Evolution* **65**: 923–931.

Hedrén, M. (1996) Genetic differentiation, polyploidization and hybridization in Northern European *Dactylorhiza* (Orchidaceae): Evidence from allozyme markers. *Plant Systematics and Evolution* **201**: 31–55.

Hedrén, M., Lorenz, R., Teppner, H. et al. (2018) Evolution and systematics of polyploid *Nigritella* (Orchidaceae). *Nordic Journal of Botany* **36**: 01539 [32 pp.].

Hedrén, M, Nordström, S., and Ståhlberg, D. (2008) Polyploid evolution and plastid DNA variation in the *Dactylorhiza incarnata/maculata* complex (Orchidaceae) in Scandinavia. *Molecular Ecology* **17**: 5075–5091.

Hedrén, M., Nordström, S, Persson, H. et al. (2007) Patterns of polyploid evolution in Greek *Dactylorhiza* (Orchidaceae) as revealed by allozymes, AFLPs and plastid DNA data. *American Journal of Botany* **94**: 1205–1218.

Heslop-Harrison, J. (1954) A synopsis of the dactylorchids of the British Isles. *Berichte der geobotanische Forschungsinstitut Rübel* **1953**: 53–82.

Hollingsworth, P. M., and 51 co-authors (2009) A DNA barcode for land plants. *Proceedings of the National Academy of Sciences USA* **106**: 12794–12797.

Hollingsworth, P. M., Squirrell, J., Hollingsworth, M. L. et al. (2006) Taxonomic complexity, conservation and recurrent origins of self-pollination in *Epipactis* (Orchidaceae). In: J. Bailey and R. G. Ellis (eds.) *Current Taxonomic Research on the British & European Flora*. BSBI, London, pp. 27–44.

Hollingsworth, P. M., Li, D-Z., van der Bank, M. et al. (2016) Telling plant species apart with DNA: From barcodes to genomes. *Philosophical Transactions of the Royal Society B* **371**: 20150338.

Jacquemyn, H., Kort, H. D., Broeck, A. V et al. (2018) Immigrant and extrinsic hybrid seed inviability contribute to reproductive isolation between forest and dune ecotypes of *Epipactis helleborine* (Orchidaceae). *Oikos* **127**: 73–84.

Jörger, K. M. and Schrödl, M. (2013) How to describe a cryptic species? Practical challenges of molecular taxonomy. *Frontiers in Zoology* **10**: 59.

Koenen, E. J. M., Ojeda, D. I., Steeves, R. et al. (2020) Large-scale genomic sequence data resolve the deepest divergences in the legume phylogeny and support a near-simultaneous evolutionary origin of all six subfamilies. *New Phytologist* **225**: 1355–1369.

Kühn, R., Pedersen, H. A., and Cribb, P. (2019) *Field Guide to the Orchids of Europe and the Mediterranean*. Royal Botanic Gardens Kew, Kew.

Larridon, I., Villaverde, T., Zuntini, A. R. et al. (2020) Tackling rapid radiations with targeted sequencing. *Frontiers in Plant Science* **10**: 1655 [17 pp.].

Lee, K.-M., Kivela, S. M., Ivanov, V. et al. (2018) Information dropout patterns in Restriction site Associated DNA phylogenomics and a comparison with multilocus Sanger data in a species-rich moth genus. *Systematic Biology* **67**: 925–939.

Li, X., Yang, Y., Henry, R. J. et al. (2015) Plant DNA barcoding: From gene to genome. *Biological Reviews* **90**: 157–166.

Liebel, H. T., Bidartondo, M. I., Preiss, K. et al. (2010) C and N stable isotope signatures reveal constraints to nutritional modes in orchids from the Mediterranean and Macronesia. *American Journal of Botany* **97**: 903–912.

Louca, S. and Pennell, M. W. (2020) Extant timetrees are consistent with a myriad of diversification histories. *Nature* **580**: 502–505.

Malmgren, S. (2008) Are there 25 or 250 *Ophrys* species? *Journal of the Hardy Orchid Society* **5**: 95–100.

Matute, D. R. and Sepulveda, D. E. (2019) Fungal species boundaries in the genomics era. *Fungal Genetics and Biology* **131**: 103249.

Mayden, R. L. (1997) A hierarchy of species concepts: The denouement in the saga of the species problem. In: M. F. Claridge, H. A. Dawah, and M. R. Wilson (eds.) *Species: The Units of Biodiversity*. Chapman & Hall, London, pp. 381–421.

Mayo, S. (2022) A review of plant taxonomic species and their role in the workflow of integrative species delimitation. *Kew Bulletin* **77**: 1–26.

Mayr, E. (1942) *Systematics and the Origin of Species*. Columbia University Press, New York.

(1982) Speciation and macroevolution. *Evolution* **36**: 1119–1132.

Minelli, A. (2020) Taxonomy needs pluralism, but a controlled and manageable one. *Megataxa* **1**: 9–18.

Morjan, C. L. and Rieseberg, L. H. (2004) How species evolve collectively: Implications of gene flow for the spread of advantageous alleles. *Molecular Ecology* **13**: 1341–1356.

Munoz-Rodriguez, P., Carruthers, T., Wood, J. R. I. et al. (2019) A taxonomic monograph of *Ipomoea* integrated across phylogenetic scales. *Nature Plants* **5**: 1136–1154.

Olson, P. D., Hughes, J., and Cotton, J. A., eds. (2016) *Next Generation Systematics*. Systematics Association Special Volume 85. Cambridge University Press, Cambridge.

Parnell, J., Rich, T., McVeigh, A. et al. (2013) The effect of preservation methods on plant morphology. *Taxon* **62**: 1259–1265.

Paulus, H. F. (2018) Pollinators as isolation mechanisms: Field observations and field experiments regarding specificity of pollinator attraction in the genus *Ophrys* (Orchidaceae und Insecta, Hymenoptera, Apoidea). *Entomologia Generalis* **37**: 261–316.

Paulus, H. F. and Gack, C. (1990) Pollinators as prepollinating isolating factors: Evolution and speciation in *Ophrys* (Orchidaceae). *Israel Journal of Botany* **39**: 43–97.

Paun, O., Bateman, R. M., Fay, M. F. et al. (2010) Stable epigenetic effects impact evolution and adaptation in allopolyploid orchids. *Molecular Biology and Evolution* **27**: 2465–2473. doi: 10.1093/molbev/msq150

Pellicer, J. and Leitch, I. J. (2020) The Plant DNA C-values database (release 7.1): An updated online repository of plant genome size data for comparative studies. *New Phytologist* **226**: 301–305.

Pérez-Escobar, O. A., Bogarin, D., Schley, R. et al. (2020) Resolving relationships in an exceedingly young orchid lineage using Genotyping-by-Sequencing data. *Molecular Phylogenetics and Evolution* **144**: 106672.

Pfenninger, M. and Schwenk, K. (2007) Cryptic animal species are homogeneously distributed among taxa and biogeographical regions. *BMC Evolutionary Biology* **7**: 121–126.

Pillon, Y., Fay, M. F., Hedrén, M. et al. (2007) Insights into the evolution and biogeography of Western European species complexes in *Dactylorhiza* (Orchidaceae). *Taxon* **56**: 1185—1208.

Pineiro-Fernandez, L., Byers, K. J. R. P., Cai, J. et al. (2019) Phylogenomic analysis of the floral transcriptomes of sexually deceptive and rewarding European orchids, *Ophrys* and *Gymnadenia*. *Frontiers in Plant Science* 01553.

Rudall, P. J. and Bateman, R. M. (2002) Roles of synorganisation, zygomorphy and heterotopy in floral evolution: The gynostemium and labellum of orchids and other lilioid monocots. *Biological Reviews* **77**: 403–441.

Schiebold, J. M. I., Bidartondo, M. I., Karasch, P. et al. (2017) You are what you get from your fungi: Nitrogen stable isotope patterns in *Epipactis* species. *Annals of Botany* **119**: 1085–1095.

Scopece, G., Cozzolino, S., Musacchio, A. et al. (2007) Patterns of reproductive isolation in Mediterranean deceptive orchids. *Evolution* **61**: 2623–2642.

Scotland, R. W. and Wood, J. R. I. (2012) Accelerating the pace of taxonomy. *Trends in Ecology & Evolution* **27**: 415–416.

Sedeek, K. E. M., Scopece, G., Staedler, Y. M. et al. (2014) Genic rather than genomewide differences between sexually deceptive *Ophrys* orchids with different pollinators. *Molecular Ecology* **23**: 6192–6205.

Sites, J. W. Jr and Marshall, J. C. (2003) Delimiting species: A Renaissance issue in systematic biology. *Trends in Ecology and Evolution* **18**: 462–470.

Sneath, P. H. A. and Sokal, R. R. (1973) *Numerical Taxonomy: The Principles and Practice of Numerical Classification.* Freeman, San Francisco.

Sramkó, G., Paun, O., Brandrud, M. K. et al. (2019) Iterative allogamy–autogamy transitions drive actual and incipient speciation during the ongoing evolutionary radiation within the orchid genus *Epipactis* (Orchidaceae). *Annals of Botany* **124**: 481–497.

Struck, T. H., Feder, J. L., Bendiksby, M. et al. (2018) Finding evolutionary processes hidden in cryptic species. *Trends in Ecology & Evolution* **33**: 153–163.

Sundermann, H. (1980) *Europäische und Mediterrane Orchideen - ein Bestimmungsflora* (3rd ed.). Schmersow, Hildesheim.

Taylor, K. (2005) Biological flora of the British Isles: *Rubus vestitus* Weihe. *Journal of Ecology* **93**: 1249–1262.

Tranchida-Lombardo, V., Cafasso, D., Cristaudo, A. et al. (2012) Phylogeographic patterns, genetic affinities and morphological differentiation between *Epipactis helleborine* and related lineages in a Mediterranean glacial refugium. *Annals of Botany*, **107**: 427–436.

Trávníček, P., Suda, J, Bateman, R. M. et al. (2012) Minority cytotypes in European populations of the *Gymnadenia conopsea* complex (Orchidaceae) greatly increase intraspecific and intrapopulation diversity. *Annals of Botany* **110**: 977–986.

Tronteij, P. and Fišer, C. (2009) Perspectives: Cryptic species diversity should not be trivialised. *Systematics and Biodiversity* **7**: 1–3.

Tyteca, D. and Dufrene, M. (1994) Biostatistical studies of western European allogamous populations of the *Epipactis helleborine* (L.) Crantz species group (Orchidaceae). *Systematic Botany* **19**: 424–442.

Valuyskikh, O. E., Shadrin, D. M., and Pylina, Y. I. (2019) Morphological variation and genetic diversity of *Gymnadenia conopsea* (L.) R. Br. (Orchidaceae) populations in the northeast of European Russia (Komi Republic). *Plant Genetics* **55**: 180–196.

Vereecken, N. J. and Francisco, A. (2014) *Ophrys* pollination: From Darwin to the present day. In: R. Edens-Meier and P. Bernhardt (eds.) *Darwin's Orchids: Then and Now*. University of Chicago Press, Chicago, pp. 47–70.

Vereecken, N. J., Streinzer, M., Ayasse, M. et al. (2011) Integrating past and present studies on *Ophrys* pollination: A comment on Bradshaw et al. *Botanical Journal of the Linnean Society* **165**: 329–335.

Wettewa, E., Bailey, N., and Wallace, L. E. (2020) Comparative analysis of genetic and morphological variation in the *Platanthera hyperborea* complex (Orchidaceae). *Systematic Botany* **45**: 767–778.

Wheeler, Q. D. and Meier, R. (2000) *Species Concepts and Phylogenetic Theory: A Debate*. Columbia University Press, New York.

Williams, B. R. M., Mitchell, T. C., Wood, J. R. I. et al. (2014) Integrating DNA barcode data in a monographic study of Convolvulus. *Taxon* **63**: 1287–1306.

Wilson, E. O. (1998) *Consilience: The Unity of Knowledge*. A. A. Knopf, New York.

Wood, J. R. I., Williams, B. R. M., Mitchell, T. C. et al. (2015) A foundation monograph of *Convolvulus* L. (Convolvulaceae). *PhytoKeys* **51**: 1–282.

Wood, J. R. I., Munoz-Roriguez, P., Degen, R. et al. (2017) New species of *Ipomoea* (Convolvulaceae) from South America. *PhytoKeys* **88**: 1–38.

Zaveska, E., Maylandt, C., Paun, O. et al. (2019) Multiple auto- and allopolyploidisations marked the Pleistocene history of the widespread Eurasian steppe plant *Astragalus onobrychis* (Fabaceae). *Molecular Phylogenetics and Evolution* **139**: 106172–106183.

4

Multilevel Organismal Diversity in an Ontogenetic Framework as a Solution for the Species Concept

Alexander Martynov and Tatiana Korshunova

Introduction

'Species' is an iconic term in biology. In addition to the countless concepts and definitions of the term in biology, in the vernacular it also has important applications, for example, in its use by custom officers, conservationists, and politicians. Obviously, it is impossible to understand, document, and study biological diversity without a theoretical framework and practical tools (classifications) to parse the overwhelming diversity of life into a smaller number of universal categories such as 'species'. Despite, or maybe because of, this, species taxa as concepts have been in persistent crisis in the absence of clear criteria to define the category 'species'. Darwin (1859) used the term 'species' as a transitional concept because constant ongoing modification was central to his concept of evolution. Following the synthetic theory, Darwin's 'modifications' are substituted by 'evolutionary changes'. For example, the polytypic species concept was suggested, which implicates broad variation over a large geographic range (Mayr 1942; Beurton 2002). This concept has been questioned, as innumerable cases of hidden ('cryptic') diversity have been documented recently, leading to many broad species being split into narrowly defined fine-scale species. Remarkably, levels of variability are commonly detected within these revised species limits that are similar to those observed in the broader taxon concepts.

One common justification of the species concept is that in prescientific times, people recognised discrete species-equivalent units within the diversity of living organisms (Mayr 1963). This tradition has continued with development of modern methods. Estimates of species numbers for major groupings of organisms vary greatly depending on the species concept and methods employed. For example, the approximate numbers of currently recognised species in major organismal groups have been estimated as: 9,110 viruses (ICTV 2020), 12,000 bacteria and archaea (Chun and Rainey 2014), 120,000 fungi (Hawksworth and Lücking 2017), 72,500 algae (including unicellular groups) (Guiry 2012), 500,000 land plants (Corlett 2016), and 1,000,000 animals (including unicellular groups) (Mora et al. 2011). Other estimates differ drastically. For example, the number of 'operational taxonomic units' (OTUs) of bacteria and archaea has been reckoned to range from a few million to several trillions (Louca et al. 2019). Similarly, estimates of diversity in multicellular organism groups are commonly ten times greater or more than the number of already described species (Scheffers et al. 2012). This implies that the number of potential species level groups has practically no upper limit, which again raises the question of what criteria should be used to delimit 'species'. Currently, a universal tool is analysis and comparison of DNA or RNA nucleotide sequences (various molecular phylogenetic analyses, delimitation methods, e.g. Yang 2014; Scornavacca et al. 2020 and many others). These tools provide an apparently universal method for comparing organisms with few, if any, morphological homologies, such as bacteria, sponges, and mammals. Quite unexpectedly, the broad application of these methods has enabled us to distinguish between taxonomic groups that can be delimited by morphological features and those that apparently can only be defined satisfactorily by using molecular data. This has led to a special emphasis on the so-called sibling or cryptic species problem (Sáez et al. 2003), which was initially proposed to deal with observed disparity between the presence or absence of distinct morphological traits and genetic isolation (Darlington 1940; Walker 1964; Knowlton 1993).

Establishing congruence between morphological and molecular (and also behavioural, ecological, physiological) data is a complex task, especially in multicellular organisms, where it is usually possible to classify using morphological data alone. Taxonomic groups where there are few morphological but considerable molecular differences (e.g. Brehm et al. 2016; Fišer et al. 2018) have been commonly referred to as 'cryptic species'. However, numerous contradictions exist in the usage of this term (Heethoff 2018; Horsáková et al. 2019; Korshunova et al. 2019). There are also species that can be distinguished using morphological data but which exhibit low genetic divergence (Mayer and von Helversen, 2001). An example is the delimitation of the genera *Homo* and *Pan* (Varki and Altheide 2005), which based on molecular data could be placed into a single genus *Homo* with *Pan* as its synonym (Diamond 1991) if classified in the same way as less well-known groups, for example, nudibranch molluscs (Korshunova et al. 2017a). These examples have direct relevance for resolving the cryptic species problem because instead of a simple cryptic *vs* non cryptic dichotomy, a general problem is revealed about the complex, multilevel structure of organismal diversity.

It should be noted that although genome-wide research programmes in a taxonomic context have been developed (Leebens-Mack et al. 2019) for the majority of multicellular organisms, only a few genes have so far been sequenced (DeSalle and Goldstein 2019). As a result, in modern taxonomic practice, 99 per cent similarity in a few genetic markers (similar to differences between *Homo* and *Pan* in selected data sets of protein-coding sequences) commonly leads to consideration as to whether individuals belong to the same species. While differences between percentages of the compared gene markers are seemingly evident, the next and most important step, species delimitation per se, is controversial and complicated (Phillips et al. 2019). This means that even though molecular data and phylogenetic analyses offer new tools for biological systematics, current molecular methods are not sufficient to properly describe the complex diversity of organisms at different scales.

Specifically, there is inconsistency in translation of molecular variation into taxonomic decisions. This is well covered in the ongoing discussion of 'cryptic species'. In this discussion, cases of minor morphological differences and significant molecular divergence have been the main focus, whilst the converse situation, smaller molecular divergence and considerable ontogenetic differences in a broad sense (morphological, behavioural, ecological, physiological), has been less well addressed. In this chapter we argue that a solution to the 'cryptic species problem' cannot be attempted without resolving the species concept problem itself (e.g. Robson and Richards 1936; Huxley 1942; Zirkle 1959; Mayr 1963; Templeton 1989; Ereshefsky 1998; Stamos 2003; Lherminier and Solignac 2005; Wilkins 2009; Richards 2010; Slater 2013; Zachos 2016; Agosta and Brooks 2020, and many others) – and, in particular, cannot be solved without addressing the general properties of living organisms. We believe that incoherence in the representation of a taxon as a classificatory unit and as a lineage within an evolutionary scenario has not been satisfactorily resolved (Wheeler and Meier 2000) and, furthermore, that biological diversity cannot be understood without a theory and practical methods that place ontogeny firmly within the central framework of taxonomy and phylogeny.

4.1 The Cryptic Species Problem

Throughout history, taxonomists have distinguished innumerable entities that they call 'species', often using subtle differences. However, prior to molecular systematics 'species' were rarely divided into 'normal' and 'difficult to distinguish' aka 'cryptic' species (for which the term 'sibling species' has also been commonly used by e.g. Mayr 1942; Knowlton 1993). An early use of the term 'cryptic species' was that of Darlington (1940: 148): "'Hidden' species were first discovered in *Drosophila* … no doubt they represent an abundant type of *cryptic* species" and Stebbins (1950: 193), " … population systems which were believed to belong to the same species until genetic evidence showed the existence of isolating mechanisms … ". Note that Darlington linked 'cryptic' and 'hidden' species. The term 'cryptic species' in this sense appeared in other publications (Walker 1964; Thorpe et al. 1978; Hauk 1995), but its usage was restricted. Currently, a popular definition of 'cryptic species' is that of Sáez et al. (2003: 7167), "These molecular differences have been taken as evidence of reproductive isolation between morphologically indistinguishable (cryptic), or only *a posteriori* distinguished species (pseudocryptic)". Sáez et al. (2003) ascribed this definition to Knowlton (1993), who, however, used the term 'sibling' species. 'Sibling' and 'cryptic' have been used interchangeably (Henry 1985). Knowlton (1986) referred to both terms but in her later work (1993) used 'sibling species' almost exclusively. The advent of molecular tools, data, and phylogenetic methods amplified this new concept, now termed 'cryptic species'. Cryptic species are generally defined as those species that show sufficient divergence in molecular characters, which cannot easily be distinguished using morphological differences.

The term 'cryptic species' is problematic on several levels (Korshunova et al. 2019), and linguistically 'cryptic' can also refer to a species that is well camouflaged on a substrate (Karp 2020). Nevertheless, the term 'cryptic species' is widely used. We summarise evidence showing that the astounding variability of taxonomic characters at different scales makes it untenable to maintain the simplistic categories of 'cryptic' and 'non cryptic' species.

4.2 Absence of Clear Distinctions within Taxa: From Individuals and Species to Genus and Family Level

The evolutionary process is traditionally considered to be continuous (Darwin 1859), a notion that prevails despite being controversial (Eldredge and Gould 1972; Coates et al. 2018; Platania et al. 2020). In contrast, the species is widely assumed to be a discrete, static unit, a view that is reflected in the botanical and zoological codes of nomenclature (ICZN 1999; Turland et al. 2018), which include the obligatory requirement to provide a diagnosis for a taxon, associated with a valid species name. This is the great tension between evolutionary processes and taxonomic practice and it has not yet been solved satisfactorily (Zachos 2018). There are no unequivocal and universal criteria for species delimitation that both reflect underlying evolutionary processes (Phillips et al. 2019; Stanton et al. 2019; Minelli 2020) and can be applied to the full range of biological diversity (Willis 2017; Reydon and Kunz 2019).

More than 30 different species concepts have already been proposed and ontology-related questions have been addressed (Wilkins 2009; Zachos 2016) but from a practical standpoint even the most widely applied criterion – reproductive isolation of interbreeding communities of organisms (Mayr 1942) – is not universally applicable, as shown by numerous hybridisation and introgression studies (Mallet 2005; Miller et al. 2012; Harrison and Larson 2014; Jančúchová-Lásková et al. 2015; Ottenburghs 2020). Hybridisation, per se, has been shown to be an important catalyst for speciation (Amaral et al. 2014; Capblancq et al. 2015; Grant and Grant 2016). For example, genomes of modern humans contain up to 3–5 per cent of *Homo neanderthalensis* genes, and molecular data unequivocally demonstrate that the 'species' *Homo sapiens* is a complex evolutionary mosaic (Jacobs et al. 2019; Durvasula and Sankararaman 2020).

At a general scale, any sexually reproducing 'species' must pass through a reticulating phase. Many accepted species share gene flow with each other, and yet maintain their identity as recognisable taxa. For example, African cichlids (Gante et al. 2016; Burress et al. 2018; Malinsky et al. 2018). The Pacific blue mussel *Mytilus trossulus* (which has invaded the Baltic Sea) genome includes mitochondrial DNA from the Atlantic *Mytilus edulis* through introgression (Kijewski et al. 2006), leading these taxa to be considered as complex, not completely isolated, species (Fraïsse et al. 2016). Among bacteria, there is evidence for existence of 'subpopulations' with stable recognisable characters inherited via epigenetic mechanisms (Casadesús and Low 2013). These observations undermine any simple understanding of the 'cryptic species' concept as well as the simple tree topology for phylogenetic relationships and challenge the assumption of distinct evolutionary lineages visualised as simple stems.

There is substantial evidence that evolution is influenced by epigenetic processes as well as classical genetic inheritance (Perez and Lehner 2019; Garcia-Dominguez et al. 2020). Cases of horizontal gene transfer have been revealed within metazoans and plants (Yoshida et al. 2019; Neelapu et al. 2019; Reiss et al. 2019), and for prokaryotes horizontal gene transfer is routine (Gogarten and Townsend 2005; Koonin and Wolf 2008). Phylogenies are not, therefore, just a branched tree but also a dynamic, reticulate network. The 'species to population continuum' has been supported by genome-wide approaches (Coates et al. 2018). Recent analyses highlight the blurred line between populations and species and confound the simple dichotomy of species *vs* not species, thus making problematic a clear discrimination between 'cryptic' and 'non cryptic' species and posing difficulties for the meaning of the term 'species' itself.

4.3 Gradations within Potential 'Cryptic' Species

Recently, it has been shown that in practice 'species' may range from entities that are similar and externally difficult to distinguish, to those that exhibit internally stable and distinct characters (Korshunova et al. 2017a, 2019). If we really wish to make a distinction between 'non-cryptic' and 'cryptic (sibling)' species we need a more elaborate terminology than this binary choice (Chenuil et al. 2019). The number of intermediate terms, however, could grow indefinitely (Korshunova et al. 2019). Knowlton (1993) already noted that usage of the term 'sibling species' (= 'cryptic species' of Sáez et al. 2003) is ambiguous and that various combinations of morphological distinctness and genetic isolation can range from 'pseudo-sibling species' (morphological differences are obvious) to 'semi-species' ('only imperfectly isolated from each other', Knowlton 1993: 190). The term 'pseudo-sibling species', renamed 'pseudocryptic' by Sáez et al. (2003), is especially controversial since it does not differ from the normal meaning of 'species' and its use is unlikely to alleviate the growing 'cryptic species' problem since many cases may represent unique combinations of genetic divergence or uniformity and morphological similarity or disparity.

4.4 Even the Apparently 'Most Cryptic' Species Still Require a Morphological Diagnosis

There are numerous examples of 'cryptic species' being proposed for multicellular organisms but not being described because of the need for a morphological diagnosis (Minelli 2020). The reasons for this are two-fold. First, nomenclature codes have a formal requirement to provide a morphological diagnosis. Second, even when species are described using mostly nucleotide data (e.g. Jorger et al. 2013), this does not preclude the later discovery of fine-scale morphological characters. This is not just a theoretical consideration. Ontogeny implies the impossibility of the existence of two organisms without genetic or epigenetic differences (Fraga et al. 2005; Casadesús and Low 2013; Bierbach et al. 2017). It is inconceivable that there are 'cryptic species' pairs without morphological differences when even individuals as closely related as twins show fine-scale molecular and morphological differences. The existence of morphological differences should be presumed at every level of taxonomic study (Korshunova et al. 2019).

Even though undiscovered fine morphological differences must exist, the notion of 'cryptic' species remains topical, and their morphological diagnoses continue to be published (Cerca et al. 2020), even for complexes in which it is apparently difficult to distinguish the species. Because discarding morphology completely from diagnoses and descriptions would be counterproductive, we anticipate the future application in taxonomy of ever more sophisticated methods to reveal fine-scale ontogenetic differences in morphology, behaviour, ecology, etc., between studied taxa. Modern technology certainly makes it possible to at least envisage the evolution of taxonomy in such a direction.

4.5 The Organism in the Ontogenetic Dimension: The 'Cryptic Species' Concept Is the Result of Failure to Assimilate Ontogeny in Taxonomy

4.5.1 Organism as Ontogeny

No single organism has ever existed without being part of a series of complex underlying biological processes (Maynard Smith and Szathmáry 1995; Calcott and Sterelny 2011), which in total form the process of *ontogeny* (Haeckel 1866). Ontogeny is not merely a developmental phase or a metamorphosis, but includes the entire life cycle, that is, the unit of organic diversity that we call an organism consists of a complete ontogeny from initial cell(s) to reproducing adult: *An organism is an ontogeny, an ontogeny is an entire life cycle.* Evolutionary changes are based on alterations in ontogenetic processes (e.g. Garstang 1922; Zimmermann 1959; Gould 1977; Hall 1999; Carroll 2008; Olsson et al. 2010; Minelli 2015), and so a plausible generalisation is that the direction of character change (the basis of the diversity of organisms) is from ontogeny to evolution. The term 'evolution' is meaningless without invoking ontogeny and it is notable that its original meaning was 'ontogeny' (Gould 1977). Since Haeckel's time, a consensus has developed that phylogeny is not produced by an abstract evolutionary process but by ontogeny, from which phylogeny results through change at the level of individual organisms. A phylogeny is a flow of modifying ontogenies in space and time, in which both complex genetic and epigenetic processes occur and result in what we know as evolution. Haeckel's (1866) motto that 'ontogeny repeats phylogeny' critically obscured the fact that ontogeny does not mechanistically accumulate evolutionary changes but instead produces them. We should therefore avoid the prevailing distinction whereby an individual ontogeny 'develops', whereas a species 'evolves'. In fact, ontogeny both develops and evolves.

After more than 150 years, the original (individual development) and secondary (historical development) meanings of the word 'evolution' can again be conjoined. Individual development *sensu stricto* and evolution in the current narrower sense are two aspects of the same ontogenetic process. Although highly conserved, ontogeny is under constant modification because each individual ontogeny is not only a product of inherited mutation and selection but is connected to both ancestral and descendant ontogenies via epigenetic inheritance and other developmental interactions that represent past adaptive responses (Müller 1991; Jablonka and Lamb 2010; Danchin et al. 2011; Chan et al. 2020).

Although individual ontogenies interact during various processes, each new ontogeny is built individually, with its own individual peculiarities. There is no population ontogeny, only individual ontogenies, each of which is the product of interaction with other ontogenies in sexual systems. The subsequent dominance of particular characters that originally arose in only a single individual and then spread throughout a group is a complex process in which genetic and epigenetic processes are involved. Living organisms began with an individual ontogeny, which was already the most complex living system ever to have hitherto appeared. An important contribution from the field of ontogeny to that of evolutionary biology would be, 'Rather than concentrating on the "survival of the fittest", evolutionary developmental biology gives us new insights into the "arrival of the fittest" . . . ' (Gilbert 2010: 683; Gefaell et al. 2020).

4.5.2 Ontogeny-Free Taxonomy and Phylogenetics

Whilst the significance of ontogeny is well understood in evolutionary developmental biology, it is still not applied to the major biological fields of taxonomy and phylogenetics. A search of publications in the Web of Science (Martynov et al. 2015; Stöhr and Martynov 2016) confirmed the absence of general ontogenetic principles in modern taxonomy. Hennig's basic argumentation schemes of phylogenetic systematics (Hennig 1966: 89, 91) evidently did not include ontogeny as a central biological process, although he included a total relationship structure termed *hologeny*. Whilst some ontogenetic traits are indicated in descriptions and general reviews, for example, seeds of plants, or distinct metamorphic stages in metazoans (e.g. Tikhomirova 1990; Bellés 2020; Costa et al. 2021), this does not amount to the consistent incorporation of ontogeny in taxonomy. There is also a growing special field of 'phylogenetic applications' of ontogeny, which, however, exists rather independently from taxonomy (e.g. Steyer 2000; Wolfe and Hegna 2013; Gee 2020).

The lack of inclusion of the ontogeny in taxonomy can lead to significant distortion in inferences of evolutionary relationships and consequent classification (Martynov 2012a). For example, without consideration of ontogenetic data, the nudibranch family Corambidae was incorrectly proposed to be the most basal dorid group, but later studies showed that corambids underwent secondary modification by the ontogenetic process of paedomorphosis (Martynov 1994; Martynov et al. 2020a). These ontogenetic data agree well with recent molecular data, in which the Corambidae never appear as a sister group to all other dorids (Korshunova et al. 2020a).

Similarly, simplified ophiuroid echinoderms were considered ancestral because of their morphology, but they share specific features with the post-larval stages of common ophiuroid families that have complex adult stages (Martynov 2009, 2012a), and so represent another example of paedomorphosis, also supported by molecular data (O'Hara et al. 2017). These two examples represent notable contributions that were first suggested on the basis of ontogenetic observations.

The premise that the same genes determine an organism's identity at the egg and adult stage of an individual, thereby obviating the need to include ontogenetic data in phylogenetic studies (Mayr 1963), ignores epigenetic and other processes beyond strict genetic control. Therefore, molecular analysis cannot substitute the ontogenetic approach.

The consistent and thorough integration of ontogenetic patterns in taxonomy and phylogeny would have many consequences. For instance, current taxon circumscriptions lack key sources of data, including at least *separate (integrated) taxonomic diagnoses of the various ontogenetic stages of a species* (Figure 4.1) on a routine basis. The taxonomic use of data from various stages within the ontogenetic cycle of the same individual will generate new arrays of informative characters. As an illustration, Figure 4.1 presents a summary of novel informative characters in the nudibranch genus *Dendronotus*. In early postlarval ontogeny the radular characters (molluscan teeth) are almost completely 'cryptic' within species shown to be different by molecular and morphological studies (Korshunova et al. 2020b). Towards adult stages, the radulae acquire specific features, but each species exhibits almost indistinguishable assemblages of characters including highly similar colour morphs (Figure 4.1). During its ontogenetic cycle, the same species may be 'cryptic' or not at different stages, depending which characters of phylotypic periods are considered (see Sections 4.5.3–4.5.5) making it meaningless to classify species as 'cryptic' or not.

The inclusion of ontogenetic data in biodiversity studies and taxonomy has several important implications: (1) the ontogenetic properties of any organism imply the impossibility of identical morphologies; (2) the same individual passes through both 'cryptic' and 'non cryptic' phases during its ontogeny; (3) the consistent application of ontogenetic information opens up the possibility of involving various genetic and epigenetic patterns and processes in taxonomic and phylogenetic studies; (4) the inclusion of ontogeny removes the 'cryptic species problem'.

4.5.3 Ontogeny Is Based on Periodic Processes: A Historical Excursion and Current Perspectives

Evolution in the narrow sense is apparently random, not because making a model of future trends is intrinsically impossible (e.g. Kupferschmidt and Cohen 2020; Pollock et al. 2020) but because we still cannot predict details of future patterns with sufficient confidence. Ontogeny, by contrast, incorporates considerable predictivity. For example, we would not expect the development of a crustacean from a molluscan egg or the emergence of a marsupial from a human uterus. Ontogeny shows clear periodic patterns, underlain by early developmental genes. Each novel ontogeny represents a condensed sequence of

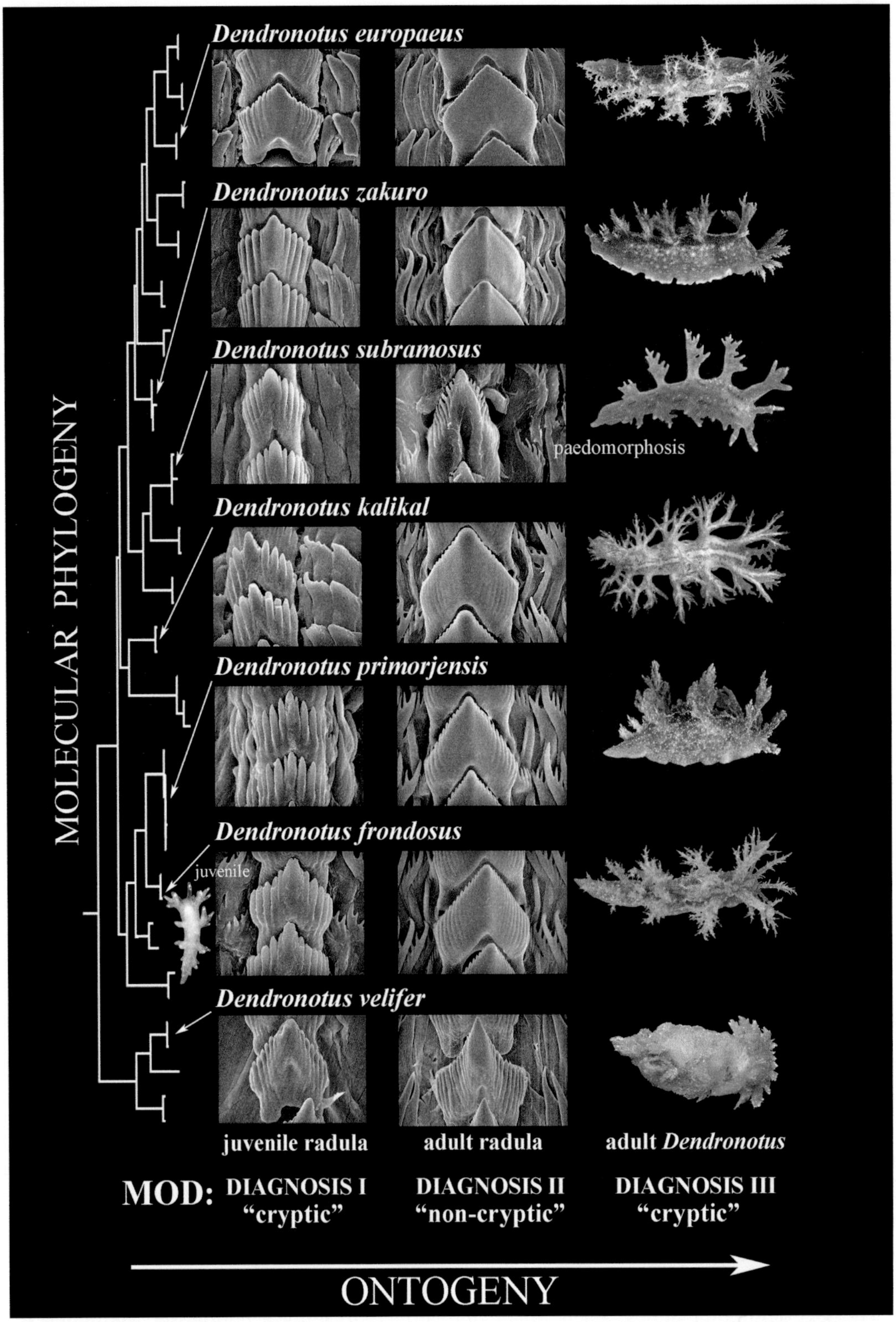

Figure 4.1 MOD in the ontogenetic framework in nudibranch genus *Dendronotus* supplemented with molecular phylogenetic analysis. See explanation in the text.

previous changes to ontogeny and a vehicle for future changes. Such changes are, however, restricted by the necessity for successful completion of the ontogeny within the limitations imposed by previous modifications.

Understanding ontogeny marks a significant boundary between pre-evolutionary and evolutionary paradigms. The pre-evolutionary paradigm recognised the recurrence of structural features within ontogenetic cycles but ignored shifts in these cycles that later would be called evolutionary changes. This can be appreciated by comparing nineteenth-century biological textbooks (Agassiz and Gould 1857; Gegenbaur 1859) to modern ones. Until the middle of the twentieth century, ontogenetic cycles remained the foundation of comparative morphology (Beklemishev 1969). Although Beklemishev worked within the evolutionary paradigm, his view of ontogenetic cycles as the basis of morphology clearly echoed his anti-evolutionary predecessors: 'Our investigations should not be limited to adult animals, but should also include the changes which they undergo during the whole course of their development' (Agassiz and Gould 1857: 31).

Astoundingly, 160 years later, understanding of ontogeny as presented in influential textbooks is still not completely evolutionary, despite the achievements of 'evo-devo'. For example, if we fail to find the same structure in the transitive pharyngeal clefts of the human embryo as in the branchial arches of adult fishes (Futuyma and Kirkpatrick 2017: 371), this is because ontogeny of descendants does not preserve all the details of ancestral adult characters. Between a particular fish-like ancestor and modern humans there is a huge ontogenetic and hence phylogenetic distance (e.g. Daeschler et al. 2006), and all the changes that occurred were not necessarily recorded in any descendant ontogenies. However, within more closely related groups, for instance, paedomorphic ophiuroids, adult shape and placement of primary plates, dental plates, and vertebrae can be well matched to homologous structures in early juveniles of related non-paedomorphic taxa (Martynov 2012a; Martynov et al. 2015). The ontogeny of modern organisms is therefore a true guide to the ontogeny of its various ancestors, and, hence, to its phylogeny. Other remarkable examples are the nudibranch molluscs that are completely shell-less at the adult stage but have shelled veliger larvae. This is *directly observable* evidence that nudibranchs originated from shelled ancestors, and is well aligned with taxonomically deployed morphological data and the results of molecular analyses (Wägele and Willan 2000; Martynov 2011a,b; Pabst and Kocot 2018).

The persistence of ancestral ontogenetic traits in the ontogeny of modern descendants is indispensable to understanding evolution. Ultimately, the fact that every person was once a small fish-like structure in the uterus of its own mother makes this evidence of evolution *central*. The fact that modern ontogenies preserve in phylotypic periods the major features of ancestral ontogenies that existed half a billion years ago is significantly undervalued by modern evolutionary biologists. Gilbert et al. (1996: 357), in a seminal article of the evo-devo synthesis, state that 'This new synthesis emphasizes three morphological areas of biology that had been marginalized by the Modern Synthesis of genetics and evolution: embryology, macroevolution, and homology'. In recent evolutionary textbooks (Futuyma and Kirkpatrick 2017: 369) the evolution and development chapters are still placed later in the book, and evolutionary developmental biology remains marginalised. Another well recognised modern textbook (Herron and Freeman 2015: 750) clearly indicates this contradiction: 'Although Darwin recognized the importance of development in understanding evolution, development was largely ignored by the architects of the 20th-century codification of evolutionary biology known as the "modern synthesis"'.

Epigenetics is closely linked to the individual events during an ontogenetic cycle and their transmission to ensuing generations. Once almost beyond the scope of science, this field is now prolific and productive (e.g. Holeski et al. 2012; Burggren 2014; Perez and Lehner 2019). Just how radically the situation has changed can be gathered from this forty-year-old quotation: 'Until the molecular revolution, which demonstrated that DNA replication is independent of the environment, there was no theoretical reason for denying the inheritance of acquired modifications' (Stebbins 1980: 140; Buss 1987: 24). The latest turn of the molecular revolution has instead brought evidence that gene expression can be dependent on the environment (Ralston and Shaw 2008) and demonstrates a mechanism for the inheritance of acquired characteristics. This does not imply an exact revival of Lamarckian evolution (Danchin et al. 2019), despite post-Soviet attempts to legitimise Lysenko (Graham 2016). Whilst individual organisms are highly constrained by their ontogenetic cycles, there is clear evidence that they are more flexible than previously accepted: for example, transgenerational odour sensitisation in mice (Dias and Ressler 2013).

Thus, an idea that was obvious to comparative morphologists of the past is still not completely incorporated into mainstream evolutionary biology, otherwise the recent statement by Gilbert (2019: 1) – 'Animals don't have life cycles; they are life cycles' – would be superfluous.

4.5.4 Conservative Periods of Ontogeny: Phylotypic Periods as a Reformulation of Haeckel's Heritage

Conservative stages appear earlier in ontogeny and can make a taxon appear 'cryptic' at certain stages, even across broad taxonomic ranks. Karl Baer stated that 'the embryo of a higher form never resembles any other form, but only its embryo' (Baer 1828–1837). Baer is widely credited with establishing the developmental evidence for evolution (Futuyma and Kirkpatrick 2017: 369, 371). However, Baer believed that the ontogenies of different taxa represented 'isolated entities' not linked by evolutionary connections. Johann Meckel (1811) instead argued that ontogenetic stages of descendants may represent adult stages of ancestors (Göbbel and Schultka 2003), with reservations when a stage of ontogeny of one organism was incorrectly considered as an adult of a different organism (Hennig 1966: 33). The juvenile stage of *one taxon* can be really similar to the adult stage of *another taxon* owing to the retention of juvenile traits into adulthood - i.e. heterochrony (including paedomorphosis) - and this is a consequence of genetically controlled differences in the timing or duration of developmental processes (Reilly et al. 1997; Martynov et al. 2020a; Lamsdell 2020). Heterochronies produce evolutionary opportunities for further changes within a highly conserved ontogenetic cycle, with mutations during early ontogenetic periods being lethal. Baer's 'laws' contradicted widely recognised heterochronies, which strongly suggested that modifications of ontogeny in one lineage can lead to the appearance of another lineage. For example, heterochronies played a significant role in the formation of modern humans (Gould 1977) and birds (Klingenberg and Marugan-Lobon 2013). From a current perspective, Baer's view can be considered anti-evolutionary and this permeated ontogeny and slowed the incorporation of developmental biology into evolutionary biology.

Early ontogenetic stages were clearly linked to the evolutionary process and were regarded as recapitulations of ancestral conditions by Müller (1864) and Haeckel (1866). This view was criticised (Mehnert 1897) and reformed (Garstang 1922; Beer de 1958; Alberch et al.1979). The term recapitulation is no longer commonly used (but continue to be discussed, Uesaka et al. 2021), and the once widely promoted 'biogenetic law' of Haeckel has been shown to be far from universal. Ontogeny, in many cases, preserves major features of an ancestral ontogenetic cycle, but not an entire sequence of evolutionary alterations. Instead, the term phylotypic stage was suggested (Slack et al. 1993; Raff 1996) and this was later altered to phylotypic period (Arthur 2002). A related 'hourglass concept' implies the occurrence of conserved phylotypic periods at the middle stages of ontogenies (Duboule 1994). However, there is usually a node of maximal conservatism close to the beginning of individual development (Arthur 2015). The graphical representation of ontogenetic conservatism during ontogeny corresponds better to an inverted cone than to an hourglass (Martynov 2014; Arthur 2015), but both models can coexist (Piasecka et al. 2013).

Divergence in ontogenetic trajectories as a basis for the formation of morphological diversity is confirmed by experimental modification of an individual's development using the thyroid hormone (Kapitanova and Shkil 2014) and by tracing of heterochronies (Lecointre et al. 2020). In angiosperms recent studies show that patterns of early ontogeny define the diversification of plant vascular systems (Claßen-Bockhoff et al. 2020). It has been proposed that phylotypic periods can be referred to different levels of ancestral ontogeny (Martynov and Korshunova 2015; Korshunova et al. 2020a). Conserved periods of ontogeny link juvenile or post-larval patterns to ancestral organisation and can approximate to different traditional ranks (phylum, classes, orders, families, and so on). It is important to note that *phylotypic period* implies not just the phylum category.

A *phylotypic period* (*phyloperiod*) can be defined as a period of ontogeny in which a specific combination of characters, patterns, and processes allows a connection to be made between a common ontogenetic period shared by related groups and their *adult ancestral organisation*. For example, in dorids, one of the earliest phylotypic periods, termed dorphylp 1 (dp 1) (Martynov and Korshunova 2015: 10), is linked to the *adult ancestral organisation* of pleurobranchids by the presence of joined rhinophores (Martynov 2011b; Figure 4.2, dp1). Subsequent dorid phylotypic periods *can be linked* to *adult organisation* of partially reduced (phanerobranchs) or evidently paedomorphic (hypobranch Corambidae) dorid taxa (Figure 4.2, dp 2). Paedomorphosis is a partial restoration of the features of ancestral ontogenetic cycles. Rather than posing a problem for phylogenetic analysis (Kluge 1985) paedomorphosis is instead strong evidence of evolution. In an essentially similar way, in vertebrates, the appearance of branchial arches is a universal phylotypic period in the early ontogeny of various major subgroups (Irie and Kuratani 2011) and corresponds to adult ancestral fish-like organisation. The possibility of this linkage is a major achievement of the evolutionary understanding of ontogeny compared to Baer's isolated entities. It is

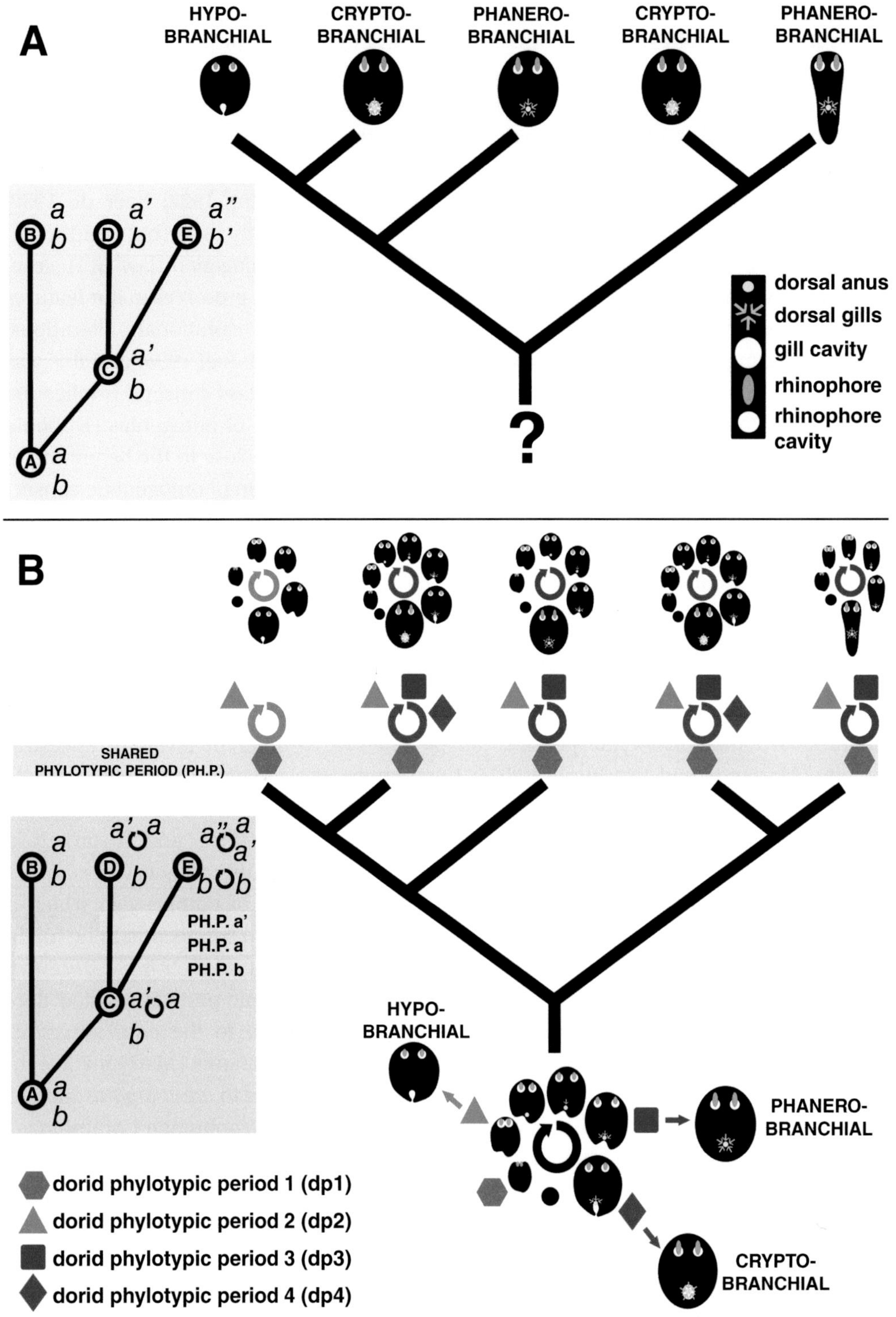

Figure 4.2 Comparison of the major features of phylogenetic (A) and ontogenetic systematics (B) exemplified by the dorid nudibranchs. Phylogenetic systematics is based on the analysis of plesiomorphies and apomorphies without consideration of the ontogeny, whereas molecular phylogenetics uses analyses of the nucleotide sequences without ontogenetic data. Hennig's original

important, therefore, to conjoin the recapitulation concept of Haekel (1866) to that of the phylotypic period approach, whereby phyloperiods partially recapitulate the ancestral ontogeny. Under this reformulation, Haeckel's 'biogenetic law' is no longer a 'mirage' (Raff and Kaufman 1983: 19) but a key pattern in an ontogeny. The shared presence of phylotypic periods in ontogeny indicates that if ontogeny and evo-devo were central to the modern synthesis, ontogenetic periodicity would play a more significant role in current taxonomy.

4.5.5 An Ontogenetic Framework Obviates the Need for Cryptic Species: The PhyloPeriodic Approach

The regular, periodic patterns observed in ontogenies suggest that a long-neglected approach, similar to the Periodic Table of Elements, could be fruitful for taxonomy and phylogeny. Whilst parallel evolution of diverse organisms has been recognised (West-Eberhard 2003), the periodic approach is not mainstream to taxonomy or phylogenetics (e.g. Schimkewitsch 1906; Vavilov 1922; Babaev 2019). The time is now ripe for the incorporation of periodicity as a fundamental component of the field.

Variation between taxa extends far beyond that of regular parallel series of characters (Bolnick et al. 2019), but in some respects it is comparable to that seen in the periodic system of chemistry. Comparisons between the periodicity of ontogeny and chemistry have previously been made (Wang and Ellington 2019). Examples include the regular diffusion patterns of interacting chemical morphogens in developing embryos, which give rise to repeated patterns of characters in distantly related organisms (Fowler et al. 1992; Kondo and Miura 2010; Futuyma and Kirkpatrick 2017; Haupaix and Manceau 2020). Bird feathers have been shown to be formed by an ontogeny of the regular periodic patterning of a coordinated reaction-diffusion-taxis system (Ho et al. 2019). Studies across a broad range of taxa, from rodents (Johnson et al. 2019), birds (Inaba and Chuong 2020), and fish (Salis et al. 2019) to flatworms (Heratha and Lobo 2020) have confirmed that such regular patterns are widespread.

Figure 4.2 (*cont.*) scheme (1966: fig. 21,3 with minor changes, yellow box) of phylogenetic character transformation. Modification of Hennig's figure 21,3 to show that phylogenetic systematics omits the important transtaxon similarity of the phylotypic periods (green box). Ontogenetic systematics reveals and compares the shared phyloperiods (highlighted by gray bars) and taxa-specific adult morphologies, and using integration with the molecular analysis targets to infer ancestral ontogenetic cycle (including ancestral adult body plan). Smaller black icons indicate developmental sequence of the dorid phyloperiods, coloured figures represent key phylotypic periods (dp 1–dp 4). Largest black icons indicate respective dorid adult body plans. Coloured arrows indicate evolutionary connections of the dorid phyloperiods to respective adult organisation. Compared to the phylogenetic systematics, ontogenetic systematics implies that phylotypic, juvenile plesiomorphies present in any subgroup of a given group as shared phylotypic periods. The juvenile plesiomorphies may independently transform to the adult apomorphies in the course of the paedomorphic evolution in different subgroups (e.g. phyloperiod dp 2 transforms to adult hypobranch organisation of corambids, see text).

We refer to these regular, repeating patterns across different characters – the product of the ontogenetic processes – as periodic patterns. Because they are shared by disparate groups, they can be a source of 'cryptic' forms. For example, similar coloration patterns in some nudibranchs and flatworms. All organisms contain repetitive (modular, Newman 2010) structures, and the phenomenon of periodicity in biological processes is widespread at molecular, cellular, and organismal levels during ontogeny (Amariei et al. 2014). Cellular mechanisms result in the appearance of segmentation-like patterns among unrelated metazoan phyla (Chipman 2020). Jung et al. (1998) concluded that 'The formation of periodic patterns is fundamental in biology'.

When the periodic ontogenetic processes that lead to the formation of regular or quasiregular morphological patterns (Haupaix and Manceau 2020) undergo evolutionary modification in various closely or distantly related taxa, a phylogenetic periodicity can emerge that can serve as a framework for taxonomic studies. Another source of periodicity in ontogenetic taxonomic characters arises from basic properties of ontogenies. Each ontogeny is a highly conserved system, so that when any organism is compared to relatives within a lineage we can expect to find a similar set of characters, structures, and processes (Holland 2013; Smaczniak et al. 2012). While not a strict rule, generally the lower the rank, the greater the similarity. Whereas evolutionary shifts in the conserved ontogeny are possible, they are constrained by the properties of complex developmental networks. This combination of a highly conserved framework with smaller-scale modifications can partly explain the appearance of periodic-like patterns during the formation of biological diversity (Figures 4.1–4.3). Conserved ontogenetic features are represented as a horizontal row of characters mapped at various ranks across taxa, the latter defined by molecular data and constituting the columns. This can be well illustrated, for example, by the families of dorid nudibranchs (Martynov and Korshunova, 2015) or the species of aeolidacean nudibranchs (Korshunova et al. 2020c). If we add to a row of ontogenetic features, conserved in many of the taxa, a column of less conserved features that specify the taxonomic characters of a particular taxonomic subgroup, the ontogenetic features will appear in a periodic-like pattern.

The periodic arrangement of conservative ontogenetic characters is supported by the concept of phylotypic periods (Arthur 2015) (see preceding section). For example, most species in the nudibranch genus *Dendronotus* (Figure 4.1) for which data are available undergo a phyloperiod characterised by strongly denticulate central radular teeth with distinct denticles. This is clear evidence that periodic ontogenetic processes directly influence taxonomic decisions. In their earlier ontogeny, the overwhelming majority of *Dendronotus* species show parallel, periodic appearance of a 'cryptic' stage, almost entirely confined to juvenile forms, characterised by distinctly denticulate radular teeth (Figure 4.1). Radular features are taxonomically important in many nudibranch taxa, including *Dendronotus*. It is, however, very difficult to distinguish species of *Dendronotus* using juvenile radulae because of their great similarity consequent on their ontogenetic periodicity, that is, at the juvenile stage the species exhibit high levels of crypticity. The juvenile teeth can partly persist in the anterior part of the adult nudibranch radulae, further

PHYLOTYPIC PERIODS

TAXONOMIC GROUPS

periods \ TAXA	CRYPTOBRANCHIAL (gill cavity; dorsal gills or ventral secondary gills)	PHANEROBRANCHIAL (no gill cavity; dorsal gills or ventral secondary gills)	HYPOBRANCHIAL (rudimental gill cavity; ventral gills)
ACTINOCYCLIDAE	all taxa		
CALYCIDORIDIDAE	all taxa		
DORIDIDAE	all taxa		
DISCODORIDIDAE	all taxa		
CADLINIDAE	all taxa		
DENDRODORIDIDAE	all taxa		
SHOWAJIDAIIDAE	all taxa		
CADLINELLIDAE	all taxa		
CHROMODORIDIDAE	all taxa		
AKIODORIDIDAE	new taxa are expected	majority of taxa	
ONCHIDORIDIDAE	*Onchimira* ✓	majority of taxa	
AEGIRIDAE	new taxa are expected	majority of taxa	
GONIODORIDIDAE	new taxa are expected	majority of taxa	new taxa are expected
HEXABRANCHIDAE	new taxa are expected	majority of taxa	
POLYCERIDAE	*Elfdaliana* ✓	majority of taxa	
GYMNODORIDIDAE	new taxa are expected	majority of taxa	
BATHYDORIDIDAE	new taxa are expected	majority of taxa	
PHYLLIDIIDAE	majority of taxa	*Fryeria*	new taxa are expected
CORAMBIDAE	*Loy* ✓		*Corambe*

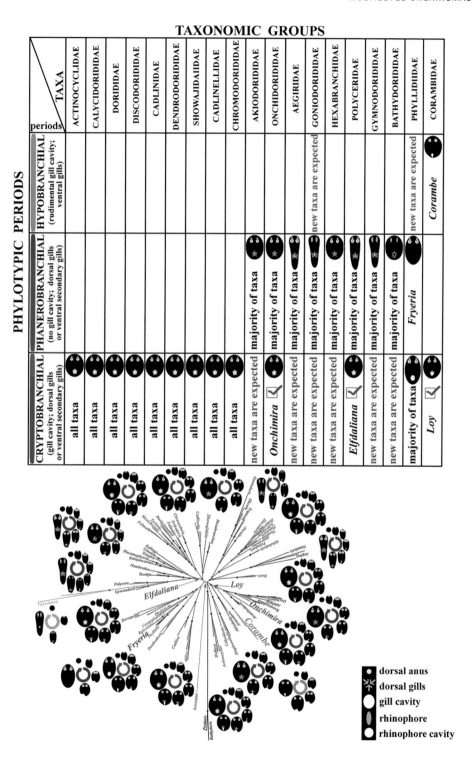

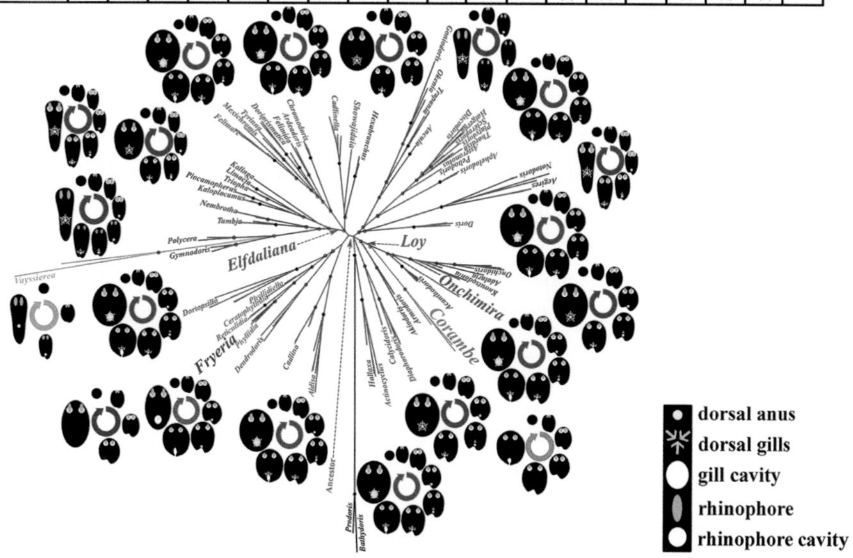

Figure 4.3 PhyloPeriodic Table of the dorid nudibranchs supplemented with dorid molecular phylogeny (based on Korshunova et al. 2020a). Three major adult body plans of the dorids (cryptobranchs, phanerobranchs, and hypobranchs) correlate with the dorid phyloperiods.

complicating hypotheses of homology. The co-occurrence of juvenile and adult radulae in mature nudibranchs has already led to considerable overestimation of adult radula variability in *Dendronotus*, which has directly influenced the delimitation of species (Robilliard 1970). While radular morphology is consistent within a species, colour is not, and so 'cryptic' species result when greater emphasis is placed on colour for their delimitation (Korshunova et al. 2017b, fig. 1). The combination of highly conserved phyloperiods of juvenile radulae and less conserved variation in colour can be considered similar to a periodic pattern (Figure 4.1), as they are underlain by a common genomic and ontogenetic basis and the processes involved are evidently periodic, as demonstrated in several disparate groups of organisms (Heratha and Lobo 2020; Inaba and Chuong 2020).

There are both differences and similarities in the usage and meaning of the term 'periodic' in the ontogenetic (biological) sense, compared to Mendeleev's Periodic Table classification of the chemical elements. Each element has discrete, relatively stable characteristics, and with increasing atomic number the chemical properties of different elements demonstrate a regular, periodic similarity in horizontal rows (*periods*) and vertical columns (*groups*) of the Table (Mendeleev 1869; Pyykkö 2019). In ontogeny, taxonomy, and phylogeny, however, morphological characters and stages, while exhibiting more continuous variation, also form repeated periodic patterns. Thus, just as the fundamental properties of chemical elements are reflected by their position within the Periodic Table, conserved *phylotypic periods* (horizontal rows) reflect fundamental ontogenetic properties, which when combined with smaller-scale differences, delimit particular *taxonomic groups* (vertical columns) (Figure 4.3).

Phyloperiods are fundamental features of the ontogeny of multicellular organisms and are basic to the construction of each respective body plan at both chemical (molecuar) and morphological levels. In unicellular organisms, complexes of DNA, proteins, and other components can be recognised as pre-phylotypic periods. Phylotypic periods can be viewed as linked to the deep homology concept (DiFrisco and Jaeger 2021). An important difference is that chemical elements have discrete unchanging properties, whereas ontogenies are able to evolve, generating considerable biological diversity, albeit within a relatively fixed framework. A biological (ontogenetic) 'Periodic Table' will therefore not be as compact and regular as the chemical one, because chemical elements possess only two main properties – atomic number and electron configuration – whereas the number of properties of ontogenies is indefinitely large. In fact, however, the periodic chemical properties of

Figure 4.3 (*cont.*) Empty cells in the table refer to yet unknown taxa with a particular combination of the shared phyloperiods and adult morphologies. In case there is a higher probability of presence of yet undiscovered taxa with particular phyloperiods and adult character combinations (because previously taxa with similar characters were discovered in the same or related families, see Martynov and Korshunova 2015), they are indicated as 'new taxa are expected'. These features constitute prognostic aspect of the PhyloPeriodic Table. See also the text.

elements have been shown more recently to be less completely regular or strictly periodic than previously thought (Scerri 2020).

Another analogy between the periodicity of ontogeny and the chemical elements is that while phylogeny is apparently continuous, an ontogeny consists of a discrete (semidiscrete) organismal unit that undergoes regular, predictable change within the periodic or quasi-periodic frame of a life cycle. When we shift our focus from the search for the source of biological diversity in continuous phylogenetic lineages to discrete, cyclically changing ontogenies, we can detect periodic ontogenetic patterns at the level of individuals and use them to classify organisms. However, ontogenies undergoing evolutionary modification display both periodic and tree-like phylogenetic patterns, and so we need to capture this combination of biological periodicity and cladogenesis by means of *PhyloPeriodic Tables* (PPT), a matrix-like representation.

An important similarity of PhyloPeriodic Tables to the Periodic Table of elements is their capacity to predict phylotypic stages in as yet unstudied subgroups. Previously, using early versions of tables (then known as 'Prognostic Table', Martynov 2011b) we predicted, by leaving a blank cell in the table, the existence of particular key features of the gill cavity in a nudibranch family Polyceridae. This was confirmed four years later by the discovery of *Elfdaliana profundimontana* (Martynov and Korshunova 2015). PhyloPeriodic Tables have a practical purpose – to incorporate ontogenetic periodicity into systematic biology. Although we have data on the early conserved radular teeth for all species included in the phylogeny of Dendronotidae (Figure 4.1), there are some species lacking data (Korshunova et al. 2020b; Martynov et al. 2020b). The PhyloPeriodic Table (Figure 4.1) can be used to predict the existence of a phylotypic period for radular teeth development (and other characters) in these species. In the same way, a PhyloPeriodic Table for dorid nudibranchs can be used to predict the existence of phyloperiods in that taxon, as well as the probable existence of particular adult characters in as yet undiscovered taxa (Figure 4.3).

PhyloPeriodicity is distinct from bioperiodicity (though partly related), which includes biological periodic patterns as circadian rhythms (Hess 2000; Packard et al. 2020). The use of phyloperiodic tables is a step towards a more complete synthesis of the terminologically distinct but intimately connected concepts of *ontogeny and phylogeny*. Every taxonomic group will have several PhyloPeriodic Tables because in each ontogeny there are phyloperiods in different hierarchies (e.g. phyla within Metazoa), the generation of which will be a fundamental task for future studies.

4.5.6 The PhyloPeriodic Approach Helps to Streamline Fine-Scale Diversity

An example of the power of the PhyloPeriodic approach to recover new 'fine-scale' character states, useful for the diagnosis of species within a complex, can be seen with respect to colour morphs, for example, in the nudibranch genus *Amphorina* (Korshunova et al. 2020c). The adult periodicity of these species is less regular compared to the phylotypic periods but can be used to recognise 'cryptic' morphological diversity in a group of studied taxa by integrating series of similar characters with molecular data in matrix-like form.

There have been some attempts, outside an ontogenetic framework, to delimit highly similar colour patterns of different taxa within the Chromodorididae nudibranchs (Rudman 1984; Padula et al. 2016). Some chromodoridids possess striped patterns that have recently been shown to be determined by ontogenetic periodicity in vertebrates (Gante et al. 2018). Currently, we have no information about the particular genes coding the appearance of stripes or spots in nudibranchs. However, a common genetic basis for these groups implies homeobox and other developmental genes and interacting chemical morphogens in developing embryos, as demonstrated in other groups (Kondo and Miura 2010). Similar modifications of these periodic and quasiperiodic patterns may appear during evolutionary shifts of the ontogeny. When similar morphs of different species are included in the same horizontal periods, differences between hitherto 'cryptic' species become apparent.

We conclude that the incorporation of ontogenetic characters in a PhyloPeriodic Table is a practical method for discovering diagnosable characters within putative cryptic species complexes as well enabling the incorporation of all phyloperiodic characters. A molecular phylogeny allows ontogenetic characters to be aligned and highlights the juvenile appearance of similar characters in clearly different species, as in the Dendronotidae (Figure 4.1). While a resolved phylogeny provides supported hypotheses of monophyly, lower molecular divergence values, per se, should not be used as the only necessary evidence for arguing that some species-level or other taxon-level groups should be merged with each other. Instead, different methods in the general ontogenetic framework can be applied to reveal multilevel fine-scale diversity. The use of a general ontogenetic framework will ensure a more rigorous and process-orientated investigation of morphology. The ontogenetic framework is not just a theoretical addition to a suite of already existing practical methods but a key element for truly integrative study. The ontogenetic framework counters the inherent bias of focussing only on adult terminal characters, and enables us to include morphological, genetic, epigenetic, behavioural, and ecological complexes of characters from different periods and levels of ontogeny in classifications.

Practical steps for using the PhyloPeriodic framework for recognising fine-scale diversity are as follows: (1) Select a taxonomic group and sample specimens; (2) Undertake the relevant morphological study of the samples; (3) Ensure that ontogenetic information is considered during taxonomic assessment, because adult diagnostic characters can be considerably transformed at different ontogenetic stages and adult paedomorphic characters can easily be confused with juvenile transitional features; (4) Undertake the appropriate bibliographic study, to evaluate the nomenclature of the target group; (5) Collect molecular data for the sample specimens; (6) Compare the results of the morphological (step 2) and taxonomic investigation (steps 3, 4) with the molecular (step 5) results; (7) Arrange diagnostic ontogenetic (including morphological) features in a PhyloPeriodic Table (Figure 4.1) to inform a further search for fine-scale differences between closely related species; (8) Where hard to distinguish ('cryptic') variants are suspected, carry out a statistical evaluation of the range of variation of potentially diagnostic features; (9) Assemble the complete study to provide fine-scale taxonomic diagnoses for the taxa at different levels (including new taxa); (10) Test the established system by further investigation with new materials and data.

4.5.7 Plesiomorphies as Integral Parts of Individuals and Their Ontogenetic Transformation into Apomorphies

Because the plesiomorphy and apomorphy framework is of key importance for phylogenetic systematics, we present here figure 21 from Hennig (1966: 89) to illustrate a significant deficiency of phylogenetic approaches that lack consistent consideration of ontogeny. Hennig distinguished the plesiomorphies *a, b* and apomorphies *a′*, *a″*, *b′* (Figure 4.2). However, the plesiomorphic states *a, b* are still part of the ontogenies (as phylotypic periods) within the exemplified lineages where apomorphic transformations *a′*, *a″*, *b′* have occurred towards adult stages. This, critically, was omitted by Hennig in his phylogenetic argumentation. For example, the apomorphic phanerobranch gill condition evolved several times independently in dorids, but the ancestral cryptobranch state can still be detected as phyloperiods in the ontogeny of dorids that have the adult phanerobranch state (Figure 4.2). Such cases are widespread in disparate groups of organisms and can be termed juvenile or *phylotypic plesiomorphies*.

Underestimation of phylotypic plesiomorphies may result in the misinterpretation of secondarily reduced or paedomorphic characters as true plesiomorphies (Figure 4.2). For example, the absence of lateral rhinophoral papillae, as well as details of the radulae teeth, are paedomorphic attributes in adult *Dendronotus subramosus* (Figure 4.1). Independent, multiple transformations of juvenile plesiomorphies into adult apomorphies in the course of paedomorphosis (which for Hennig 1966: 228–230 were just particular cases) have repeatedly been demonstrated in nudibranchs (Korshunova et al. 2018, 2019, 2020a, b; Korshunova and Martynov 2020) and ophiuroid echinoderms (Martynov 2009; Stöhr and Martynov 2016). Common phylotypic stages (*juvenile plesiomorphies*) that are transitional in the ontogeny of all taxa can be transformed into *adult apomorphies* as a consequence of heterochronic shifts in their evolution (Figure 4.2). The fundamental importance of this distinction is confirmed by the fact that a recent phylogenetic study especially mentioned the 'inability of the analyses to differentiate retained plesiomorphies from juvenile features' (Gee 2020: 79), despite that ontogeny has been discussed both in Hennig (1966) and also later (Kluge 1985).

Thus, ancestral plesiomorphic states transformed or reduced at adult stage are still integral parts of the ontogenies of crown taxa. This makes a fundamental difference between phylogenetic and ontogenetic systematics. Organisms must be compared at equivalent stages in their ontogenetic cycle to attain maximum coherence between taxonomic arrangement and underlying genetic and epigenetic systems. However, ontogeny is not a combination of semaphoronts (Sharma et al. 2017) but a complex natural phenomenon that far exceeds the simple binary choice of plesiomorphy-apomorphy. The continued use of the terms 'plesiomorphic-apomorphic' is possible only with special modifications (i.e. phylotypic, juvenile, and adult plesiomorphies-apomorphies) (Figure 4.2).

4.5.8 The Ability to Incorporate Any Respective Biological Features and Processes into Classification

An ontogenetic framework highlights the presence in biological organisms of multiple levels and numerous elements underlain by various processes and allows inclusion of many more biological features. Real biological diversity far exceeds that reflected in current classifications. For example, capturing inter-taxon 'cryptic' ontogenetic similarity expressed in the phylotypic periods of every individual is not part of current systematic methods, but nevertheless significantly influences both the taxonomic process and resulting groupings. Such a concept of an inter-kingdom organism – the holobiont – cannot be expressed (Jaspers et al. 2020; Vannier et al. 2019) within current systems of nomenclature. Individuals of two 'cryptic species' should not only possess finer morphological differences but also differ in their chemical composition and microbiomes. After drafting of this chapter, we received remarkable confirmation of this statement (Evans et al. 2021).

Attempts to replace strict isolation species recognition criteria in favour of mutual recognition (Paterson 1985) or cohesion (Templeton 1989) have already been made. The ontogeny-related framework, however, escapes from the need to adhere to concepts such as 'species', 'reproductively isolated breeding communities', or 'separate lineages' by relying on the ontogenetic cycle of each individual to classify biological properties.

Horizontal gene transfer occurs among all organisms. This is not reflected in current systems of nomenclature and classification but crosses barriers between closely and distantly related organisms (Jablonka and Lamb 2010); for example, metazoan bdelloid rotifers (phylum Rotifera) are able to borrow genes from other kingdoms of organisms (Gladyshev et al. 2008; Yoshida et al. 2019).

Immune systems are well documented properties of all organisms, both unicellular and multicellular (Pradeu 2012). Every organism can be expected to be immunologically most similar to itself or to its clonal descendants, according to what may be called the 'immune (protection)' principle, although this often operates within a fragile balance of cooperation and conflict between genes and cells (Carmel and Shavit 2020; Simpson et al. 2020; Blackstone 2021). We can estimate that maximally related individuals at any level of the evolutionary hierarchy should show an increasing similarity of immune and protection properties from the most general to the most narrowly defined groups. For example, in monozygotic twins, transplants can be carried out successfully without immunosuppression (Jorgensen et al. 2020). For increasingly complex organismal systems at various levels, immune-protection systems are a better foundation for recognising biological diversity than attempts to categorise all levels of biological organisation into 'species' with the addendum of 'cryptic' or 'non cryptic'. Our proposal is in line with previous conceptualisations in that a species concept should not be centred exclusively on the isolation of groups from each other, but rather on underlying biological traits that allow recognition and communication (in the broadest sense, including language-related phenomena) at different organismal levels (Paterson 1985; Templeton 1989). For discussion and continuity purposes, in a transitional period, we mention here 'species' and 'taxa', but all levels of biological diversity should be incorporated into a classification – and we term this biological concept *multilevel (fine-scale) organismal diversity (MOD).*

4.6 Organism-Centred Classification

4.6.1 Classification and Species Concepts Must Consider the Various Ontogenetic Properties of Organisms and Diverse Evolutionary Properties at Different Levels of Complexity

To encompass a wide variety of scenarios, eight ontogenetic modes of reproduction and structure that contribute to the framework of constraints that an ontogeny imposes on any potential 'species' are presented in Figure 4.4. Ontogeny 0 denotes viruses, which are non-cellular, that is, with only a rudimentary ontogeny, implying either an ancestral condition for all life, or a secondary simplification, and hence 'Species 0'. Ontogeny I ('Species I') consists of bacteria and archaea, that is, single-celled prokaryotic organisms dominated by clonal reproduction and horizontal gene transfer and lacking true sexual reproduction. Ontogeny II ('Species II') includes unicellular protists, microalgae, and a few fungi, that is, single-celled eukaryotic organisms with well-defined clonal reproduction and with or without sexual reproduction. Ontogeny III ('Species III') consists of multicellular organisms with 'pre-tissues' (animals) or specific plant tissue types, lacking clearly defined organs (animals) or internal organs (plants and fungi) – clonal reproduction prevails alongside the capacity for sexual reproduction (e.g. sponges). Ontogeny IV ('Species IV') consists of multicellular organisms with less differentiated diploblastic tissues and differentiated simple internal organs – clonal reproduction prevails alongside the capacity for sexual reproduction (e.g. cnidarians). Ontogeny V ('Species V') consists of multicellular organisms with differentiated triploblastic tissues and complex differentiated internal organs – clonal reproduction prevails alongside the capacity for sexual reproduction (e.g. various marine metazoan bilaterian phyla). Ontogeny VI ('Species VI') comprises multicellular organisms with differentiated triploblastic tissues and complex differentiated internal organs – clonal reproduction is lacking, and sexual reproduction prevails, sometimes reduced to parthenogenesis (various bilaterian metazoan phyla). Ontogeny VII ('Species VII') consists of multicellular organisms with differentiated complex triploblastic tissues and differentiated internal organs – clonal reproduction is lacking, and reproduction is exclusively sexual, without parthenogenesis (e.g. mammals). In reality, more levels of structural complexity and ontogenetic patterns could be recognised. For example, we have united ontogenetic categories of some metazoans and plants due to their similarity and the prevalence of clonal reproduction, although this is somewhat artificial.

Because there is no universally accepted species concept, this preliminary ontogenetic classification could be of considerable consequence for resolving the 'species problem' and associated 'cryptic species subproblem'. For example, organisms from approximately 20 Metazoan phyla reproduce partly or exclusively by cloning (Minelli 2009; Martynov

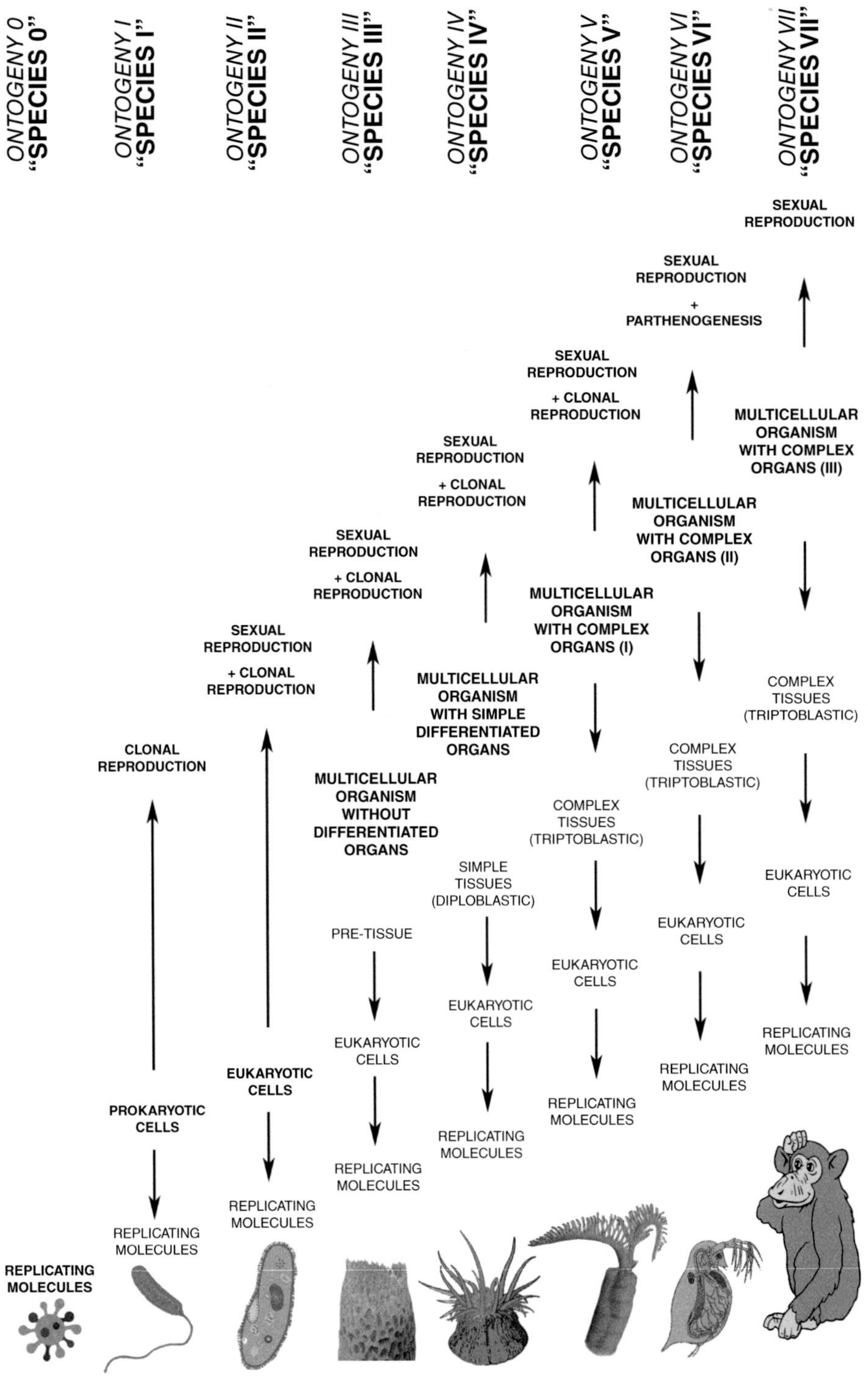

Figure 4.4 Eight different modes of reproduction that form particular variants of ontogeny (0–VII) and respective 'species' (0–VII) as a basis of MOD. See explanation in the text.

2013; Hiebert et al. 2020). This greatly limits the applicability of reproductive isolation as a criterion of species definition, since it depends on sexual reproduction. Clonal reproduction, clonality, is distinct from 'asexual reproduction'. In its original sense, clonality involved only somatic tissue, whereas 'asexual' includes sexual reproduction without fertilisation (parthenogenesis, apomixis).

In a process related to clonal reproduction, organisms with highly developed clonal reproduction (e.g. cnidarian metazoans) can change their body plan drastically by modifications of some homeobox genes (Jakob and Schierwater 2007) or during regeneration (Reitzel et al. 2007), whereas exclusively sexual organisms are unable to do so. Clonal reproduction is more widespread in plants than metazoans (Stuefer et al. 2004; Vannier et al. 2019). The impact of ontogeny on evolutionary processes (and thus on 'species' properties) is considerably different in clonal and sexual organisms. Despite the potential disadvantages of long-term clonality for the genomic system, clonal organisms have more adaptive flexibility (e.g. ability to adapt quicker) because they do not depend on another individual for reproduction.

Importantly, clonally reproducing organisms do not need any population processes to form a clonal lineage (characterised by a particular combination of genetic and epigenetic characters). A group of individuals resulting from clonal reproduction will be closely related and have identical immune defence systems so that genetic conflict within clusters of uniclonal individuals is minimal (Ratcliff et al. 2012). Spatially and morphologically distinct groups of clonal individuals are well known, for example, in cnidarian actinias (Glon et al. 2019).

This framework implies that ontogenetic properties (and hence 'species') differ considerably between clonal and sexual organisms. Problems with the strict application of the biological species concept are widely known (Paterson 1985), yet it remains generally accepted. The biological species concept arose from an overestimation of importance of isolation criteria for species definition (Mayr 1942). For clonal taxa, however, there is no barrier to the spread of characters through a population. Clonal reproduction provides a pathway for the production of a population of morphologically similar individuals (and allows rapid colonisation of new habitats) that could be considered distinct species using morphological criteria. For organisms with simpler tissues (Ontogeny III, IV), clonal reproduction can enable changes in basic body plan that would be impossible under the strong constraints of sexual reproduction (Figure 4.4). All of today's higher taxonomic ranks, such as phyla, must have originated as small clonal groups in which drastically different body plans (i.e. bilaterian phyla) could arise significantly quicker than by transformation under sexual reproduction. Sexual reproduction can be viewed as an exchange of genetic information between initially clonal groups (e.g. horizontal gene transfer among bacteria), and it is therefore a secondary process in evolutionary terms. This is supported by the broad phylogenetic distribution of clonal organisms. From this we can infer that any group that can be called a 'species' inevitably shows some features of clones.

Clones accumulate epigenetic (Fraga et al. 2005; Bierbach et al. 2017) and genetic changes (Herrera 2009). The currently recognised 'species' taxa include a vast diversity of life strategies. Even predominantly clonal organisms that have occasional sexual reproduction remain as clone-like groups. Clonal or quasi-clonal groups are able to interact with each other to a degree that depends on the level of cooperation and conflict relations (Queller and Strassmann 2009). Such interactions may include horizontal gene transfer or sexual reproduction through which they can exchange recently modified variants of characters. The ability of sexually reproducing groups to hybridise and produce fertile offspring crosses the boundaries of species that are well characterised phenotypically or genetically (Harrison and Larson 2014; Crossman et al. 2016; Jacobs et al. 2019; Ottenburghs 2020) and this networking process of sexual reproduction contributes further to the formation of multilevel organismal diversity.

4.6.2 Biological (Ontogenetic) Properties Shift the 'Species Concept'

Within advanced multicellular organisms there are numerous cases where females can reproduce parthenogenetically in metazoans or apomictically in angiosperms (León-Martínez and Vielle-Calzada 2019). The impact of asexual reproduction on the genome is not immediately disadvantageous (Pellino et al. 2013), but it results in an absence of genome-wide recombination in persistently asexual lineages (Glémin et al. 2019; Jaron et al. 2021). The underlying reasons for persistence of sexual reproduction are not completely explained (Sharp and Otto 2016; Neiman et al. 2017; Hörandl et al. 2020) by the need to reduce the load of deleterious mutations (Kondrashov 2018). Recent studies have failed to demonstrate the accumulation of deleterious mutations across a broad range of parthenogenetic insects (Brandt et al. 2019) or in clonal fish hybrids (Koĉi et al. 2020).

True clonality is the most ancient mechanism of reproduction in organisms (Hörandl et al. 2020) (Figure 4.4). Evolutionary radiation and biological diversity have been driven in many prokaryotic and eukaryotic organisms by clonal processes, and sexual reproduction cannot be fundamental to organismal evolution. The broad distribution of clonality and parthenogenesis (apomixis), even within multicellular organisms, makes the species criterion of reproductive isolation difficult to be apply and undermines the necessity for species to embody population structure and gene flow. This observation shifts the focus of the main evolutionary forces from groups of organisms to the interacting organism, *per se*. An additional challenge for a conventional species concept arises where the gradation from individual organism to a population or species is even more fine scale. For example, plants can undergo complex interactions between phases of clonal and sexual reproduction in the same organism (Vannier et al. 2019). The terms 'broader species' and 'narrow species' (comparable to small groups of clones) are more common in botany ('linneons', 'jordanons', 'microspecies' etc., Lotsy 1906; Mishler and Budd 1990; Majeský et al. 2015; Sharma and Bhat 2020). Among vertebrates a substantial number of species reproduce unisexually (Laskowski et al. 2019). Obligate clonal reproduction has also been demonstrated for animals in which genome-wide studies have not been able to verify a full complement of meiotic genes (Wilding et al. 2020). Numerous epigenetic studies (e.g. Holliday 2006; Burton and Metcalfe 2014; Yoshikawa et al. 2020) suggest that evolution occurs at the level of the individual. By organism level we refer not only to a canonical separate organism with dominance of sexual reproduction but to a continuum from clonal groups of individuals to true colonies. This has serious consequences for understanding species concepts, including the 'unified species concept' (De Queiroz 2007), which depends on gene flow as evidence for lineages. The properties (constraints) of ontogeny, and hence of evolution, will differ according to combinations of features (Figure 4.4). For example, a clonal organism can reproduce characters without the requirement for a complex population process. The species concept would therefore need to be different in groups with very divergent ontogenetic patterns.

Current codes of biological nomenclature only partially consider major differences in ontogenetic, and hence evolutionary, properties at different levels of organisation (Figure 4.4). In the code of nomenclature for algae, fungi, and plants (Turland et al. 2018) it is possible to recognise the ranks of variety or subform, but in the zoological code only a vague concept of subspecies is recognised. This difference in nomenclatural codes may reflect a greater degree of plasticity in plants, possibly linked to shifts between asexual (in broad sense) and sexual reproduction. It might also be because botanists are less attached to the biological species concept. Animals, however, exhibit variation below the rank of species that is worth recognising and fixing nomenclaturally. Virus nomenclature is governed by the International Committee on Taxonomy of Viruses (Adams et al. 2017; ICTV 2020), which partly allows non-binomial names. The second Baltimore classification (Simmonds and Aiewsakun 2018) is more similar to a chemical nomenclature than a biological classification. This reflects the fact that viruses can be described as chemical units whilst being able to replicate, a feature of living organisms. Despite the flexibility of biological nomenclature, the term 'species' continues to be used without any clear definition or delimitation.

There have been proposals to place clonal and unisexual organisms in a special category, as 'agamospecies' or 'uniparental' species (Willmann and Meier 2000). This category should include several million bacterial taxa, species from ca. 400 plant genera, members of at least 18 phyla of metazoans, and several hundred species of vertebrates (Fei et al. 2019; Laskovsky et al. 2019). Considering that clonal and sexual reproduction can be intertwined (León-Martínez and Vielle-Calzada 2019; Koĉi et al. 2020), it follows that an organism can be classified under different species concepts at different stages in their life cycle (ontogeny). For example, the cladoceran arthropod *Daphnia* alternates parthenogenetic and sexual phases during a single year. Hennig (1966: 209), however, refused to consider parthenogenetic organisms as 'species'. Because cyclic parthenogenesis occurs in at least 15,000 'species' across several animal groups (Thielsch et al. 2009), a significant amount of diversity would thereby be excluded from that covered by the term 'species'. This seriously undermines the ability of both 'species concepts' and phylogenetic systematics to accommodate real patterns of natural diversity. We propose that instead of trying to unite strongly contrasting ontogenetic modes under a universal term by adding prefixes to the term 'species' (including secondary term 'cryptic'), we should instead revise our expectations of the species concept.

Any individual organism, including clones, is unique in its patterns and processes. This statement should be central in revising our understanding of the 'species' concept. The individual organism, perceived within a multilevel framework, should be the focus for evolutionary and taxonomic study. By removing 'species' from evolutionary concepts we can focus on real natural phenomena of multilevel organismal diversity (MOD), instead of constraining the comprehension of evolution by premising that 'evolution is the origin of species'.

The idea of the 'individuality' of higher categories appears to be an old one (Hennig 1966: 221), but species rather than individuals became main focus for phylogenetic systematics. Ghiselin (1974) considered species to be individuals by analogy with commercial companies and without an emphasis on the individual organism. Within systematic biology it has occasionally been proposed that only individuals can be real terminals (Vrana and Wheeler 1992). Adopting the *ontogenetic systematics* approach advocated here would require connecting taxonomic nomenclature to the MOD concept and implies a potential ranking at the level of the individual, because populations do not have an ontogeny. Any individual, or clonal group is, per se, an extremely complex multilevel system at the molecular, cell, tissue, and macromorphological levels, composed of trillions of elements encompassed within a particular ontogenetic cycle. With increasing complexity in successive ontogenetic frames, the autonomisation of particular individuals (Rosslenbroich 2014) and the constraints of sexual reproduction increase, whereas the clonal component decreases (Figure 4.4). All biological organisms, even bacteria, exhibit individuality, and every individual, even as a transient phase during clonal reproduction, bears a complex of energy-dependent structures that support feeding, growth, defense, and reproduction.

We do not propose to simply substitute 'species' by 'individuals' or deny the existence of a 'species-related' phenomenon (Gregg 1950). What we propose is that our understanding and study of organisms should shift its focus from the terminals in an analysis to interacting individuals as the central phenomenon in evolution. Each individual organism encapsulates its evolutionary history through the inherited genetic material manifested in its phylotypic periods, and epigenetic processes modify rigid ontogenetic periods at the fine scale. Thus, an organism can be reimagined as a replicating supermolecule structured into numerous levels and exhibiting various capabilities to interact with other such 'supermolecules'.

4.7 Ontogenetic Systematics

Hennig (1966) was largely responsible for the present-day predominance of phylogenetics as a basis of classification. Establishing robust hypotheses of phylogenetic lineages, however, omits the underlying dynamics of ontogenetic cycles, including heterochronic and heterotopic modifications that can inform phylogenetic inference (Figure 4.2). Although Hennig presented a broader view of phylogeny, ' ... perhaps better called hologenetic' (Hennig 1966: 31), he did not make clear that 'hologeny' is ontogeny in its true sense (a life cycle) and stated, 'Phylogenetic kinship exists only between species and species groups, not between larval and imaginal stages of these species' (Hennig 1966: 122). Inter-taxon relations can, however, be established via phyloperiods (Figure 4.2). This may be the reason systematic biology after 60 years remains largely ontogeny-free.

Achievements in evolutionary developmental biology and epigenetics make it clear that, rather than an immobile black box on which blind evolutionary forces act to produce a phylogeny, individual organisms are complex molecular-morphological systems that exist as ontogenetic cycles (ontogeneses) with a considerable capacity for adaptation. Adaptations acquired during an individual's life can be transferred to subsequent generations through epigenetic and genomic events (Yehuda et al. 2016; Stenz et al 2018; Sweatt 2019). We propose the term *ontogenetic systematics* for a wider development of systematics to integrate classical taxonomy, molecular phylogenetics, evolutionary developmental biology, and epigenetics.

Previous attempts to incorporate or discuss ontogenetic principles into taxonomy (e.g. Orton 1955; Kluge and Strauss 1985; Ho 1992) did not result in consistent taxonomic applications. Within the field of 'evo-devo', taxonomy has rarely been mentioned (Hawkins 2002; Minelli 2007). Whilst there are precedents for coding ontogenetic stages (e.g. Wolfe and Hegna 2013), taxonomy has remained essentially ontogeny-free. For example, the paper by Albert et al. (1998) entitled 'Ontogenetic systematics, molecular developmental genetics, and the angiosperm petal' is an important attempt to link ontogenetic patterns with character evolution, but it has only limited application to taxonomy and phylogeny. The term *ontogenetic systematics* was independently proposed and unambiguously applied to the field of taxonomy by Martynov (2009), unaware of Albert et al.'s (1998) previous and different use of the term. Ontogenetic systematics in our sense (Martynov 2012a; Martynov et al. 2020a), that is, as *ontogenetic taxonomy*, has been included in a recent review of invertebrate developmental biology (Minelli 2015). The need to integrate evolutionary developmental biology, phylogenetics, and morphology has been mentioned by Richter and Wirkner (2014), and Wanninger (2015). Faria et al. (2020) highlighted a 'shortfall' of ontogenetic data.

The failure to incorporate ontogenetic data into systematic biology limits the precision of taxonomic representations of natural patterns. Ontogenetic systematics is based not only on formalised molecular genealogies of taxa but on analysis and description of the major structural modifications and transformations of organisms (as expressed in phylotypic periods), which lead to character patterns that systematists use to construct taxa. Seen from the viewpoint of the multilevel organismal diversity (MOD) approach, the problem of 'cryptic' species would not arise in ontogenetic systematics. Because living organisms are ontogeneses it is justified to decipher MOD also as the multilevel ontogenetic diversity.

The ontogenetic challenge for taxonomy is comparable with those that preceded the acceptance of the evolutionary paradigm, and consistent application of ontogeny in taxonomy opens up new scientific horizons, just as occurred previously with the implementation of evolutionary ideas in systematics.

4.8 Practical Workflow of Ontogenetic Systematics

I Selection of a taxonomic group. Our suggestion is a group classified traditionally as an order or class.

II Primary taxonomic evaluation based mostly on adult characters. Estimation of the group's diversity in the form of major subgroups and number of traditionally recognised species, etc.

III Primary ontogenetic investigation. Study of a selection of individuals belonging to different subgroups (families, suborders, etc.) and recognition of the main stages of larval and postlarval ontogeny.

IV Evaluation of the conservatism and variability of corresponding stages of ontogeny between various subgroups.

V Recognition and identification of similar phylotypic periods by detection of similar conserved developmental periods across various taxa of a studied group.

VI Formulation of a preliminary model of the ontogenetic cycle. The model should express a higher probability of finding similar conservative ('cryptic') elements among the subgroups in their phylotypic periods. Potentially, the presence of certain phylotypic periods can be predicted in subgroups for which individual development details are as yet unknown. More general examinations of ontogeneses are made possible by this method since it is not necessary to study every individual developmental pattern within the group.

VII Random and non-random similarities should be recognised in phylotypic periods of subgroups and adult periods of other, distantly or closely related taxonomic groups.

VIII Using data from VI and VII, a model of the ancestral ontogenetic cycle (including potential initial variability) should be constructed. This implies a particular sequence of evolutionary modifications in character homologies and a particular phylogenetic topology of related groups showing the ancestor–descendant relationships. Ontogenesis, expressed and conceived within the framework of phylotypic periods, can be construed as a more direct observation of evolutionary succession than phylogenetic reconstruction using solely molecular data.

IX All available methods of phylogenetic reconstruction (e.g. morphological and molecular phylogenetics, omics, paleontological data, etc.) should be applied and the results compared with the predictions of the ontogenetic cycle model. Inconsistencies should be detected and discussed. Further testing of the ontogenetic model should be proposed and carried out.

The principles and the workflow are mostly applicable to multicellular groups, but common morphological and molecular patterns can be found in unicellular prokaryotes and eukaryotes.

4.9 Phylotypic Periods as Trans Taxon (Inter-taxon) Nodes of Similarity and Relationship

The term phylotypic period (phyloperiod) implies the presence of similar, homologous, conservative periods of ontogeny shared by various groups with disparate adult morphology. In various taxonomic groups, these phyloperiods form *trans taxon (inter-taxon) nodes of similarity* (Figure 4.2), which can be 'cryptic', even between phylogenetically distant groupings. The early appearance of conservative, 'cryptic' phylotypic periods followed by transitions to 'non-cryptic' divergent stages in adult morphology has been demonstrated for large taxonomic groups both in morphology (Richards 2009) and by expression of developmental genes. It has been observed in plants (Quint et al. 2012) and metazoans (Piasecka et al. 2013). Phyloperiods and their underlying transcriptomic activity have been documented in phyla as disparate as nematodes, arthropods, and chordates (Domazet-Lošo and Tautz 2010; Kalinka et al. 2010; Ninova et al. 2014), but detailed linkages to body-plans have yet to be undertaken (Irie and Kuratani 2014). There is no agreed nomenclature for phylotypic periods. For example, whilst phyloperiods (as stages) for all Bilateria (Irie and Sehara-Fujisawa 2007; Levin et al. 2012) and for dorid nudibranchs (as periods, Martynov and Korshunova 2015) have been proposed, no equivalent terms have been suggested for the majority of phyla.

The hidden conservatism of phylotypic periods is underestimated by modern taxonomic methods because they are based exclusively on adult stages, although taxonomists are aware of ontogenetic conservatism (Raso et al. 2014). Systematic biology significantly underestimates independent reductions of initially complex characters in early ontogeny, and this may result in the incorrect inference of a simpler ancestral organisation (Martynov et al. 2020a). A best approximation of evolutionary scenarios can be attained only when framed by a well-defined ontogenetic model. It has already been shown (Korshunova et al. 2017a, 2020a) that wherever an ontogenetic model is missing, it is easy to misinterpret molecular phylogenies. The relevance of ontogenetic models to phylogenetic reconstruction is most relevant at higher ranks where molecular data can be ambiguous (Kumar et al. 2012). For example, placement of the phylum Ctenophora (comb-jellies) as the basal-most clade of multicellular animals (Metazoa), lower than phylum Porifera (sponges) (Moroz et al. 2014), is inconsistent with known morphological and physiological data and implies an independent origin of complex nervous systems (Jékely et al. 2015). More recent studies have demonstrated that this placement was due to errors in the phylogenetic methods used, and that Ctenophorans are unlikely to be the most basal metazoans (Redmond and McLysaght 2021: 1). Within an ontogenetic framework it appears evident that sponges (Martynov 2012b) are the only metazoans that share a similar organisation with the pre-metazoan colonial protists, Choanoflagellatea (Ros-Rocher et al. 2021). Ferretti et al. (2020) when evaluating conserved phylotypic periods of Bilateria, state 'We argue that evolutionary biologists should return from a purely gene-centric view of evolution and place more focus on analysing and defining conserved developmental processes and periods' (Ferretti et al. 2020: 1). This call confirms the need for *ontogenetic systematics*. With this approach, we have already put forward a phylotypic period-based model of dorid evolution that contributed to the resolution of previous incongruence in their taxonomy and phylogeny (Martynov and Korshunova 2015; Korshunova et al. 2020a).

Taxonomy is commonly viewed as a utilitarian tool for communication between various scientific and social fields. However, the application of the ontogenetic approach to classical taxonomy and modern phylogenetics reveals their excessive simplification of the natural complexity of organisms. Persistent issues with species definition, especially the current 'cryptic species' problem, indicate that before taxonomy can become really effective for communication, the natural phenomena used for a classification need to be much more clearly defined and understood. This will in turn radically change our understanding of the 'species' as a concept.

4.10 Multilevel Organismal Diversity (MOD) Within the Ontogenetic Framework

4.10.1 MOD and Species

MOD is an alternative to the highly controversial 'species concept', with a broader biological basis. As outlined earlier in this chapter, the term 'species' is used to refer to groups with fundamentally different ontogenetic properties. The comparison of 'species' of *Streptococcus* (bacterium) and *Dendronotus* (a nudibranch) at the same rank is meaningless because their basic ontogenetic features are very different (Figure 4.4). Furthermore, small groups of near-identical individuals, or groups of less closely related individuals (Figure 4.1), or even the equivalent of genera and families may all be hidden under the term species, giving an erroneous impression of equivalence (Korshunova et al. 2017a,b, 2018, 2020a–c, 2021). For these reasons, although Reydon and Kunz (2019: 12) argued that 'a straightforward realism about species is not an option', we do not support the 'weak realism' approach to retaining the 'species' as a universally applicable concept. Furthermore, the potential 'unified species concept' (De Queiroz 2007) does not reflect the great variety of the ontogenetic patterns and processes or that 'separately evolving metapopulations' may interact before and after 'separation' in many ways that are inconsistent with this proposal.

A next step would be to consider that no 'species' definition can satisfactorily describe the multilevel nature of organism (ontogenetic) diversity as it is now understood. We conclude that there needs to be a shift of focus away from populations to individual organisms' responses to environmental challenges. Society is conservative in this respect and favours the term 'species' over potential replacements (Burma 1954; Wilkins 2009). However, already by the end of the nineteenth century, accumulated biological data had shown that it was no longer possible to categorise all biological diversity within the framework of a single basic term 'species' (Lotsy 1906) and concern reappeared a century later with development of phylogenetic methodology (Plejel 1999; Vrana and Wheeler 1992; Hey 2001; Mishler 2010). In addition, the original Latin root ('*specere*') of the word 'species' – 'to look at' – has been noted to introduce an element of uncertainty into the concept itself (Zirkle 1959: 638). The popular markers used in molecular systematics to confirm the existence of 'species' (for example, COI gene in metazoans) do not reveal in all cases considerable differences since the effectiveness of their usage depends on the taxa. For some authors, different species can have values of interspecies genetic difference lower than 2 per cent, whereas others have argued that values higher than 2 per cent are not enough for species delimitation (Jirapatrasilp et al. 2019). Genome-wide coverage makes it possible to model allele distribution in and between recognised species, but does not provide better definition of the term 'species' itself (e.g. Coates et al. 2018). The conventional view that 'species' comprise millions of genetically homogenous individuals is not even valid for one of the most common species, codfish (Árnason and Halldórsdóttir 2019).

We believe that the continued usage of the term 'species' can only be justified by the need to communicate information on diversity between the scientific community and general society. Doing so, however, obscures the fact that organismal diversity is *multilevel*. One of the best examples to highlight the multilevel nature of biological reality in opposition to the species-centred paradigm is the fundamentally uncertain taxonomy of the genus *Homo*. Denisovans do not have official separate species status, but differ from both *H. sapiens* and *H. neanderthalensis*, whereas the latter can be considered as either a separate species or a subspecies of *H. sapiens*. To further confuse not only a traditional taxonomist but also a strict phylogeneticist, *H. sapiens* is not a monophyletic taxon because it contains some features of the closely related *H. neanderthalensis*, Denisovians (Pääbo 2014), and a distantly related archaic hominin (Durvasula and Sankararaman 2020). These facts contradict the still very widely accepted Mayrian biological species concept and Hennigian phylogenetic systematics because, 'This stem species must have monophyletic groups in common with no other group in the system'. (Hennig 1966: 209). In ontogenetic systematics this is instead a normal situation of MOD, because the focus is on intrinsic features of every individual, rather than artificial 'species divisions'.

Before Wallace and Darwin, the ability of species to change was largely denied, even if taxonomists were not strict essentialists (Winsor 2006). Despite many subsequent attempts to employ evolutionary principles in taxonomy, this only became general after 1966 with the English translation of Hennig's book. To make an analogy with physics, before Poincaré, Lorentz, and then the *annus mirabilis* of Einstein, understanding of main properties of space and time was very different. This *status quo* was reflected in Lord Kelvin's statement that there were just two clouds in the almost perfect sky of contemporary physics (at the beginning of the twentieth century) (Thomson 1900). Ironically, the relativity and quantum mechanics hurricane soon fundamentally changed the general landscape of physics.

Now, after decades of ongoing biological revolution, it has become clear that restricting systematics to just the consideration of one or more phylogenies is insufficient. An organism has an extremely complex dynamic structure, where genetic, epigenetic, and other processes occur that, in total, form the process of ontogeny. All this requires changing our understanding of taxonomic entities. The term 'species' is well suited for a pre-evolutionary counting of entities but is a poorer fit to the phylogenetic framework, and is not at all adequate for the ontogenetic understanding of organisms. We can continue to maintain the 'species' *status quo* and omit the complex dynamics of the organism, but then real biology will present us with problems such as the 'cryptic species', just as real physics presented pre-relativity astronomers with such problems as the anomaly of Mercury's orbit. Conventions are necessary for communication, but they must be changed when evidence is found that they seriously distort understanding of underlying natural phenomena.

4.10.2 The Billions of Years of Evolutionary Modification of Ontogeny Are Manifested at Many Levels (MOD) in Every Particular Currently Existing Individual

The MOD approach incorporates the ontogenetic properties of living organisms and it overcomes shadows of racism because it leads to the conclusion that *any human individual* not only contains features of both closely and distantly related ancestors but also the phylotypic periods of remote fish-like ancestors and the DNA complex of maximally distant unicellular ancestors. Whilst this diversity can be very roughly translated into a system of binomial names, in reality it needs further modification if it is to reflect the immensely complex, dynamic, and multilevel nature of organismal diversity. An ontogenetic perspective in taxonomy would require changing the nomenclatural system from one in which ranks are externally imposed to one in which they are derived from internal ontogenetic patterns.

4.10.3 MOD and Integrative Taxonomy

Ontogenetic systematics fundamentally encompasses 'static' taxonomic assessments of biodiversity and dynamic evolutionary aspects. We therefore argue that the 'integrative taxonomy' approach (e.g. Dayrat 2005; Schlick-Steiner et al. 2010; Gómez Daglio and Dawson 2019) has emerged as a reaction to neglect of morphology in molecular analysis. However, it remains within the already ontogeny-free framework of phylogenetic systematics. If ontogeny were genuinely central to taxonomy, its integrative power, per se, would obviate calls for integration. *Ontogenetic systematics* in its current reformulation is based on MOD and should be considered the foundation of taxonomy because ontogeny is the most direct and natural integrator of molecules and macro morphology and can be used to classify organisms. By contrast, current understanding of 'integrative taxonomy' tends to mean the auxiliary addition of morphological data to molecular analysis and to a great extent is just a part of the much more fundamental ontogenetic approach. MOD, which is all-encompassing, cannot be regarded as equivalent to the OTUs recognised in current studies, which are mostly based only on molecular differences (Raphaël et al. 2016).

At a broad scale the most complex multicellular eukaryotes represent phylogenetically a single branch of the Archaea radiation (Williams et al. 2020). We could unite *all* organismal diversity into a single super-genus and then number the endless subgroups and so avoid the challenges of how to document the fine-scale phenotypic differences among them. This would, however, mask an immense organismal diversity. The main goal of current taxonomy should therefore be to consistently integrate morphological data, molecular phylogenetics, and taxon delimitation methods. Partition of all organismal diversity into a single vast set of species, delimited in some artificial molecular way to generate a phylogenetic hierarchy and then discard everything between species level and the overall taxon Life would also just throw away a huge array of information on diversity. This is a practical problem. With the increasing recent rate of discovery of complexes of similar but genetically distinct species, the need for finer scale genus- and family-level taxa has become evident (Korshunova et al. 2017a,c, 2020a,b, 2021, e.g. over lumped *Tenellia*) and modern taxonomy offers no alternative to fine-scale splitting.

Ontogenetic systematics, compared to 'integrative taxonomy', is in principle a different understanding of 'species', which aims to integrate classical taxonomy, molecular phylogenetics, and evolutionary developmental biology through the MOD. In this, MOD differs from any 'typological' approach. For a transitional period, when MOD nomenclature will be developed, we propose following practical rules for integrating molecular and morphological data as a step towards the application of ontogenetic data to recover more consistent classifications: (1) Morphological and molecular data should be given equal weight, as opposed to allowing molecular data to dominate, and in conflicting cases data from different sources should be carefully evaluated. (2) Morphologically highly divergent taxa nested inside a broader molecular taxon should be kept distinct to reflect their significant morphological differences. (3) Where molecular data persistently indicate heterogeneity (para- or polyphyletic) in a taxon with homogeneous morphology, the latter should be divided accordingly into several taxa at the same rank. (4) The recognition of species-rich genera should be avoided because they obscure morphological and molecular diversity and hamper recognition of hidden diversity.

4.10.4 Nomenclatural Consequences of MOD

Within a hierarchical system in traditional taxonomy and in phylogenetic systematics, the role of the 'species' rank and an individual organism are very dependent. The externally imposed taxonomic ranks create problems and paradoxes such as the reality of 'higher taxa', the need for sub ranks, the separation of taxon characters and composition, and at a fundamental level, the incongruence between taxonomy and phylogeny. Proposals for a consistent phylogenetic nomenclature (Phylocode, Cantino and Queiroz 2010) that aimed to overcome the pitfalls of rank-based taxonomy have, in practice, created further complex problems, that is, the need to change a taxon name after every modification of accepted phylogenetic relationships. Perhaps for these reasons the Phylocode has yet to be embraced by the taxonomic community (Rieppel 2006).

In ontogenetic systematics and MOD, the individual organism becomes the focus of classification. Every organism encompasses key ontogenetic features from remote to closely related ancestors; these features are *intrinsic* to each individual rather than characteristics of externally ranked taxa, which are only loosely connected to actual ontogenies. According to these principles, we propose individual, intrinsic MOD formulae instead of external hierarchies. Every organism will still belong to a few, broad ancestral levels of ontogenetic change. Thus, the most general categories, encompassing trillions of individuals, will not be lost. Instead of the current external taxonomic ranks, which are hierarchical approximations of common phyloperiods, the intrinsic MOD identifiers will provide more flexibility at the organismal and individual levels. The MOD formula will free individuals from the constraints of external ranks by characterising them instead by internal phyloperiods shared with other individuals (Figure 4.2).

The elaboration of MOD as a methodological approach for classifying biodiversity requires development of its main principles and a new code of nomenclature, the basic outlines of which are given in this section. This future nomenclature is conceived as a formula for any individual, in which its phylotypic periods are manifested by particular

Table 4.1 Acronyms of the extant metazoan phyla as kernels for the MOD formula

Acanthocephala **Act**	Ctenophora **Cte**	Loricifera **Lor**	Placozoa **Pla**
Acoelomorpha **Aco**	Cycliophora **Cyc**	Micrognathozoa **Mic**	Platyhelminthes **Plt**
Annelida **Ann**	Dicyemida **Dic**	Mollusca **Mol**	Porifera **Por**
Arthropoda **Art**	Echinodermata **Ech**	Nematoda **Nem**	Priapulida **Pri**
Brachiopoda **Bra**	Enteropneusta **Enp**	Nematomorpha **Nmt**	Pterobranchia **Pte**
Bryozoa **Bry**	Entoprocta **Ent**	Nemertea **Nta**	Rotifera **Rot**
Chaetognatha **Cha**	Gastrotricha **Gst**	Onychophora **Ony**	Sipuncula **Sip**
Chordata **Cho**	Gnathostomulida **Gna**	Orthonectida **Ort**	Tardigrada **Tar**
Cnidaria **Cni**	Kinorhyncha **Kin**	Phoronida **Pho**	Xenoturbellida **Xen**

morphological traits. The evolutionary sequence of phyloperiods in ontogenetic systematics implies a hierarchy-based classification; however, the main problem of any hierarchical system of high-level taxa (both classical and phylogenetic) is instability. The long-term goal of ontogenetic MOD taxonomy is to infer the exact evolutionary order of the ontogenetic formation of phylotypic periods. At present, only approximations can be reconstructed. In order for an MOD formula to be stable, we propose to anchor the trillions of existing ontogenies using a few universal ontogenetic patterns as the main kernels: for example, metazoan phyla (Table 4.1), the divisions in plants, and universal molecular-morphological properties in unicellular organisms. The current 'order-family-genus' external ranks will still be detectable as intrinsic levels in other respective phyloperiods, as in dorid nudibranchs (Figures 4.2 and 4.3).

Phylotypic periods are natural nodes that link the molecular basis of development to macromorphology. The integration of molecules and morphology is important for the development of MOD nomenclature. For example, in viruses that lack an ontogeny (Ontogeny 0, Figure 4.4) or prokaryotes with a simple ontogeny (Ontogeny I, Figure 4.4), the entire organism is close to the molecular level. While morphological traits exist in viruses, bacteria, and archaea, they are of much less importance than in multicellular organisms. Since phyloperiods link molecular and morphological levels, we propose that for multicellular organisms a formula-based classification would efficiently describe the major features of all the individuals attributable to any particular taxon.

When compared to 'external' taxonomic ranks, a formula can be understood as intrinsic, underlining the fact that each individual exhibits a particular set of previously acquired or modified characters, based on phyloperiods shared with other individuals. In this way ontogenetic systematics would incorporate phyloperiods as a major criterion of classification and provide universal indicators of relationship across millions of groups while also supporting the delimitation of finer scale taxonomic groups. Here we outline the major features of formulae using the dorid nudibranch mollusc *Elfdaliana profundimontana* as an example. It is possible to indicate in the formula only a few identifiers. Firmly established phylotypic periods, corresponding to Phylum, Class, and Order (Mollusca, Gastropoda, Doridida, = Mol-Gastr-Dorid) provide the main identifiers of the MOD formula.

The lower taxonomic ranks are often less stable but could be indicated by formula extensions that include disputed taxa such as Subclass Heterobranchia or Opisthobranchia, and the family Polyceridae, that is, *Elfdaliana profundimontana* (Mol-Gastr-Heter-Opisth-Dorid-Pol-*Elfd-profund*) Martynov and Korshunova 2015. By using special identifier indexes, MOD formulae could incorporate other details representing fine-scale gradations within the individual – 'species' continuum. For example, if it is necessary to describe differences such as coloration, these can be incorporated into the formula: *Elfdaliana profundimontana* (Mol-Gastr-Heter-Opisth-Dorid-Pol-*Elfd-profund*WR) Martynov and Korshunova 2015, as opposed to *Elfdaliana profundimontana* (Mol-Gastr-Heter-Opisth-Dorid-Pol-*Elfd-profund*W); WR indicating white and red general coloration and W indicating white. Other identifiers could be used, that is, those related to the holobiont. Potentially, genetic or molecular information could also be incorporated into the formula, as was already proposed for the molecular approach (Raphaël et al. 2016). If there is a need for a decision on conservation or other practical applications, a format such as Mol. . .Dor..*Elfd-profund* should be considered, and by this reduce the problem of extensive polynomials. A major challenge compared to the traditional or phylogenetic 'species concept' is that in the framework of MOD, each individual or clonal continuum represents a novel entity. During a transitional period, the terms 'species' and 'taxon', including 'sp.nov. . . taxon nov.' taxonomic indicators could be used with reservations in parallel with MOD nomenclature. With the further development of MOD, traditional usage would be decreased and replaced by various identifiers that would describe a particular organism in a unique and precise way, importantly, linked through identifiers to other organisms, and form potential multilevel networks. The traditional 'new species' could then be understood as a broader 'novelty' within the framework of MOD, whereas other organismal novelties would not be diminished. In moving towards MOD, traditional species should be kept as a maximally narrow-defined entity.

We expect ontogenetic nomenclature to be more stable than traditional or molecular phylogenetic taxonomy. The greater stability of MOD nomenclature will be derived from the fact that the main identifiers will rely on universal shared features of ontogenies, including phylotypic periods. Phyloperiods reflect past evolutionary sequences of the characters and do not necessarily require a name change with every new analysis, as implied by the PhyloCode. That is, it is not necessary to show all the details of evolutionary sequence or hierarchy to indicate the taxonomic position of an individual organism. A few main identifiers (kernels) would be sufficient to indicate the organism's main features. For example, millions of potential species of Metazoa can be accommodated within less than 40 main ontogenetic systems, including their respective phyloperiods, represented by the currently recognised metazoan phyla (Table 4.1). The ontogenetic kernels of the phyla are highly conserved and so very stable in a taxonomic sense, and the discovery of a completely new phylum or a change to its phylogenetic position is a rare event. The status of the majority of metazoan phyla has not been challenged by the most recent molecular data (Laumer et al. 2019) and disparities in the evolutionary history of a phylum (Deline et al. 2020), although in a few cases there are minor differences in their treatments. For example, if treated as subphyla, no change to the main picture is involved. The metazoan phyla listed in Table 4.1 represent specific patterns of early developmental phylotypic periods and adult morphologies, and can be used to provide short acronyms as identifiers of their respective ancestral evolutionary patterns, for example, Art for Arthropoda, Mol for Mollusca, and Xen for Xenoturbellida. We propose acronyms of the commonly recognised bilaterian phyla as a base for the MOD formula (Table 4.1). Supraphyletic phyloperiods, which include both recent and potential fossil taxa, for example, for Bilateria (Levin et al. 2012) can also be inferred.

The main identifier closest to the individual level would be a Latin binomial name, the retention of which would enable the integration of 300 years of taxonomic observation and publication into current nomenclatural codes. The use of the Latin binomial as a unique identifier allows continuity between ontogenetic nomenclature, its application in taxonomy, and the communication, documentation, and management of diversity. Although formally maintaining their traditional format, binomial identifiers do not refer to ranks in the framework of ontogenetic taxonomy. To keep the binomial MOD identifier as close as possible to the individual level, the narrowest defined genus names of classical taxonomy should be used. This is because MOD targets individuals that are maximally similar in their morphological and molecular patterns (ontogenetic in the broad sense).

Thus, instead of an externally imposed, complex, and unstable hierarchy, we can denote in a short MOD formula the phylum-related ontogenetic identifiers and binomial name as the main kernels, and these will diagnose a particular small MOD group of individuals. For example, Mol-*Elfd-profund.* Using the less stable inferred 'intermediate' levels is not necessary where there is no universal agreement on the delimitation of the taxa concerned. Traditional taxon names can be used during the transition to MOD, to aid in the switch from ranks to ontogenetic taxonomy.

In addition to the stability of the main identifiers, ontogenetic nomenclature offers greater flexibility than the current systems. The MOD formula offers the possibility of using a few stable phylotypic periods (as acronyms) to denote a particular individual, for example Mol-Dor 1 ... Mol- Dor N, with the potential to use digital identifiers at both phylum and individual ontogenetic levels. Ontogenetic taxonomy can incorporate various patterns and processes within specific ontogenies or holobionts (e.g. Gilbert 2019), with the addition of more identifiers as required. Another important advantage of ontogenetic systematics is that it can accommodate paraphyletic groups, which, though excluded in a strict phylogenetic classification, are not prohibited by the nomenclatural codes. From an ontogenetic perspective, paraphyletic groups are not fundamentally different from monophyletic ones and frequently present modifications of the shared ontogenetic cycle of their closest ancestors. The unique similarities represented by a major phylotypic period are no less important for a classification than the smaller ontogenetic modifications. If paraphyly emerges from the application of ontogenetic and molecular analyses, the formula will need to be updated. For example, although numerous analyses (using molecular data) have been carried out that support the monophyly of the phylum Mollusca, class Gastropoda, or order (suborder) Doridida (Laumer et al. 2019; Kocot et al. 2020; Korshunova et al. 2020a), in the unlikely scenario that molluscan or dorid organisation was shown to have originated independently more than once, the taxonomic formula could be modified: for example, Mol1-Gastr-Dorid, or Mol-Gastr-Dorid1, etc., depending on which taxon originated twice. The creation of a database of potential identifiers for each particular group would be required and we give here, as an example, a list of kernel identifiers at phylum level for Metazoa (Table 4.1).

Our proposed nomenclature differs from 'formal-logical' taxon formulae that have appeared sporadically during the history of taxonomy (ICZN 1999), because it is based on real properties - the ontogenies - of the organisms. Molecular-based formulae (Raphaël et al. 2016; Minelli 2020), though not yet conceived in the ontogenetic sense, indicate that such approaches are slowly beginning to affect the core of traditional taxonomy.

4.10.5 Future of Organismal Research within the Framework of MOD

We believe that the adoption of Ontogenetic Systematics by the taxonomic community will enable the synthesis of an enormous amount of current and future biological data. We conclude that our proposal to rely on naturally existing nodes that connect molecules with macro morphology (phylotypic periods) is consistent with the need to update taxonomy and one that will provide new opportunities for an increased understanding of biological diversity. A comprehensive, multilevel understanding of organismal diversity within a consistent ontogenetic framework needs to be defined. Currently, there is a chaotic mixture of rigid, pre-evolutionary Linnean nomenclature and persistent but not very successful attempts to make taxonomy more phylogenetic. A big array of morphological and molecular methods are now available, but taxonomists adhere to basically pre-evolutionary 'species concepts'. Modern methods allow the investigation of organismal diversity to unprecedented depths, but without a renewed general taxonomic framework new data will be just a collection of loosely connected facts. We advocate MOD as a starting point for further constructive discussion towards an ontogenetic nomenclature and classification. Ontogeny is a primary driving force of evolution, and phylogeny and should become a central part of systematics just as phylogeny has during the past 60 years. The effective integration of ontogeny will allow classification to reflect as many biological levels as possible and this may well be more complex than the incorporation of evolutionary ideas into the taxonomy of an earlier epoch.

Conclusions

Currently, the species is the central focus for evolutionary analysis. Systematists use imprecise definitions to define and delimit species. This can be problematic as species are important for resource management, conservation, and other purposes. Concurrently, theoreticians continue to discuss the 'species problem' with a wide range of opinions and little agreement, and this is leading to a shift in our view of diversity away from a reliance on widely distributed, polytypic, and morphologically variable species to more narrowly defined entities. The issue of cryptic species demonstrates this well. Despite these calls for change, the use of the 'species' concept has remained almost unmodified. Our hope is that this poorly defined but rigidly applied 'species concept' can finally be discarded in favour of an ontogenetic systematics and multilevel organismal diversity (MOD) that considers any organism as the product of different levels of ontogenetic complexity and reproduction pattern (Figure 4.4).

Change is needed if MOD, based on the ontogenetic cycle of interacting individuals instead of a focus on population properties, is to become central to the theory of evolution. The main purpose of this chapter is to highlight that biology is on the verge of a breakdown in the meaning of traditional tenets such as 'species', and that to better promote communication on biodiversity, the development of a different approach is necessary, including a different nomenclatural system. Currently, a huge variety of entities and underlying processes are obscured by the use of the term 'species'. Our proposal is a practical programme to challenge the predominant notion of the species concept and to urge a major shift towards understanding organisms within the framework of *multilevel organismal diversity* (MOD), a more comprehensive biological concept. The terminology used in science for sharing ideas and discoveries, and the concepts they denote, should be consistent with modern progress.

Acknowledgements

This study was supported by the research project of MSU Zoological Museum (18-1-21 №121032300105-0). The work of TK was conducted under the IDB RAS Government basic research program in 2021 №0088-2021-0008. Electron Microscopy Laboratory MSU is thanked for support with electron microscopy. Special thanks to Alexandre Monro and Simon Mayo (Royal Botanic Gardens, Kew) for great editorial work and valuable comments, which have helped to improve the English language text.

References

Adams, M. J. et al. (2017) 50 years of the international committee on taxonomy of viruses: Progress and prospects. *Archives of Virology* 162: 1441–1446.

Agassiz, A. A. and Gould, A. A. (1857) *Principles of Zoölogy*. Gould and Lincoln, Boston, 250 pp.

Agosta, S. J and Brooks, D. R. (2020) *The Major Metaphors of Evolution*. Springer Intern. Publishing, Berlin, 273 pp.

Alberch, P. S., Gould, S. J., Oster G. F., and Wake, D. B. (1979) Size and shape in ontogeny and phylogeny. *Paleobiology* 5: 296–317.

Albert, V. A., Gustafsson., M. H. G., and Di Laurenzio, L. (1998) Ontogenetic systematics, molecular developmental genetics, and the angiosperm petal. In: P. S. Soltis, D. E. Soltis, and J. J. Doyle (eds.) *Molecular Systematics of Plants II*. Kluwer Academic Publishing, Boston, pp. 349–374.

Amaral, A. R., Lovewell, G., Coelho, M. M., Amato, G., and Rosenbaum, H. C. (2014) Hybrid speciation in a marine mammal: The Clymene dolphin (*Stenella clymene*). *PLoS ONE* 9(1): e83645.

Amariei, C., Tomita, M., and Murray, D. B. (2014) Quantifying periodicity in omics data. *Frontiers in Cell and Developmental Biology* 2: 1–9.

Árnason, E. and Halldórsdóttir, K. (2019) Codweb: Whole-genome sequencing uncovers extensive reticulations fueling adaptation among Atlantic, Arctic, and Pacific gadids. *Science Advance* 5: eaat8788.

Arthur, W. (2002) The emerging conceptual framework of evolutionary developmental biology. *Nature* 415: 757–764.

(2015) Internal factors in evolution: The morphogenetic tree, developmental bias, and some thoughts on the conceptual structure of evo-devo. In: Alan C. Love (ed.) *Conceptual Change in Biology: Scientific and Philosophical Perspectives on Evolution and Development.* Springer Science & Business Media, Dordrecht, pp. 343–363.

Babaev, E. (2019) Periodic law in chemistry and other sciences. *Pure and Applied Chemistry* 91: 2023–2035.

Baer, K. E., von (1828–1837) *Entwickelungsgeschichte der Thiere: Beobachtung und Reflexion.* 2 Bd. Bornträger, Königsberg, 264 S.

Beer, G. R. de (1958) *Embryos and Ancestors*, 3rd ed. Oxford University Press, Oxford, 197 pp.

Beklemishev, V. N. (1969) *Principles of Comparative Anatomy of Invertebrates*, Vol 1. University of Chicago Press, Chicago, 490 pp.

Bellés, X. (2020) *Insect Metamorphosis: From Natural History to Regulation of Development and Evolution.* Elsevier, Academic Press, Amsterdam and New York, 304 pp.

Beurton, P. J. (2002) Ernst Mayr through time on the biological species concept: A conceptual analysis. *Theory in Biosciences* 121: 81–98.

Bierbach, D., Laskowski, K., and Wolf, M. (2017) Behavioural individuality in clonal fish arises despite near-identical rearing conditions. *Nature Communications* 8: 15361.

Blackstone, N. W. (2021) Evolutionary conflict and coloniality in animals. *Journal of Experimental Zoology* 336: 212–220.

Bolnick, D. I., Barrett, R. D. H., Oke, K. B., Rennison, D. J., and Stuart, Y. E. (2019) (Non)parallel evolution. *Annual Review of Ecology, Evolution, and Systematics* 49: 303–330.

Brandt, A. et al. (2019) No signal of deleterious mutation accumulation in conserved gene sequences of extant asexual hexapods. *Scientific Reports* 9: 5338.

Brehm, G. et al. (2016) Turning up the heat on a hotspot: DNA barcodes reveal 80% more species of geometrid moths along an Andean elevational gradient. *PLoS ONE* 11: e0150327.

Burggren, W. W. (2014) Epigenetics as a source of variation in comparative animal physiology – or – Lamarck is lookin' pretty good these days. *Journal of Experimental Biology* 217: 682–689.

Burma, B. H. (1954) Reality, existence, and classification: A discussion of the species problem. *Madroño* 12: 193–209.

Burress, E. D., Alda, F., Duarte, A. et al. (2018) Phylogenomics of pike cichlids (Cichlidae: Crenicichla): The rapid ecological speciation of an incipient species flock. *Journal of Evolutionary Biology* 31: 14–30.

Burton, T. and Metcalfe, N. B. (2014) Can environmental conditions experienced in early life influence future generations? *Proceedings of the Royal Society B* 281: 20140311.

Buss, L. (1987) *The Evolution of Individuality.* Princeton University Press, Princeton, NJ, 201 p.

Calcott, B. and Sterelny, K. (2011) *The Major Transitions in Evolution Revisited.* Vienna Series in Theoretical Biology, MIT Press, Cambridge, MA, 319 pp.

Cantino, P. D. and Queiroz, K. de (2010) *International Code of Phylogenetic Nomenclature*, Version 4c, 102 pp.

Capblancq, T., Després, L., Rioux, D., and Mavárez, J. (2015) Hybridization promotes speciation in *Coenonympha* butterflies. *Molecular Ecology* 24: 6209–6222.

Carmel, Y. and Shavit, A. (2020) Operationalizing evolutionary transitions in individuality. *Proceedings of the Royal Society B* 287: 20192805.

Carroll, S. B. (2008) Evo-Devo and an expanding evolutionary synthesis: A genetic theory of morphological evolution. *Cell* 134: 25–36.

Casadesús, J. and Low, D. A. (2013) Programmed heterogeneity: Epigenetic mechanisms in Bacteria. *Journal of Biological Chemistry* 288: 13929–13935.

Cerca, J. Meyer, C., Purschke, G., and Struck, T. H. (2020) Delimitation of cryptic species drastically reduces the geographical ranges of marine interstitial ghost-worms (Stygocapitella, Annelida, Sedentaria). *Molecular Phylogenetics and Evolution* 143: 106663.

Chan, J. C. et al. (2020) Reproductive tract extracellular vesicles are sufficient to transmit intergenerational stress and program neurodevelopment. *Nature Communications* 11: 1499.

Chenuil, A., Cahill, A. E., Délémontey, N. et al. (2019) Problems and questions posed by cryptic species: A framework to guide future studies. In: E. Casetta, M. J. da Silva, and D. Vecchi (eds.) *From Assessing to Conserving Biodiversity*. Springer Publishing, New York, pp. 77–107.

Chipman, A. D. (ed.) (2020) *Cellular Processes in Segmentation*. CRC Press, Taylor & Francis, Boca Raton, FL, 299 pp.

Chun, J. and Rainey, F. A. (2014) Integrating genomics into the taxonomy and systematics of the Bacteria and Archaea. *International Journal of Systematic Evolutionary Microbiology* 64: 316–324.

Claßen-Bockhoff, R., Franke, D., and Krähmer, H. (2020) Early ontogeny defines the diversification of primary vascular bundle systems in angiosperms. *Botanical Journal of the Linnean Society* 195: 281–307.

Coates, D. J., Byrne, M., and Moritz, C. (2018) Genetic diversity and conservation units: Dealing with the species-population continuum in the age of genomics. *Frontiers Ecology and Evolution* 6: 165.

Corlett, R. T. (2016) Plant diversity in a changing world: Status, trends, and conservation needs. *Plant Diversity* 38: 10–16.

Costa, S. G. D. S., Welbourn, C., Klimov, P., and Pepato, A. R. (2021). Integrating phylogeny, ontogeny and systematics of the mite family Smarididae (Prostigmata, Parasitengona): Classification, identification key, and description of new taxa. *Systematic and Applied Acarology* 26: 85–123.

Crossman, C. A., Taylor, E. B., and Barrett-Lennard, L. G. (2016) Hybridization in the Cetacea: Widespread occurrence and associated morphological, behavioral, and ecological factors. *Ecology and Evolution* 6: 1293–1303.

Daeschler, E., Shubin, N., and Jenkins, F. (2006) A Devonian tetrapod-like fish and the evolution of the tetrapod body plan. *Nature* 440: 757–763.

Danchin, E., Charmantier, A., and Champagne, F. C. et al. (2011) Beyond DNA: Integrating inclusive inheritance into an extended theory of evolution. *Nature Revue Genetics* 12: 475–486.

Danchin, E., Pocheville, A., Rey, O., Pujol, B., and Blanchet, S. (2019) Epigenetically facilitated mutational assimilation: Epigenetics as a hub within the inclusive evolutionary synthesis. *Biological Reviews* 94: 259–282.

Darlington, C. D. (1940) Taxonomic species and genetic systems. In J. Huxley (ed.) *The New Systematics*. Oxford University Press, Oxford, pp. 137–160.

Darwin, C. (1859) *On the Origin of Species by Means of Natural Selection*. John Murray, London.

Dayrat, B. (2005) Toward integrative taxonomy. *Biological Journal of the Linnean Society* 85: 407–415.

Deline, B. et al. (2020) Evolution and development at the origin of a phylum. *Current Biology* 30: 1672–1679.

DeSalle, R. and Goldstein, P. (2019) Review and interpretation of trends in DNA Barcoding. *Frontiers Ecology and Evolution* 7: 302.

De Queiroz, K. (2007) Species concepts and species delimitation. *Systematic Biology* 56: 879–886.

Diamond, J. (1991) *The Rise and Fall of the Third Chimpanzee: How Our Animal Heritage Affects the Way We Live*. Hutchinson Radius, London, 364 pp.

Dias, B. G. and Ressler, K. J. (2013) Parental olfactory experience influences behavior and neural structure in subsequent generations. *Nature Neuroscience* 17: 89–96.

DiFrisco, J. and Jaeger, J. (2021) Homology of process: Developmental dynamics in comparative biology. *Interface Focus* 11: 20210007.

Domazet-Lošo, T. and Tautz, D. (2010) A phylogenetically based transcriptome age index mirrors ontogenetic divergence patterns. *Nature* 468: 815–818.

Duboule, D. (1994) Temporal colinearity and the phylotypic progression: A basis for the stability of a vertebrate Bauplan and the evolution of morphologies through heterochrony. *Development* Suppl: 135–142.

Durvasula, A. and Sankararaman, S. (2020) Recovering signals of ghost archaic introgression in African populations. *Science Advance* 6: eaax5097.

Eldredge, N. and Gould, S. J. (1972) Punctuated equilibria: An alternative to phyletic gradualism. In: T. J. M. Schopf (ed.) *Models in Paleobiology*. Freeman Cooper, San Francisco, pp. 82–115.

Ereshefsky, M. (1998) Species pluralism and 'anti-realism'. *Philosophy of Science* 65: 103–120.

Evans, J. S., Erwin, P. M., Sihaloho, H. F., and López- Legentil, S. (2021) Cryptic genetic lineages of a colonial ascidian host distinct microbiomes. *Zoologica Scripta*, https://doi .org/10.1111/zsc.12482.

Faria, L. R. R., Pie, M. R., Salles, F. F., and Soares, E. D. G. (2020) The Haeckelian shortfall or the tale of the missing semaphoronts. *Journal of Zoologcial Systematics and Evolutionary Research*, 1–11.

Fei, X., Shi, J., Liu, Y., Niu, J., and Wei, A. (2019) The steps from sexual reproduction to apomixis. *Planta* 249: 1715–1730.

Ferretti, L., Krämer-Eis, A., and Schifferet, P. H. (2020) Conserved patterns in developmental processes and phases, rather than genes, unite the highly divergent Bilateria. Life *(Basel)* 10: 182.

Fišer, C., Robinson, C. T., and Malard, F. (2018) Cryptic species as a window into the paradigm shift of the species concept. *Molecular Ecology* 27: 613–635.

Fowler, D. R., Meinhardt, H., and Prusinkiewicz, P. (1992) Modeling seashells. Proceedings SIGGRAPH'92. *Computer Graphics* 26: 379–387.

Fraga, M. F. et al. (2005) Epigenetic differences arise during the lifetime of monozygotic twins. *PNAS* 26: 10604–10609.

Fraïsse, C., Belkhir, K., Welch, J. J., and Bierne, N. (2016) Local interspecies introgression is the main cause of extreme levels of intraspecific differentiation in mussels. *Molecular Ecology and Evolution* 25: 269–286.

Futuyma, D. J. and Kirkpatrick, M. (2017) *Evolution*. Fourth Edition. Sinauer, Associates, Inc., Publishers, Sunderland, MA, 599 pp.

Gante, H. F. (2018) How fish get their stripes-again and again. *Science* 362: 396–397.

Gante, H. F., Matschiner, M., Malmstrøm, M. et al. (2016) Genomics of speciation and introgression in Princess cichlid fishes from Lake Tanganyika. *Molecular Ecology* 25: 6143–6161.

Garcia-Dominguez, X., Marco-Jiménez, F., Peñaranda, D. S. et al. (2020) Long-term and transgenerational phenotypic, transcriptional and metabolic effects in rabbit males born following vitrified embryo transfer. *Scientific Reports* 10: 11313.

Garstang, W. (1922) The theory of recapitulation: A critical restatement of the biogenetic law. *Proceeding of the Linnean Society, Zoology* 35: 81–101.

Gegenbaur, C. (1859) *Grundzüge der vergleichenden Anatomie*. Wilhelm Engelmann Verlag, Leipzig.

Gefaell, J., Varela, N., and Rolán-Alvarez, E. (2020) Comparing shape along growth trajectories in two marine snail ecotypes of *Littorina saxatilis*: A test of evolution by paedomorphosis. *Journal of Molluscan Studies* 86: 382–388.

Gee, B. M. (2020) Size matters: The effects of ontogenetic disparity on the phylogeny of Trematopidae (Amphibia: Temnospondyli). *Zoological Journal of the Linnean Society* 190: 79–113.

Ghiselin, M. (1974) A radical solution to the species problem. *Systematic Zoology* 23: 536–544.

Gilbert, S. F. (2010) *Developmental Biology*. Ninth ed. Sinauer Associates, Sunderland, MA, 711 pp.

(2019) Evolutionary transitions revisited: Holobiont evo-devo. *Journal of Experimental Zoology*: 1–8.

Gilbert S. F., Opitz J. M., and Raff R. A. (1996) Resynthesizing evolutionary and developmental biology. *Developmental Biology* 173: 357–372.

Gladyshev, E. A., Meselson, M., and Arkhipova, I. R. (2008) Massive horizontal gene transfer in bdelloid rotifers. *Science* 320: 1210–1213.

Glémin, S., François, C. M., and Galtier, N. (2019) Genome evolution in outcrossing vs. selfing vs. asexual species. In: M. Anisimova (ed.) *Evolutionary Genomics: Methods in Molecular Biology*, vol. 1910. Humana, New York.

Glon, H., Haruka, Ya, Daly, M. et al. (2019) Temperature and salinity survival limits of the fluffy sea anemone, *Metridium senile* (L.), in Japan. *Hydrobiologia* 830: 303–315.

Göbbel, L. and Schultka, R. (2003) Meckel the Younger and his epistemology of organic form: morphology in the pre-Gegenbaurian age. *Theory Bioscience* 122: 127–141.

Gogarten, J. P. and Townsend, J. P. (2005) Horizontal gene transfer, genome innovation and evolution. *Nature Reviews Microbiology* 3: 679–687.

Gómez Daglio, L. and Dawson, M. N. (2019). Integrative taxonomy: Ghosts of past, present and future. *Journal of Marine Biological Association of United Kingdom* 99: 1237–1246.

Gould, S. J. (1977) *Ontogeny and Phylogeny*. Harvard University Press, Cambridge, MA, 501 pp.

Graham, L. (2016) *Lysenko's Ghost: Epigenetics and Russia*. Harvard University Press, Cambridge, MA, 224 pp.

Grant, P. R. and Grant B. R. (2016) Introgressive hybridization and natural selection in Darwin's finches. *Biological Journal of the Linnean Society* 117: 812–822.

Gregg, J. R. (1950) Taxonomy, language and reality. *American Naturalist* 84: 419–435.

Guiry, M. D. (2012) How many species of algae are there? *Journal of Phycology* 48: 1057–1063.

Haeckel, E. (1866) *Generelle Morphologie der Organismen*. Bd. 1–2. G. Reimer, Berlin, 574 S. 462 S.

Hall, B. K. (1999) *Evolutionary Developmental Biology* (2nd ed.). Kluwer Academic Publishers, Dordrecht, 491 pp.

Harrison, R. G. and Larson, E. L. (2014) Hybridization, introgression, and the nature of species boundaries. *Journal of Heredity* 105: 795–809.

Hauk, W. D. (1995) A molecular assessment of relationships among cryptic species of *Botrychium* subgenus *Botrychium* (Ophioglossaceae). *American Fern Journal* 85: 375–394.

Haupaix, N. and Manceau, M. (2020) The embryonic origin of periodic color patterns. *Developmental Biology* 460: 70–76.

Hawkins, J. (2002) Evolutionary developmental biology: Impact on systematic theory and practice, and the contribution of systematics. In: *Developmental Genetics and Plant Evolution*. Taylor and Francis, London, pp. 32–51.

Hawksworth, D. L. and Lücking, R. (2017) Fungal diversity revisited: 2.2 to 3.8 million species. *Microbiology Spectrum* 5: FUNK-0052-2016.

Henry, C. S. (1985) Sibling species, call differences, and speciation in green lacewings (Neuroptera: Chrysopidae: Chrysoperla). *Evolution* 39: 965–984.

Heethoff, M. (2018) Cryptic species: Conceptual or terminological chaos? *Trends in Ecology & Evolution* 33: 310.

Hennig, W. (1966) *Phylogenetic Systematics*. University of Illinois Press, Urbana, 263 pp.

Heratha, S. and Lobo, D. (2020) Cross-inhibition of Turing patterns explains the self-organized regulatory mechanism of planarian fission. *Journal of Theoretical Biology* 485: 110042.

Herrera, C. M. (2009) *Multiplicity in Unity*. University Chicago Press. 448 pp.

Herron, J. C. and Freeman S. (2015) *Evolutionary Analysis*. Fifth Edition. Pearson Education Limited, London, 864 pp.

Hess, B. (2000) Periodic patterns in biology. *Naturwissenschaften* 87: 199–211.

Hey, J. (2001) The mind of the species problem. *Trends in Ecology & Evolution* 16: 326–329.

Hiebert, L. S., Simpson, C., and Tiozzo, S. (2020) Coloniality, clonality, and modularity in animals: The elephant in the room. *Journal of Experimental Zoology Part B*: 1–14.

Ho, M. W. (1992) Development, rational taxonomy and systematics. *Rivista di Biologia-Biology Forum* 85: 193–211.

Ho, W. K. W et al. (2019) Feather arrays are patterned by interacting signalling and cell density waves. *PLoS Biology* 17(2): e3000132.

Holeski, L. M., Jander, G., and Agrawal, A. A. (2012) Transgenerational defense induction and epigenetic inheritance in plants. *Trends in Ecology & Evolution* 27: 618–626.

Holland, P. W. H. (2013) Evolution of homeobox genes. *WIREs Developmental Biology* 2: 31–45.

Holliday, R. (2006) Epigenetics: A historical overview. *Epigenetics* 1: 76–80.

Horsáková, V., Nekola, J. C., and Horsák, M. (2019) When is a 'cryptic' species not a cryptic species: A consideration from the Holarctic micro-land snail genus *Euconulus* (Gastropoda: Stylommatophora). *Molecular Phylogenetics and Evolution* 132: 307–320.

Hörandl, E. et al. (2020) Genome evolution of asexual organisms and paradox of sex in eukaryotes. In: P. Pontarotti (eds.) *Evolutionary Biology: A Transdisciplinary Approach*. Springer, Champaign, IL, pp. 133–167.

Huxley, J. (1942) *Evolution: The Modern Synthesis*. George Allen & Unwin Ltd., London, 645 pp.

ICTV. (2020) *International Committee on Taxonomy of Viruses*. www.ictvonline.org

ICZN. (1999) *International Code of Zoological Nomenclature*. The International Trust for Zoological Nomenclature, London.

Inaba, M. and Chuong, C. M. (2020) Avian pigment pattern formation: Developmental control of macro- (across the body) and micro- (within a feather) level of pigment patterns. *Frontiers in Cellular and Developmental Biology* 8: 620.

Irie, N. and Kuratani, S. (2011) Comparative transcriptome analysis reveals vertebrate phylotypic period during organogenesis. *Nature Communications* 2: 248.

(2014) The developmental hourglass model: A predictor of the basic body plan? *Development* 141: 4649–4655.

Irie, N. and Sehara-Fujisawa, A. (2007) The vertebrate phylotypic stage and an early bilaterian-related stage in mouse embryogenesis defined by genomic information. *BMC Biology* 5: 1.

Jablonka, E. and Lamb, M. J. (2010) Transgenerational epigenetic inheritance. In: M. Pigliucci and G. B. Müller (eds.) *Evolution, the Extended Synthesis*. The MIT Press, Cambridge, MA, pp. 137–174.

Jacobs, G. S. et al. (2019) Multiple deeply divergent denisovan ancestries in Papuans. *Cell* 177: 1010–1021.

Jakob, W. and Schierwater, B. (2007) Changing hydrozoan bauplans by silencing hox-like genes. *PLoS ONE* 2(8): e69.

Jančúchová-Lásková, J., Landová, E., and Frynta, D. (2015) Experimental crossing of two distinct species of leopard geckos, *Eublepharis angramainyu* and *E. macularius*: Viability, fertility and phenotypic variation of the hybrids. *PLoS ONE* 10(12): e0143630.

Jaron, K. S., Bast, J., Nowell ,R. W. et al. (2021) Genomic features of parthenogenetic animals. *Journal of Heredity* 112: 19–33.

Jaspers, C. et al. (2020) Resolving structure and function of metaorganisms through a holistic framework combining reductionist and integrative approaches. *Zoology* 133: 81–87.

Jékely, G., Paps, J., and Nielsen, C. (2015) The phylogenetic position of ctenophores and the origin(s) of nervous systems. *EvoDevo* 6: 1.

Jirapatrasilp P. et al. (2019) Untangling a mess of worms: Species delimitations reveal morphological crypsis and variability in Southeast Asian semi-aquatic earthworms. *Molecular Phylogenetics and Evolution* 139: 1–20.

Johnson, M. R., Barsh, G. S., and Mallarino, R. (2019) Periodic patterns in Rodentia: Development and evolution. *Experimental Dermatology* 28: 509–551.

Jorgensen, D. R., Wu, C. M., and Hariharan, S. (2020) Epidemiology of end-stage renal failure among twins and diagnosis, management, and current outcomes of kidney transplantation between identical twins. *American Journal of Transplantation* 20: 761–768.

Jörger, K. M. and Schrödl, M. (2013) How to describe a cryptic species? Practical challenges of molecular taxonomy. *Frontiers in Zoology* 10: 59.

Jung, H. S. et al. (1998) Local inhibitory action of BMPs and their relationships with activators in feather formation: Implications for periodic patterning. *Developmental Biology* 196: 11–23.

Kalinka, A. T., Varga, K. M., Gerrard, D. T. et al. (2010) Gene expression divergence recapitulates the developmental hourglass model. *Nature* 468: 811–814.

Kapitanova, D. V. and Shkil, F. N. (2014) Effects of thyroid hormone level alterations on the development of supraneural series in zebrafish, *Danio rerio*. *Journal of Applied Ichthyology* 30: 821–824.

Karp, D. (2020) Detecting small and cryptic animals by combining thermography and a wildlife detection dog. *Scientific Reports* 10: 5220.

Kijewski, T. K., Zbawicka, M., Väinölä, R., and Wenne, R. (2006) Introgression and mitochondrial DNA heteroplasmy in the Baltic populations of mussels *Mytilus trossulus* and *M. edulis*. *Marine Biology* 149: 1371–1385.

Klingenberg, C. P. and Marugán-Lobón, J. (2013) Evolutionary covariation in geometric morphometric data: analyzing integration, modularity, and allometry in a phylogenetic context. *Systematic Biology* 62: 591–610.

Kluge A. G. (1985) Ontogeny and phylogenetic systematics. *Cladistics* 1: 13–27.

Kluge, A. G. and Strauss, R. E. (1985) Ontogeny and systematics. *Annual Review of Ecology Systematics* 16: 247–268.

Knowlton, N. (1986) Cryptic and sibling species among the decapod Crustacea. *Journal of Crustacean Biology* 6: 356–363.

(1993) Sibling species in the sea. *Annual Review of Ecology and Systematics* 24: 189–216.

Koĉi, J. et al. (2020) No evidence for accumulation of deleterious mutations and fitness degradation in clonal fish hybrids: Abandoning sex without regrets. *Molecular Ecology* 29: 3038–3055.

Kocot, K. M., Poustka, A. J., Stöger,I., Halanych, K. M., and Schrödl, M. (2020) New data from Monoplacophora and a carefully-curated dataset resolve molluscan relationships. *Scientific Reports* 10: 101.

Kondo, S. and Miura, T. (2010) Reaction-diffusion model as a framework for understanding biological pattern formation. *Science* 329: 1616–1620.

Kondrashov A. S. (2018) Through sex, nature is telling us something important. *Trends in Genetics* 34: 352–361.

Koonin, E. V. and Wolf, Yu. I. (2008) Genomics of bacteria and archaea: The emerging dynamic view of the prokaryotic world. *Nucleic Acids Research* 36: 6688–6719.

Korshunova, T. A. and Martynov A. V. (2020) Consolidated data on the phylogeny and evolution of the family Tritoniidae (Gastropoda: Nudibranchia) contribute to genera reassessment and clarify the taxonomic status of the neuroscience models *Tritonia* and *Tochuina*. *PLoS ONE* 15: e0242103.

Korshunova, T. A., Martynov, A. V., and Picton, B. E (2017a) Ontogeny as an important part of integrative taxonomy in tergipedid aeolidaceans (Gastropoda: Nudibranchia) with a description of a new genus and species from the Barents Sea. *Zootaxa* 4324: 1–22.

Korshunova, T. A., Martynov, A. V., Bakken, T., and Picton, B. E. (2017b) External diversity is restrained by internal conservatism: New nudibranch mollusc contributes to the cryptic species problem. *Zoologica Scripta* 46: 683–692.

Korshunova, T. A. et al. (2017c) Polyphyly of the traditional family Flabellinidae affects a major group of Nudibranchia: aeolidacean taxonomic reassessment with descriptions of several new families, genera, and species (Mollusca, Gastropoda). *ZooKeys* 717: 1–139.

Korshunova, T. A., Lundin, K., Malmberg, K., Picton, B., and Martynov, A. V. (2018) First true brackish water nudibranch mollusc provides new insights for phylogeny and biogeography and reveals paedomorphosis-driven evolution. *PLoS ONE* 13(3): e0192177.

Korshunova, T. A. et al. (2019) Multilevel fine-scale diversity challenges the 'cryptic species' concept. *Scientific Reports* 9: 1–23.

Korshunova, T. A. et al. (2020a) The Emperor *Cadlina*, hidden diversity and gill cavity evolution: new insights for taxonomy and phylogeny of dorid nudibranchs (Mollusca: Gastropoda). *Zoological Journal of the Linnean Society* 189: 762–827.

Korshunova, T. A., Bakken, T., Grøtan, V. et al. (2020b). A synoptic review of the family Dendronotidae (Mollusca: Nudibranchia): A multilevel organismal diversity approach. *Contribution to Zoology* 90: 93–153.

Korshunova, T. A. et al. (2020c) Fine-scale species delimitation: Speciation in process and periodic patterns in nudibranch diversity. *ZooKeys* 917: 15–50.

Korshunova, T. A., Sanamyan, N. P., Sanamyan, K. E. et al. (2021) Biodiversity hotspot in cold waters: A review of the genus *Cuthonella* with descriptions of seven new species (Mollusca, Nudibranchia). *Contributions to Zoology*: 1–68.

Kumar, S., Filipski, A. J., Battistuzzi, F. U., Kosakovsky Pond, S. J., and Tamura, K. (2012) Statistics and truth in phylogenomics. *Molecular Biology and Evolution* 29: 457–472.

Kupferschmidt, K. and Cohen, J. (2020) Will novel virus go pandemic or be contained? *Science* 367: 610–611.

Lamsdell, J. C. (2020) A new method for quantifying heterochrony in evolutionary lineages. *Paleobiology*: 1–22.

Laskowski, K. L., Doran, C., Bierbach, D., Krause, J., and Wolf, M. (2019) Naturally clonal vertebrates are an untapped resource in ecology and evolution research. *Nature Ecology and Evolution* 3: 161–169.

Laumer, C. E. et al. (2019) Revisiting metazoan phylogeny with genomic sampling of all phyla. *Proceedings of the Royal Society B* 286: 20190831.

Lecointre, G., Schnell, N. K., and Teletchea, F. (2020) Hierarchical analysis of ontogenetic time to describe heterochrony and taxonomy of developmental stages. *Scientific Reports* 10: 19732.

Leebens-Mack, J. H. et al. (2019) One thousand plant transcriptomes and phylogenomics of green plants. *Nature* 574: 679–698.

León-Martínez, G. and Vielle-Calzada, J. P. (2019) Apomixis in flowering plants: Developmental and evolutionary considerations. *Current Topics in Developmental Biology* 131: 565–604.

Levin, M., Hashimshony, T., Wagner, F., and Yanai I. (2012) Developmental milestones punctuate gene expression in the *Caenorhabditis* embryo. *Developmental Cell* 22: 1101–1108.

Lherminier, P. and Solignac, M. (2005) *De l'espèce. Syllepse.* Paris, 694 pp.

Louca, S., Mazel, F., Doebeli, M., and Parfrey, L. W. (2019) A census-based estimate of Earth's bacterial and archaeal diversity. *PLoS Biology* 17: e3000106.

Lotsy, J. P. (1906) *Vorlesungen uber Deszendenztheorien mit besonderer Berücksichtigung der Botanischen Seite der Frage gehalten an der Reichsuniversität zu Leiden.* Gustav Fischer, Jena, 384 S.

Majeský, L., Vašut R. J., and Kitner, M. (2015) Genotypic diversity of apomictic microspecies of *Taraxacum scanicum* group (*Taraxacum* sect. *Erythrosperma*). *Plant Systematics and Evolution* 301: 2105–2124.

Malinsky, M., Svardal, H., Tyers, A. et al. (2018) Whole-genome sequences of Malawi cichlids reveal multiple radiations interconnected by gene flow. *Nature Ecology and Evolution* 2: 1940–1955.

Mallet, J. (2005) Hybridization as an invasion of the genome. *Trends Ecology and Evolution* 20: 229–237.

Martynov, A. V. (1994) Materials for revision of nudibranch molluscs of the family Corambidae (Gastropoda, Opisthobranchia). Part II. Origin. *Zoologichesky Zhurnal* 73: 36–43.

(2009) From ontogeny to evolution: An expectation for changing current systematic paradigm. *Trudy Zoologicheskogo Muzeya Moskovskogo Gosudarstvennogo Universiteta* 50: 145–229.

(2011a) *Ontogenetic Systematics and a New Model of Evolution of Bilateria*. KMK, Moscow, 286 pp.

(2011b) From 'tree-thinking' to 'cycle-thinking': Ontogenetic systematics of nudibranch molluscs. *Thalassas* 27: 193–224.

(2012a) Ontogenetic systematics: The synthesis of taxonomy, phylogenetics, and evolutionary developmental biology. *Paleontological Journal* 46: 833–864.

(2012b) Ontogeny, systematics, and phylogenetics: Perspectives of future synthesis and a new model of evolution of Bilateria. *Biological Bulletin* 39: 393–401.

(2013) Evolutionary history of Metazoa, ancestral status of Bilateria clonal reproduction, and semicolonial origin of Mollusca. *Zhurnal Obshei Biologii* 74: 201–240.

(2014) *Ontogeny as a Central Paradigm of Biology: Declarative Importance and Practical Underestimation*. 3rd International Congress on Invertebrate Morphology, Berlin. Program and Abstract Book, pp. 128.

Martynov, A. V. and Korshunova, T. A. (2015) A new deep-sea genus of the family Polyceridae (Nudibranchia) possesses a gill cavity, with implications for cryptobranch condition and a 'Periodic Table' approach to taxonomy. *Journal of Molluscan Studies* 81: 365–379.

Martynov, A. V., Ishida, Y., Irimura, S. et al. (2015) When ontogeny matters: A new Japanese species of brittle star illustrates importance of considering both adult and juvenile characters in taxonomic practice. *PLoS ONE* 10: e0139463.

Martynov, A. V., Lundin, K., Picton, B. et al. (2020a) Multiple paedomorphic lineages of soft-substrate burrowing invertebrates: Parallels in the origin of *Xenocratena* and *Xenoturbella*. *PLoS ONE* 15(1): e0227173.

Martynov, A. V. et al. (2020b) Three new species of the genus *Dendronotus* from Japan and Russia (Mollusca, Nudibranchia). *Zootaxa* 4747: 495–513.

Mayer, F. and von Helversen, O. (2001) Cryptic diversity in European bats. *Proceedings of the Royal Society London B* 268: 1825–1832.

Maynard Smith, J. and Szathmáry, E. (1995) *The Major Transitions in Evolution*. Freeman, Oxford, 343 pp.

Mayr, E. (1942) *Systematics and the Origin of Species*. Columbia University Press, New York, 372 pp.

(1963) *Animal Species and Evolution*. Harvard University Press, Cambridge, MA, 797 pp.

Meckel, J. F. (1811) *Entwurf einer Darstellung der zwischen dem Embryozustände der höheren Tiere und dem permanenten der niederen stattfindenden Parallele*. Carl Heinrich Reclam, Leipzig, pp. 1–60.

Mehnert, E. (1897) Kainogenese. *Morphologische Arbeiten* 7: 1–156.

Mendeleev, D. (1869) Über die Beziehungen der Eigenschaften zu den Atomgewichten der Elemente. *Zeitschrift für Chemie* 12: 405–406.

Miller, W. et al. (2012) Polar and brown bear genomes reveal ancient admixture and demographic footprints of past climate change. *PNAS* 109: E2382–E2390.

Minelli, A. (2007) Invertebrate taxonomy and evolutionary developmental biology. *Zootaxa* 1668: 55–60.

(2009) *Perspectives in Animal Phylogeny and Evolution*. Oxford University Press, Oxford, 336 pp.
(2015) EvoDevo and its significance for animal evolution and phylogeny. In: A. Wanninger (ed.) *Evolutionary Developmental Biology of Invertebrates*, Vol. 1. Introduction, Non-Bilateria, Acoelomorpha, Xenoturbellida, Chaetognatha. Springer-Verlag, Vienna.
(2020) Taxonomy needs pluralism, but a controlled and manageable one. *Megataxa* 1: 9–18.
Mishler, B. D. (2010) Species are not uniquely real biological entities. In: F. Ayala and R. Arp (eds.) *Contemporary Debates in Philosophy of Biology*. Wiley Blackwell, Weinheim, pp. 110–122.
Mishler, B. D. and Budd, A. F. (1990) Species and evolution in clonal organisms: Introduction. *Systematic Botany* 15: 79–85.
Mora, C., Tittensor, D. P., Adl, S. et al. (2011) How many species are there on Earth and in the Ocean? *PLoS Biology* 9: e1001127.
Moroz, L. et al. (2014) The ctenophore genome and the evolutionary origins of neural systems. *Nature* 510: 109–114.
Müller, F. (1864) *Für Darwin*. Verlag von Wilhelm Engelmann, Leipzig.
Müller, G. B. (1991) Experimental strategies in evolutionary embryology. *American Zoologist* 31: 605–615.
Neelapu, N. R. R., Dutta, T., and Challaet, S. (2019) Role of horizontal gene transfer in evolution of the plant genome. In: T. G. Villa and M. Viñas (eds.) *Horizontal Gene Transfer*. Springer Publishing, Champaign, IL, pp. 425.
Neiman, M., Lively, C. M., and Meirmans, S. (2017) Why sex? A pluralist approach revisited. *Trends in Ecology and Evolution* 32: 589–600.
Newman, S. A. (2010) Dynamical patterning modules. In: M. Pigliucci and G. B. Müller (eds.) *Evolution, the Extended Synthesis*. The MIT Press, Cambridge, MA, pp. 281–306.
Ninova, M., Ronshaugen, M., and Griffiths-Jones, S. (2014) Conserved temporal patterns of microRNA expression in *Drosophila* support a developmental hourglass model. *Genome Biology and Evolution* 6: 2459–2467.
O'Hara, T. D., Hugall A. F., Thuy B., Stöhr S., and Martynov, A. V. (2017) Restructuring higher taxonomy using broadscale phylogenomics: The living Ophiuroidea. *Molecular Phylogenetic and Evolution* 107: 415–430.
Olsson, L., Levit, G. S., and Hossfeld, U. (2010) Evolutionary developmental biology: Its concepts and history with a focus on Russian and German contributions. *Naturwissenschaften* 97: 951–969.
Orton, G. R. (1955) The role of ontogeny in systematics and evolution. *Evolution* 9: 75–83.
Ottenburghs, J. (2020) Ghost introgression: Spooky gene flow in the distant past. *BioEssays* 42: 2000012.
Pääbo, S. (2014). *Neanderthal Man: In Search of Lost Genomes*. Basic Books, New York, 275 pp.
Pabst, E. A. and Kocot, K. M. (2018) Phylogenomics confirms monophyly of Nudipleura (Gastropoda: Heterobranchia). *Journal of Molluscan Studies* 84: 259–265.
Packard, A., Smotherman, C., and Jovanovic, N. (2020) Effect of circadian rhythm on the pain associated with preventive onabotulinumtoxinA injections for migraines. *Chronobiology International* 37: 1766–1771.
Padula, V., Bahia, J., Stöger, I. et al. (2016) A test of color-based taxonomy in nudibranchs: Molecular phylogeny and species delimitation of the *Felimida clenchi* (Mollusca: Chromodorididae) species complex. *Molecular Phylogenetics and Evolution* 103: 215–229.
Paterson, H. E. H. (1985) The recognition concept of species. In: E. S. Vrba (ed.) *Species and Speciation*. Transvaal Museum Monograph No. 4., pp. 21–29.
Pellino, M. et al. (2013) Asexual genome evolution in apomictic *Ranunculus auricomus* complex: Examining the effects of hybridization and mutation accumulation. *Molecular Ecology* 22: 5908–5921.

Perez, M. and Lehner, B. (2019) Intergenerational and transgenerational epigenetic inheritance in animals. *Nature Cell Biology* 21: 143–151.

Phillips, J. D., Gillis, D. J., and Hanner, R. H. (2019) Incomplete estimates of genetic diversity within species: Implications for DNA barcoding. *Ecology and Evolution* 9: 2996–3010.

Piasecka, B. et al. (2013) The hourglass and the early conservation models: Co-existing patterns of developmental constraints in vertebrates. *PLoS Genetics* 9 (4): e1003476.

Platania, L., Vodă, R., Dincă, V. et al. (2020) Integrative analyses on Western Palearctic *Lasiommata* reveal a mosaic of nascent butterfly species. *Journal of Zoological Systematics and Evolutionary Research* 58: 809–822.

Pleijel, F. (1999) Phylogenetic taxonomy, a farewell to species, and a revision of *Heteropodarke* (Hesionidae, Polychaeta, Annelida). *Systematic Biology* 48: 755–789.

Pollock, L. J., O'Connor, M. J., Mokany, K. et al. (2020) Protecting biodiversity (in all its complexity): New models and methods. *Trends in Ecology and Evolution* 35: 1119–1128.

Pradeu, T. (2012) *The Limits of the Self: Immunology and Biological Identity*. Oxford University Press, Oxford, 302 pp.

Pyykkö, P. (2019) An essay on periodic tables. *Pure and Applied Chemistry* 91: 1–10.

Queller, D. C. and Strassmann, J. E. (2009) Beyond society: The evolution of organismality. *Philosophical Transactions of the Royal Society B* 364: 3143–3155.

Quint, M. et al. (2012) A transcriptomic hourglass in plant embryogenesis. *Nature* 490: 98–101.

Raff, R. A. (1996) *The Shape of Life*. University of Chicago Press, Chicago, 520 pp.

Raff, R. A. and Kaufman, T. C. (1983) *Embryos, Genes, and Evolution*. Indiana University Press, Bloomington, 395 pp.

Ralston, A. and Shaw, K. (2008) Environment controls gene expression: Sex determination and the onset of genetic disorders. *Nature Education* 1: 203.

Raphaël, M. et al. (2016) Nomenclature for the nameless: A proposal for an integrative molecular taxonomy of cryptic diversity exemplified by planktonic Foraminifera. *Systematic Biology* 65: 925–940.

Raso, L. et al. (2014) Molecular identification of adult and juvenile linyphiid and theridiid spiders in Alpine glacier foreland communities. *PLoS ONE* 9(7): e101755.

Ratcliff, W. C., Denison, R. F., Borrello, M., and Travisano, M. (2012) Experimental evolution of multicellularity. *PNAS* 109: 1595–1600.

Redmond, A. K. and McLysaght, A. (2021) Evidence for sponges as sister to all other animals from partitioned phylogenomics with mixture models and recoding. *Nature Communications* 12: 1783.

Reilly, S. M., Wiley, E. O., and Meinhardt, D. J. (1997) An integrative approach to heterochrony: The distinction between interspecific and intraspecific phenomena. *Biological Journal of the Linnean Society* 60: 119–143.

Reiss, D. et al. (2019) Global survey of mobile DNA horizontal transfer in arthropods reveals Lepidoptera as a prime hotspot. *PLoS Genetics* 15(2): e1007965.

Reitzel, A. M., Burton, P. M., Krone, C., and Finnerty, J. R. (2007) Comparison of developmental trajectories in the starlet sea anemone *Nematostella vectensis*: Embryogenesis, regeneration, and two forms of asexual fission. *Invertebrate Biology* 126: 99–112.

Reydon, T. A. C. and Kunz, W. (2019) Species as natural entities, instrumental units and ranked taxa: New perspectives on the grouping and ranking problems. *Biological Journal of the Linnean Society* 126: 623–636.

Richards, R. A. (2010) *The Species Problem. A Philosophical Analysis*. Cambridge University Press, Cambridge, 236 pp.

Richards, R. J. (2009) Haeckel's embryos: Fraud not proven. *Biological Philosophy* 24: 147–154.

Richter, S. and Wirkner, C. S. (2014). A research program for evolutionary morphology. *Journal of Zoological Systematics and Evolutionary Research* 52: 338–350.

Rieppel, O. (2006) The PhyloCode: A critical discussion of its theoretical foundation. *Cladistics* 22: 186–197.

Robilliard, G. A. (1970) The systematics and some aspects of the ecology of the genus *Dendronotus* (Gastropoda: Nudibranchia). *The Veliger* 12: 433–479.

Robson, G. C. and Richards, O. W. (1936) *The Variation of Animals in Nature*. Longmans, Green and Co, London, 425 pp.

Ros-Rocher, N., Pérez-Posada, A., Leger, M. M., and Ruiz-Trillo, I. (2021) The origin of animals: An ancestral reconstruction of the unicellular-to-multicellular transition. *Open Biology* 11: 200359.

Rosslenbroich, B. (2014) *On the Origin of Autonomy: A New Look at the Major Transitions in Evolution*. Springer International Publishing, Switzerland, 302 pp.

Rudman, W. B. (1984) The Chromodorididae (Opisthobranchia: Mollusca) of the Indo-West Pacific: A review of the genera. *Zoological Journal of the Linnean Society* 81: 115–273.

Sáez, A. G., Probert, I., Geisen, M. et al. (2003) Pseudo-cryptic speciation in coccolithophores. *PNAS* 100: 7163–7168.

Salis, P., Lorin, T., Laudet, V., and Frédérich, B. (2019) Magic traits in magic fish: Understanding color pattern evolution using reef fish. *Trends in Genetics* 35: 265–278.

Scerri, E. (2020) Recent attempts to change the periodic table. *Philosophical Transactions of the Royal Society A* 378: 20190300

Scheffers, B. R. et al. (2012) What we know and don't know about Earth's missing biodiversity. *Trends in Ecology and Evolution*, 27: 501–510.

Schimkewitsch, W. (1906) Über die Periodizität in den System der Pantopoda. *Zoologischer Anzeiger* 30: 1–22.

Schlick-Steiner, B. C., Steiner, F. M., Seifert, B. et al. (2010) Integrative taxonomy: A multisource approach to exploring biodiversity. *Annual Review of Entomology* 55: 421–438.

Scornavacca, C., Delsuc, F., and Galtier, N. (eds.) (2020) *Phylogenetics in the Genomic Era*. Authors open access book, hal-02535070, 568 pp.

Sharma, P. P., Clouse, R. M., and Wheeler, W. C. (2017) Hennig's semaphoront concept and the use of ontogenetic stages in phylogenetic reconstruction. *Cladistics* 33: 93–108.

Sharma, R. and Bhat, V. (2020) Role of apomixis in perpetuation of flowering plants: ecological perspective. In: R. Tandon, K. Shivanna, and M. Koul (eds.) *Reproductive Ecology of Flowering Plants: Patterns and Processes*. Springer, Singapore.

Sharp, N. P. and Otto, S. P. (2016) Evolution of sex: Using experimental genomics to select among competing theories. *Bioessays* 38: 751–757.

Simmonds, P. and Aiewsakun, P. (2018) Virus classification: Where do you draw the line? *Archives of Virology* 163: 2037–2046.

Simpson, C., Herrera-Cubilla, A., and Jackson, J. B. C. (2020) How colonial animals evolve. *Science Advances* 6: eaaw9530.

Slack, J. M. W., Holland, P. W. H., and Graham, C. F. (1993) The zootype and the phylotypic stage. *Nature* 361: 490–492.

Slater, M. (2013) *Are Species Real? An Essay on the Metaphysics of Species*. Palgrave Macmillan, London, 214 pp.

Smaczniak, C., Immink, R. G. H., and Angenent, G. C. (2012) Developmental and evolutionary diversity of plant MADS-domain factors: Insights from recent studies. *Development* 139: 3081–3098.

Stanton, D. W. G. et al. (2019) More grist for the mill? Species delimitation in the genomic era and its implications for conservation. *Conservation Genetics* 20: 101–113.

Stamos, D. N. (2003). *The Species Problem: Biological Species, Ontology, and the Metaphysics of Biology*. Lexington Press, 380 pp.

Stebbins, G. L. (1950) *Variation and Evolution in Plants*. Columbia University Press, New York, 643 pp.

(1980) Botany and the synthetic theory of evolution. In: E. Mayr and W. B. Provine, *The Evolutionary Synthesis: Perspectives on the Unification of Biology*. Harvard University Press, Cambridge, MA, pp. 139–152.

Stenz, L., Schechter, D. S., Serpa, S. R. et al. (2018) Intergenerational transmission of DNA methylation signatures associated with early life stress. *Current Genomics* 19: 665–675.

Steyer, J. S. (2000) Ontogeny and phylogeny in temnospondyls: A new method of analysis. *Zoological Journal of the Linnean Society* 130: 449–467.

Stöhr, S. and Martynov, A. V. (2016) Paedomorphosis as an evolutionary driving force: Insights from deep-sea brittle stars. *PLOS ONE* 11: e0164562.

Stuefer, J. F., Gómez, S., and Van Mölken, T. (2004) Clonal integration beyond resource sharing: Implications for defence signalling and disease transmission in clonal plant networks. *Evolution and Ecology* 18: 647–667.

Sweatt, J. D. (2019) The epigenetic basis of individuality. *Current Opinion in Behavioral Sciences* 25: 51–56.

Templeton, A. R. (1989) The meaning of species and speciation: A genetic perspective. In: D. Otte and J. A. Endler (eds.) *Speciation and Its Consequences*. Sinauer Associates, Sunderland, MA, pp. 3–27.

Thielsch, A., Brede, N., Petrusek, A., de Meester, L., and Schwenk, K. (2009) Contribution of cyclic parthenogenesis and colonization history to population structure in *Daphnia*. *Molecular Ecology* 18: 1616–1628.

Thomson, W. (1st Baron Kelvin) (1900) Nineteenth century clouds over the dynamical theory of heat and light. *Notices of Proceedings at Meetings of Members of Royal Institution of Great Britain* 16: 363–397.

Thorpe, J. P., Beardmore, J. A., and Ryland, J. S. (1978) Genetic evidence for cryptic speciation in the marine bryozoan *Alcyonidium gelatinosum*. *Marine Biology* 49: 27–32.

Tikhomirova, A. L. (1990) *Alteration of Ontogeny as Mechanism of Insect Evolution*. Nauka, Moscow, 168 pp.

Turland, N. J. et al. (eds.) (2018) *International Code of Nomenclature for Algae, Fungi, and Plants (Shenzhen Code)*. Regnum Vegetabile 159, Koeltz Botanical Books, Glashütten.

Uesaka M., Kuratani, S., and Irie, N. (2021) The developmental hourglass model and recapitulation: An attempt to integrate the two models. *Journal of Experimental Zoology*: 1–11.

Vannier, N., Mony, C., Bittebiere, A. K. et al. (2019) Clonal plants as meta-holobionts. *mSystems* 4: e00213-18.

Varki, A. and Altheide, T. K. (2005) Comparing the human and chimpanzee genomes: Searching for needles in a haystack. *Genome Research* 151: 746–1758.

Vavilov, N. I. (1922) The law of homologous series in variation. *Journal of Genetics* 12: 47–89.

Vrba, E. (ed.) (1985) Species and speciation. *Transvaal Museum Monograph* No. 4: 73 pp.

Vrana, P. and Wheeler, W. (1992) Individual organisms as terminal entities: Laying the species problem to rest. *Cladistics* 8: 67–72.

Wägele, H. and Willan, R. C. (2000) Phylogeny of the Nudibranchia. *Zoological Journal of the Linnean Society* 130: 83–181.

Walker, T. J. (1964) Cryptic species among sound-producing ensiferan Orthoptera (Gryllidae and Tettigoniidae). *Quarterly Review of Biology* 39: 345–355.

Wang, S. S. and Ellington, A. D. (2019) Pattern generation with nucleic acid chemical reaction networks. *Chemical Reviewers* 119: 6370–6383.

Wanninger, A. (2015) Morphology is dead – long live morphology! Integrating MorphoEvoDevo into molecular EvoDevo and phylogenomics. *Frontiers of Ecology and Evolution* 3: 54.

West-Eberhard, M. J. (2003) *Developmental Plasticity and Evolution*. Oxford University Press, New York.

Wheeler, Q. D. and Meier, R. (eds.) (2000) *Species Concepts and Phylogenetic Theory:*

A Debate. Columbia University Press, New York, 230 pp.

Willmann, R. and Meier, R. (2000) A critique from the Hennigian species concept perspective. In: Q. D. Wheeler and R. Meier (eds.) *Species Concept and Phylogenetic Theory: A Debate*. Columbia University Press, New York, pp. 101–118.

Wilding, C. S., Fletcher, N., Smith, E. et al. (2020) The genome of the sea anemone *Actinia equina* (L.): Meiotic toolkit genes and the question of sexual reproduction. *Marine Genomics* 53: 100753.

Williams, T. A., Cox, C. J., Foster, P. G., Szöllősi, G. J., and Embley, T. M. (2020) Phylogenomics provides robust support for a two-domains tree of life. *Nature Ecology & Evolution* 4: 138–147.

Wilkins, J. S. (2009) *Species: A History of the Idea*. University of California Press, Berkeley.

Willis, S. C. (2017) One species or four? Yes!... and, no. Or, arbitrary assignment of lineage to species obscures the diversification processes of Neotropical fishes. *PLoS ONE* 12: e0172349.

Winsor, M. P. (2006) Linnaeus's biology was not essentialist. *Annals of the Missouri Botanical Garden* 93: 2–7.

Wolfe, J. M. and Hegna, T. A. (2013) Testing the phylogenetic position of Cambrian pancrustacean larval fossils by coding ontogenetic stages. *Cladistics* 30: 366–390.

Yang, Z. (2014). *Molecular Evolution: A Statistical Approach*. Oxford University Press, Oxford, 492 pp.

Yehuda, R. et al. (2016) Holocaust exposure induced intergenerational effects on FKBP5 methylation. *Biological Psychiatry* 80: 372–380.

Yoshida, Yu., Nowell, R. W., Arakawa, K., and Blaxter, M. (2019) Horizontal gene transfer in Metazoa: examples and methods. In: T. G. Villa and M. Viñas (eds.) *Horizontal Gene Transfer*, Springer Nature, Switzerland, Cham, 220 pp.

Yoshikawa, H. et al. (2020) Effects of the global coronavirus disease-2019 pandemic on early childhood development: Short- and long-term risks and mitigating program and policy actions. *The Journal of Pediatrics* 223: 188–193.

Zachos, F. E. (2016) *Species concepts in biology. Historical Development, Theoretical Foundations and Practical Relevance*. Springer Nature, Switzerland, Cham, 220 pp.

(2018) (New) Species concepts, species delimitation and the inherent limitations of taxonomy. *Journal of Genetics* 97: 811–815.

Zimmermann, W. (1959) *Die Phylogenie der Pflanzen*. Gustav Fischer Verlag, Stuttgart, 777 S.

Zirkle, C. (1959) Species before Darwin. *Proceedings of the American Philosophical Society* 103: 636–644.

5

Diagnosability and Cryptic Nodes in Angiosperms

A Case Study from Ipomoea

Pablo Muñoz-Rodríguez, John R. I. Wood, and Robert W. Scotland

Three important properties associated with a classification of any group of organisms are diagnosability, monophyly and resolution. In this chapter we explore the interrelationships between these three properties in the context of cryptic taxa, here defined as a *clade with no obvious diagnostic morphological support.* We present the view that the number of nodes on a phylogenetic tree of all flowering plants that have morphological diagnostic support is less than five percent; as such, cryptic nodes are much more common than non-cryptic nodes. Because of this, we suggest that the phrase 'cryptic nodes' is a preferable description as opposed to cryptic taxa because taxa in the sense of traditional classifications are generally diagnostic. By reference to a global taxonomic study of the genus *Ipomoea*, we discuss the role of diagnosability at various scales including major infrageneric clade, genus and species. We demonstrate that the level of diagnosability for *Ipomoea* is relatively low, therefore making cryptic nodes the rule and not the exception. We provide several examples of such cryptic nodes, detail how we discovered them and place them in a wider conceptual framework of diagnosability in angiosperms.

5.1 Diagnosability, Monophyly, and Resolution in Angiosperms

Classifications of organisms into groups at various hierarchical levels – species, genera, families – are most readily understood when the taxon is a natural group and has at least one unique diagnostic character. A taxon such as the angiosperms, for example, is distinguished (diagnosability) from other organisms based on a unique morphological character: the carpel. A key feature in this example is that the unique character resolves and diagnoses the Angiosperms clade (monophyletic group). In this sense, biological classifications are

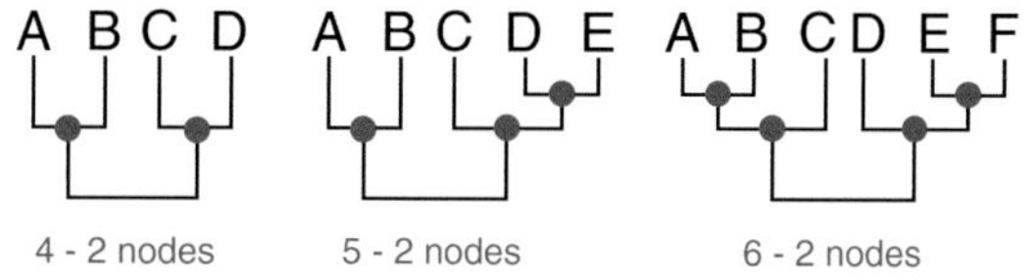

Figure 5.1 The number of nodes in a fully resolved phylogeny is the number of tips minus 2. A 4-taxon phylogeny has two nodes, a 5-taxon phylogeny has 3 nodes, a 6-taxon phylogeny has 4 nodes, and so on. Given 361,750 accepted species (tips), a fully resolved phylogeny including all of them contains 361,748 nodes, i.e., 361,750 minus 2.

structured hierarchies of monophyletic taxa that have diagnostic features. These three properties of a classification – diagnosability, monophyly, and resolution – are the hallmarks of classification (Carine and Scotland 2002). One question that is relevant to the subject of this volume is, how much resolution is there across the tree of life?

In the case of Angiosperms, for example, a current classification, APG4, arranges a total of 361,750 species into 57 orders, 446 families, and approximately 13,208 genera (Stevens 2020). How much diagnosability, monophyly, and resolution does this classification contain compared to what is theoretically possible? A fully resolved bifurcating phylogeny including all 361,750 accepted species would contain 361,750 minus 2 internal nodes (Figure 5.1). The information content (Mickevich and Platnick 1989) of this fully resolved tree, terminal species notwithstanding, is the number of nodes in the tree that show relationships between species. Thus, the *theoretically* possible information content of this phylogeny is 361,748 units of information. We can contrast this theoretical value with a very rough estimate of the actual value based on the current classification of flowering plants. This is achieved by assuming that all taxa at ranks above the species level in this classification represent a node that has been resolved and is therefore monophyletic and – presumably to some degree – morphologically diagnosable. Given that each rank is equivalent to a node on a tree, we can calculate a rough and ready approximation of diagnostic nodes by adding up the number of recognised taxa at each rank. This means that there are 57 (orders) + 446 (families) + 13,208 (genera) = 13,711 units of information that represent diagnostic nodes on the angiosperm tree. Thus, the current classification contains approximately 3.8 per cent of the theoretically possible amount of information (13,711 out of 361,748 nodes). In other words, the current classification of angiosperms into genera, families, and orders recognises less than 4 per cent of all possible nodes in a fully resolved phylogeny.

Given the increasing amount of molecular sequence data being gathered for more and more species, and recent calls to sequence all species on Earth (Callaway 2018), the possibility of obtaining a fully resolved phylogeny for all eukaryotes seems to be within reach. In other words, resolution and monophyly are two increasingly achievable goals for biological classifications. What then of diagnosability, the third desired property of a classification? One possible view of future research into the classification of the tree of life is that, similarly to resolution and monophyly, increasing levels of diagnosability are possible and desirable. However, we have seen that the number of diagnosable nodes on the tree of life is limited. Therefore, although the main aim of a classification is to recognise diagnostic monophyly, it must also account for the fact that the number of diagnosable nodes on the tree of life is limited.

5.2 Diagnosability is Limited by the Nature of Character Evolution

Throughout the history of systematics, the delimitation of taxa has been based on characters such as the carpel for angiosperms, the legume for Fabaceae, the capitulum for Asteraceae, and as THE spiny pollen for *Ipomoea*. These examples demonstrate the close relationship between diagnostic morphological characters and the monophyletic taxa they characterize. Diagnostic characters play a central role in the recognition and discovery of biological taxa at all ranks of classical taxonomy. Patterson (1982) argued that 'to discover a character is to discover a group', reflecting his view of the close relationship between characters and taxa – a property that Patterson referred to as *taxic homology*. However, these types of character, which define and diagnose monophyletic groups (synapomorphies *sensu* Hennig 1966), are relatively few in number compared to the vast majority of characters, which are highly homoplastic (Scotland 2010, 2011). In other words, the achievable goals for an increased level of *diagnosable* monophyly are highly constrained by the nature of character evolution across angiosperms.

Individual morphological characters and combinations of morphological characters are used to diagnose monothetic and polythetic groups, respectively (Sneath and Sokal 1973). Monothetic groups have at least one diagnostic character necessary for group membership, for instance all flowering plants have carpel(s) that contain seeds and later develop into the fruit. Characters of monothetic groups are broadly equivalent to the phylogenetic term 'synapomorphy' (evolutionary novelty) and were termed conservative or *good* characters in the pre-phylogenetics era. These often unique and conserved characters are thus very useful for defining and diagnosing taxa, for example, carpel for flowering plants, vertebrae for vertebrates, amnion for amniotes, and nucleic acid for life.

On the other hand, polythetic groups – which make up most taxa across the tree of life – are diagnosed and recognised by combinations of characters, none of which is universally present in all members of the group. In other words, in a polythetic group no character is shared by all members of the group. Thus, for example, species belonging to the family Lamiaceae tend to have square stems, verticillate inflorescences, four nutlets, gynobasic styles, and opposite leaves, but none of these characters is found in all species belonging to the family. Instead, various combinations of these characters are needed to diagnose membership of the polythetic Lamiaceae (Stevens 1984).

The ratio of polythetic/monothetic groups in taxonomy is more or less proportional to the extent of homoplasy or iterative evolution: characters evolve at one phylogenetic level but change/reverse at another, such that all descendants of a particular node on the tree do not all share any one character. Systematics sometimes uses a pragmatic solution when diagnosability is absent. The clade Eudicots (originally termed tricolpates), for example, is supposedly diagnosed, as first proposed, by the presence of tricolpate pollen, even though a large number of taxa within it do not have tricolpate pollen. This pragmatic solution was accepted because the plesiomorphic condition within Eudicots is considered to be tricolpate pollen and the character state is optimised intuitively at the Eudicots node such that it is treated as the evolutionary novelty of the group.

It is also the case that, in some parts of the tree of life, diagnosability is emphasised at the expense of monophyly, such that groups can be non-monophyletic and diagnosed by a large number of plesiomorphic characters. This is the case for prokaryotes – which feature in biology textbooks and biology courses despite their non-monophyletic status. This non-monophyletic taxon (prokaryotes) persists simply because of the phenetic distance (i.e. number of plesiomorphic characters) shared between prokaryotes but absent from eukaryotes, which is viewed as a basic division in the way life is organised.

Contemporary classifications of angiosperms such as the Angiosperm Phylogeny Group IV system (APG4) increasingly reflect monophyletic groups, inferred using DNA sequence data; these groups are either monothetic or, more usually, polythetic, and therefore have discernible diagnostic morphological characters or diagnostic combinations of morphological characters. Consequently, for any node on a tree it is possible to list the characters or combinations of characters for the terminal taxa included, but this does not necessarily achieve diagnosability. Our central message in this chapter is that the extent of diagnosability across the angiosperm phylogeny seems minimal at best and is certainly less than 10 per cent of all possible nodes, perhaps as low as 5 per cent. In modern systematics, the diagnostic traits of clades are often decided in one of two ways: intuitively – as a result of phylogenetic analysis and morphological observation – or by using transformational models of character evolution – which attempt to optimise homoplastic character states at the appropriate node on a tree (e.g. Sauquet et al. 2017; Stevens 2020). However, when there are appreciable levels of homoplasy neither method is likely to provide a satisfactory level of diagnosability (e.g. Sauquet et al. 2017).

It is surprising that, given the importance of diagnosability for classifications and the tree of life, there has been little or no theoretical treatment of diagnosability within systematics. The lack of discussion in systematics about diagnosability is especially surprising given the tension and discrepancy that exists between increasing levels of monophyletic resolution and decreasing levels of diagnostic resolution. This tension has also been explored relative to the PhyloCode (Wojciechowski 2013). It is against this background of phylogenetic resolution of nodes with no obvious morphological support that the phenomenon of cryptic taxa or cyptic nodes has emerged.

It is evident that cryptic nodes – *clades with no obvious diagnostic morphological support* – should be the rule rather than the exception across the tree of life. This is based on the fact that the extent of unique diagnostic synapomorphy is minimal and that the vast majority of morphological tree space is characterised by homoplasy, that is, the iterative re-invention of the same evolutionary strategies.

An example of the issues around diagnosability can be found in the megadiverse plant genus *Ipomoea*, object of a monographic study by our research group (Muñoz-Rodríguez et al. 2019; Wood et al. 2020). The densely sampled phylogenies in our analyses revealed the existence of several major clades in *Ipomoea* that are strongly supported in all methods of phylogenetic inference but have no diagnostic morphological characters. These clades conflict with all previous infrageneric classifications of the genus based on perceived morphological distinctions, which should consequently be abandoned. Furthermore, several smaller clades containing well-known species (e.g. *Ipomoea pes-*

caprae (L.) R.Br. and allies) also lack diagnostic morphological characters, reflecting the difficulty of finding monophyletic, diagnostic, and mutually exclusive groups (Carine and Scotland 2002).

Furthermore, Linnean classifications assume and demand that all species are accounted for at a given rank (completeness). This places a further constraint on the recognition of diagnostic monophyly as it is not satisfactory to classify some of the taxa that may have diagnostic characters whilst leaving others unaccounted for. The need to achieve this fourth property, completeness, when combined with diagnosability and monophyly, has meant that many large genera of plants that are similar to *Ipomoea* (e.g. *Begonia* (Moonlight et al. 2018), *Solanum* (Peralta and Spooner 2000)) have not been split up into smaller diagnostic monophyletic groups. In the following section we demonstrate that it is not possible to exhaustively classify all species diversity within *Ipomoea* into mutually exclusive and morphologically distinct taxa of equal rank as in a traditional classification system.

5.3 Diagnosability in Convolvulaceae and *Ipomoeeae*

Convolvulaceae Juss. is, with c. 2,000 species described, a large family of Angiosperms (Royal Botanic Gardens, Kew 2019). The taxonomic history of Convolvulaceae is similar to many families in that several attempts to reconcile the properties of diagnosability, monophyly, and resolution have been made over time. As in many other plant families, these three properties have been applied in an inconsistent way that partially reflects changing concepts of underlying methodology but also the lack of sampling across the entire geographical range and phylogenetic diversity of the family (Choisy 1838, 1845; van Ooststroom 1953; Austin 1975; Austin 1979; Stefanovic et al. 2002, 2003; Eserman et al. 2014).

Several proposals to classify the diversity within Convolvulaceae in formal taxonomic ranks have been presented since the nineteenth century, but nowadays most studies follow the phylogenetic approach proposed by Stefanovic et al. (2002, 2003). In their work, Stefanovic and colleagues redefined the tribes recognised by Austin (1973), which they found to be non-monophyletic. They recognised 12 tribes corresponding to as many monophyletic lineages (see figure 1 in Stefanovic et al. (2003) for lineages and a comparison with previous works). An alternative proposal in APG4 recognises six subfamilies and several tribes (Stevens 2020) but does not include all tribes recognised by Stefanovic and colleagues. The identification of diagnostic characters for lower ranks in Convolvulaceae has often been challenging, especially in the tribe Ipomoeeae but also in other tribes such as Merremieae and Convolvuleae (Verdcourt 1957; Austin 1975; McDonald 1991; Wood et al. 2015; Simões et al. 2015; Simões and Staples 2017; Wood et al. 2020).

Ipomoeeae, the largest group in the family, includes all members of Convolvulaceae with echinate, pantoporate pollen. This pollen character was first used by Hallier to establish this group (Hallier 1893, 1894), and subsequent molecular analyses have confirmed Ipomoeeae as a monophyletic group.

Until recently, the tribe Ipomoeeae was generally divided into ten genera: *Ipomoea* L., with over 500 species, and nine smaller genera described to accommodate morphological variations within the tribe: *Argyreia* Lour., *Astripomoea* A.Meeuse, *Blinkworthia* Choisy,

Lepistemon Blume, *Lepistemonopsis* Dammer, *Paralepistemon* Lejoly & Lisowski, *Rivea* Choisy, *Stictocardia* Hallier f. and *Turbina* Raf. *Ipomoea* was further divided into three subgenera and a variable number of sections and series depending on the author (House 1908; van Ooststroom 1953; Verdcourt 1957; Austin 1979, 1980; Austin and Huáman 1996; Miller et al. 2004).

In the last three decades, cladistic and molecular studies using limited data showed that some of these genera were not monophyletic (Miller et al. 1999; Wilkin 1999; Stefanovic et al. 2003; Eserman et al. 2014). We confirmed this through genomic studies using extensive taxon sampling from across all genera in Ipomoeeae based on both nuclear and chloroplast genomic data (Muñoz-Rodríguez et al. 2019). We showed that all these smaller genera are nested inside *Ipomoea* and only *Astripomoea* forms a monophyletic group of any size within it (Figure 5.2). Similarly, the three previously recognised subgenera and almost all series and sections within *Ipomoea* are not monophyletic (Figure 5.3).

In our phylogenies, the tribe Ipomoeeae is divided into two large clades dominated by taxa from different geographical regions but including a significant number of representatives from elsewhere (Figure 5.4). The first clade is dominated by taxa from the Palaeotropical region (hereinafter Old-World Clade, OWC) and includes all smaller genera formerly recognised, whereas the second clade includes a small paraphyletic grade of mainly African taxa and a large clade dominated by species from the Neotropics (New-World Clade, NWC). OWC and NWC lack morphological synapomorphies and are identified from DNA data only but are strongly supported by our molecular data, regardless of the method of phylogenetic inference.

In light of these results, we redefined the genus *Ipomoea* and included all the smaller genera to make *Ipomoea* monophyletic. We consider this is the most logical decision because (1) all smaller genera intermingle with species of *Ipomoea*; (2) of the hitherto recognised genera of any size only *Astripomoea* is monophyletic; and (3) there are no clear diagnostic morphological characters to redefine these smaller genera.

The newly delimited *Ipomoea* includes all members of Convolvulaceae with echinate pantoporate pollen, a morphological synapomorphy within the family, and it is the only genus in the tribe *Ipomoeeae*. It includes around 800 currently accepted species. Now for the first time there is a well sampled phylogeny (**resolution**) that contains well-supported robust clades (**monophyly**) but very few of these have morphological **diagnostic** characters and are therefore cryptic (Muñoz-Rodríguez et al. 2019; Wood et al. 2020).

5.4 Cryptic Taxa, Rapid Radiations, and Diagnosability in *Ipomoea*

Cryptic taxa are often discussed and studied as potential new entities separate from existing ones to which they were thought to belong. In preparing our monograph of *Ipomoea* in the New World (Wood et al. 2020), for example, we discovered several new entities that could fit the definition of *cryptic species*. On several occasions, our DNA studies revealed that new collections thought to belong to species previously described were only distantly related. These results prompted the re-examination of morphological data and the subsequent recognition of several new species with robust evidence.

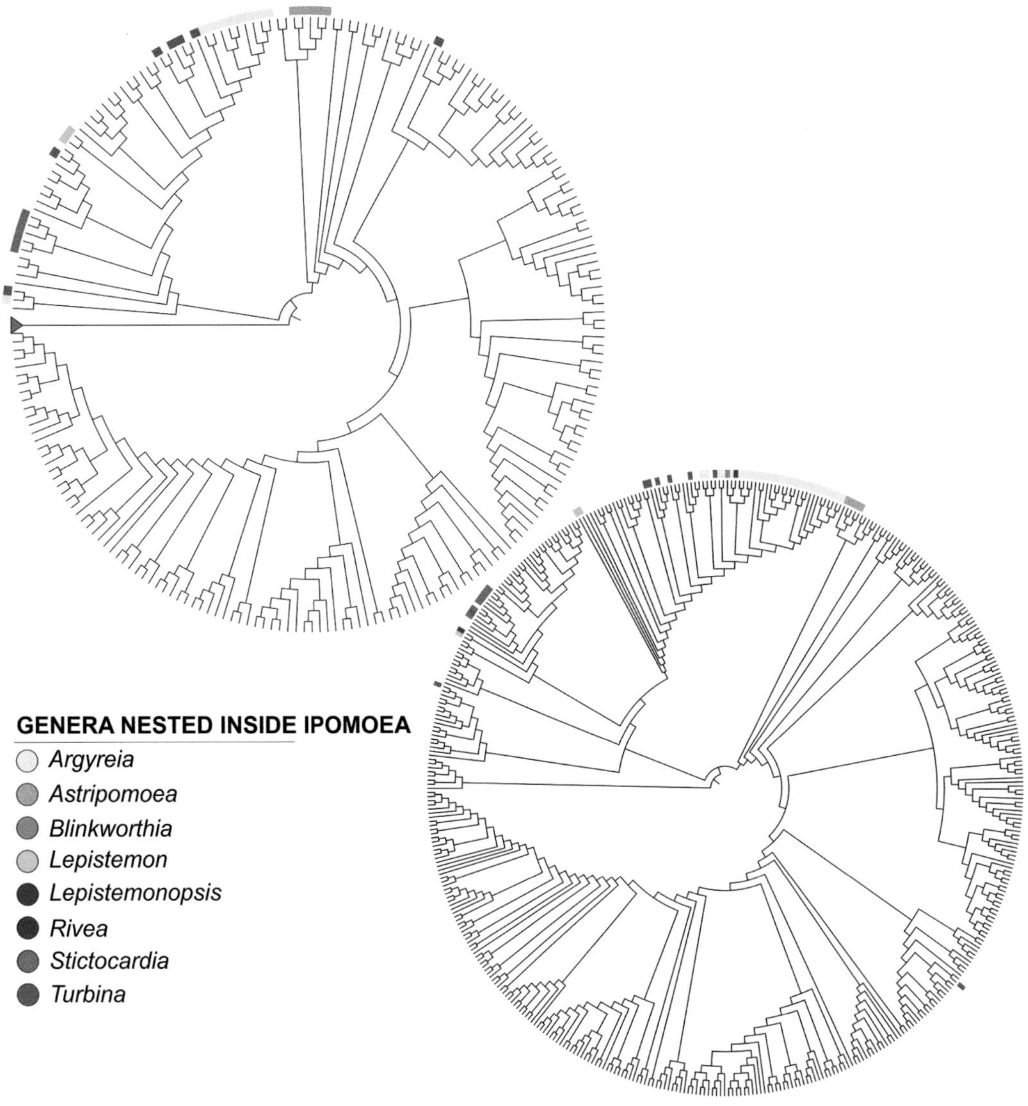

Figure 5.2 Summary phylogenies of the tribe Ipomoeeae using (A) nuclear coding DNA regions and (B) ITS sequences. Colours in the outer circle indicate the position of specimens belonging to species from the various genera now included in Ipomoea. All branches without colours are Ipomoea specimens. Segregate genera have some degree of diagnosability, generally through a combination of non-exclusive morphological characters but most are not monophyletic. The distribution of these segregate genera and the relatively few species they contain as well as their lack of monophyly demonstrate the futility of trying to split Ipomoea into a smaller number of diagnostic monophyletic taxa at the same rank.

An example of this is *Ipomoea cryptica* J. R. I. Wood and Scotland, a species described from Bolivia that is almost identical morphologically to two other species widespread in the Americas, *I. squamosa* Choisy and *I. anisomeres* B. L. Rob. and Bartlett (see Supplementary

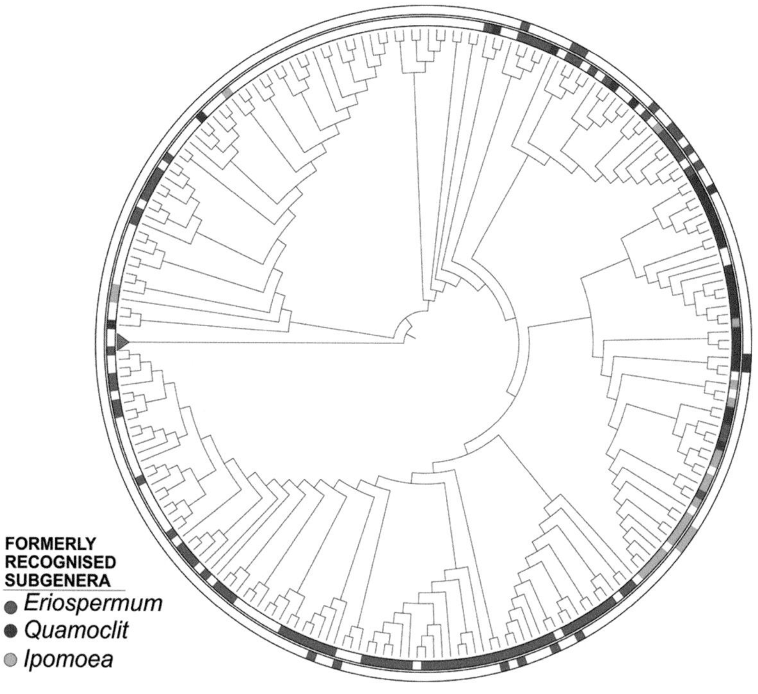

Figure 5.3 Nuclear phylogeny of Ipomoea showing the position of the species in the subgenera recognised by Austin (1977–1996) (inner circle), and Miller (1999) (outer circle). Not all species in the phylogeny were assigned to a specific subgenus by these authors, hence the lack of colour in some tips.

Information in Muñoz-Rodríguez et al. 2019 for a detailed account of this discovery). One other example is *Ipomoea australis* (O'Donell) J. R. I. Wood and P. Muñoz, a South American close relative of the sweet potato previously treated as a variety of *I. cordatotriloba* Dennst., which we showed represents a different evolutionary lineage restricted to North America. Other examples of species recently described that are morphologically very similar to species only distantly related are *I. longibarbis* J. R. I. Wood and Scotland vs *I. rubens* O'Donell; *I. graniticola* J. R. I. Wood vs *I. subrevoluta* Choisy; or *I. psammophila* J. R. I. Wood and Scotland vs *I. nitida* Griseb. In some cases, the pairs of species are morphologically identical or almost identical and very difficult to distinguish based on morphology only. However, in most cases the new species had gone unnoticed because of the limited knowledge of already described species. A better understanding of *Ipomoea* diversity, together with the close morphological study of many specimens, revealed clear differences supporting the molecular results.

A different kind of cryptic taxon also occurs. A comprehensive species-level taxonomy of *Ipomoea* revealed the existence of two very diverse clades within the genus, namely Clade A1 and Clade A2 (part of the bigger NWC mentioned above) (Figure 5.4). Species in each of these clades have overlapping morphologies and, at least from DNA barcode phylogenies (we only have genomic data from one sample per species in these clades), their relationships are poorly resolved. Clade A1, with c. 130 species, is distributed throughout the

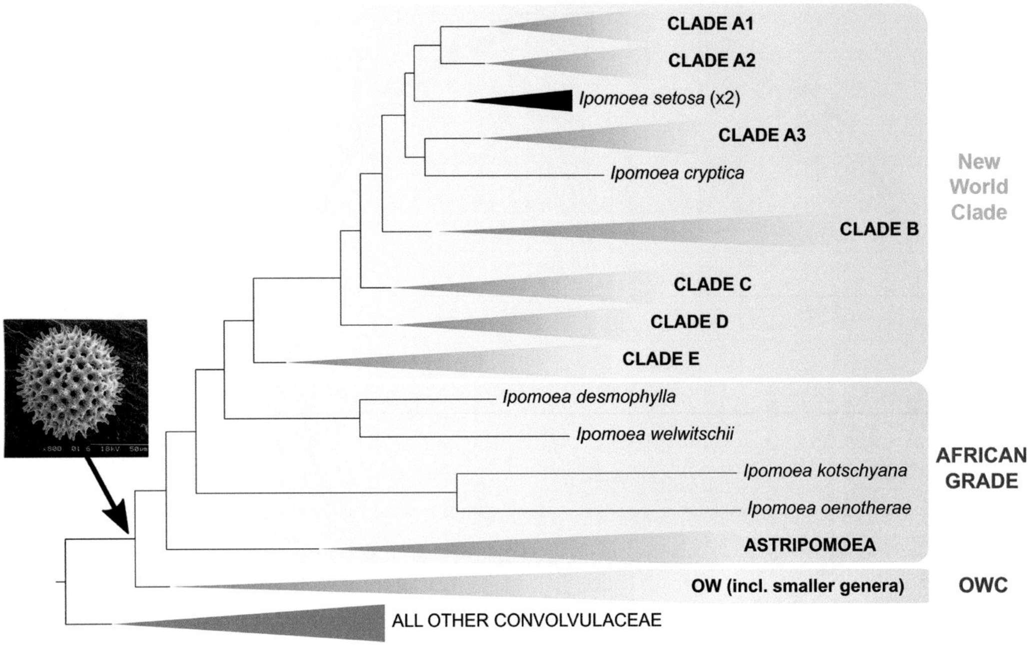

Figure 5.4 Nuclear phylogeny of Ipomoea inferred using non-coding single copy nuclear regions. All branches have 100 per cent support. See Muñoz-Rodríguez et al. (2019) for details on the methodology. The clade indicated by the black arrow includes all members of Convolvulaceae with spiny pollen, a synapomorphy of the tribe Ipomoeeae and the expanded genus Ipomoea. NWC, New World Clade; OWC, Old World Clade.

Americas but is clearly centred on Paraguay, southeast Bolivia, southwest Brazil, and northern Argentina. Clade A2, on the other hand, is well represented in tropical America with c. 90 species, but the number of species is proportionately highest in the Caribbean region. Both clades are closely related and part of the bigger Clade A, which also includes the sweet potato and its wild relatives that constitute Clade A3 (Figure 5.4).

Diversification rates are relatively constant in most of the genus but change in the part of the phylogeny including the two clades A1 and A2. In this part of the tree, we found a greater than 5.5-fold increase in net diversification (from 0.127 to 0.719 species Myr^{-1}) as the result of increased speciation rates (Muñoz-Rodríguez et al. 2019; Carruthers et al. 2020). Extinction rates, on the other hand, remained relatively constant. This increase in diversification rates probably occurred in the early Miocene, although more recent phenomena such as the establishment of the Cerrado biome may have contributed to the diversification dynamics of these species (Muñoz-Rodríguez et al. 2019). Of particular interest is that these rapid radiations, unlike other radiations previously described, show several shifts between growth habits and biomes.

This aspect of *Ipomoea* evolution – increased diversification rate in one part of the phylogeny – clearly demonstrates the close relationship between resolution, monophyly, and diagnosability in taxonomic and phylogenetic studies. First, in this part of the phylogeny all the branch lengths are relatively short, providing minimal resolution. Second, the monophyly of the radiation is relatively clear although there are differences for nuclear and chloroplast genome data. And finally, the radiation itself has no diagnostic characters and some species within the radiation are difficult to diagnose. In short, the taxonomy of these clades is more difficult than for the rest of the genus because the underlying tree model has not resulted in clear-cut boundaries between species nor between the radiation and the rest of *Ipomoea*. Further studies using high-throughput sequencing and including other still unsampled species in these clades will help assess how far genetic data alone can help reveal the true levels of species diversity in this part of the phylogeny.

5.5 Diagnosability of Infrageneric Ranks in *Ipomoea*

The majority of *Ipomoea* species can be readily identified if adequate material is available. A host of morphological characters have proved useful including variations in the sepals, corolla, and seeds, as well as significant differences in habit and leaf shape. Species can generally be diagnosed by a combination of hard/qualitative characters, such as corolla and seed indumentum or sepal morphology, supported by softer, often quantitative characters, such as corolla and leaf dimensions, together with distinct geographical distribution or ecological requirements. Most species are shown to be monophyletic when sequenced for DNA barcodes.

This assertion may be regarded as unnecessarily bold by some researchers, but diagnosability at species level is only problematic in certain species clusters - notably some relatives of the well-known species *I. batatas* and *I. hederifolia* L., and to a degree in Clade A1 and Clade A2, where many species form largely undifferentiated polytomies. In many cases, additional collections can be expected to resolve uncertainties. However, some species that can only be distinguished by soft characters may need review as intermediates suggest that introgression or hybridisation is taking place.

Although the situation at the species level is relatively straightforward, diagnosability at higher levels is problematic. As explained before, an expanded *Ipomoea* divides in two large clades, dominated by species from the Palaeotropics (OWC) and the Neotropics (NWC), respectively. However, there is no morphological character or combination of characters by which OWC can be distinguished from NWC. In addition, NWC is further divided into five strongly supported clades A–E (Figure 5.4), but there is no morphological character or combination of characters that can be used to define them.

Only at the level of smaller clades of less than 15 species does diagnosability sometimes become possible. One example, the former genus *Astripomoea*, has been referred to earlier in the chapter. This African clade can be distinguished by the presence of stellate hairs and a somewhat elongate stigma. However, although these characters serve to distinguish the clade in Africa, both are also found in a number of American species, although not in combination. A second example in OWC is the former genus *Stictocardia*, whose members intermingle with several *Ipomoea* species in a clade that is identifiable by the glandular abaxial surface of the leaves and the accrescent sepals, but neither character is unique to this clade.

Similar cases appear in NWC; for instance, the clade that includes species formerly included in the subgenus *Quamoclit* is diagnosable by the subapically awned sepals and 4-locular ovary and is perhaps the best-defined clade of all. Very small clades of c. four species in NWC, such as the *Ipomoea alba* or *I. asarifolia* alliances, are also quite frequently discernible. Clade A is divisible into three smaller clades, one of which includes the sweet potato and its close relatives, A3. Clade A3 can be diagnosed with difficulty and only by a combination of rather imprecise characters including the mucronate, thin, sometimes papery sepals, the small corolla, and often annual habit and hirsute ovary. All these characters are found in other clades or occur inconsistently, so species from this clade are often wrongly placed. Clades A1 and A2 are even more cryptic. Clade A1 is usually recognised by the pubescent exterior of the corolla and relatively soft pubescent sepals, but there are several exceptions, and these characters appear elsewhere in the genus, for example in the OWC species *Ipomoea rubens* Choisy. Conversely, clade A2 is usually recognisable by the glabrous corolla and rigid, concave sepals but again there are exceptions including species found in Clade A1.

The lack of morphological diagnostic characters except for very small clades renders an infrageneric classification based on morphology impossible in our current state of knowledge. Although we have no intrinsic objection to the naming of cryptic taxa, there seems to be no or little utility or societal benefit in establishing a complex named hierarchy of non-diagnosable taxa. Consequently, we did not attempt any formal recognition of infrageneric ranks, although this has been attempted in other megadiverse genera such as *Begonia* (Moonlight et al. 2018). The problem with infrageneric ranks is that, in order to meet the requirement of completeness, it is necessary to place all species in a taxon at each rank. Given the complex tree model of *Ipomoea* and the low level of diagnosability, this is virtually impossible if the wish is to maintain both monophyletic and diagnostic taxa.

Conclusions

Thanks to comprehensive taxon sampling, and the combination of morphological and molecular data, we showed that most taxonomic groups traditionally recognised in the tribe Ipomoeeae and the genus *Ipomoea* do not represent distinct evolutionary lineages. We also showed that the morphological diagnostic characters traditionally used do not identify monophyletic entities and that, instead, cryptic nodes are numerous. In this situation there are two solutions: the recognition of an expanded monophyletic *Ipomoea* that includes all other genera that are nested within or, alternatively, the division of *Ipomoea* into a series of smaller, monophyletic groups. Our research clearly demonstrates that the latter solution of splitting *Ipomoea* into smaller clades is unrealistic, as many parts of the *Ipomoea* phylogeny have few or no diagnostic characters and would produce a taxonomy that would be unusable outside the laboratory. Further, this approach would certainly leave many species unplaced, as their position could not be resolved. In other words, such a classification would not be exhaustive.

In conclusion, we advocate an expanded, monophyletic *Ipomoea* to include all the smaller, formerly recognised genera. In our view, the lack of diagnosability argues against splitting Ipomoeeae into smaller genera and instead supports sinking all these genera into a broader, monophyletic genus. An expanded monophyletic *Ipomoea* makes the tribe Ipomoeeae monogeneric and the spiny pollen a synapomorphy of the group within Convolvulaceae.

References

Austin, D. F. (1973) The American Erycibeae (Convolvulaceae): Maripa, Dicranostyles, and Lysiostyles I. systematics. *Annals of the Missouri Botanical Garden* 60(2): 306.

(1975) Typification of the New World subdivisions of *Ipomoea* L. (Convolvulaceae). *Taxon* 24(1): 107–110.

(1979) An infrageneric classification for *Ipomoea* (Convolvulaceae). *Taxon* 28(4): 359–361.

(1980) Additional comments on infrageneric taxa in *Ipomoea* (Convolvulaceae). *Taxon* 29 (4): 501–502.

Austin, D. F. and Huáman, Z. (1996) A synopsis of *Ipomoea* (Convolvulaceae) in the Americas. *Taxon* 45(1): 3–38.

Callaway, E. (2018) 'Why not sequence everything?' A plan to decode every complex species on Earth. *Nature*: d41586-018-07279-z.

Carine, M.A. and Scotland, R. W. (2002) Classification of Strobilanthinae (Acanthaceae): Trying to classify the unclassifiable? *Taxon* 51(2): 259.

Carruthers, T., Muñoz-Rodríguez, P., Wood, J. R. I., and Scotland R. W. 2020) The temporal dynamics of evolutionary diversification in *Ipomoea. Molecular Phylogenetics and Evolution* 146: 106768.

Choisy, J.D. (1838) De Convolvulaceis. In: *Memoires de la Société de Physique et d'Histoire Naturelle de Genève [Internet]*. Vol. 8. Librairie d'Abraham Cherbuliez, Geneva, pp. 43–86. www.biodiversitylibrary.org/item/39750

(1845) Convolvulaceae. In: A-L-P-P de Candolle (ed.) *Prodromus Systematis Naturalis Regni Vegetabilis* [Internet]. Vol. 9. Sumptibus Sociorum Treuttel et Würtz, Parisii, pp. 323–462. http://bibdigital.rjb.csic.es/ing/Libro.php?Libro=6160

Eserman, L. A., Tiley, G. P., Jarret, R. L., Leebens-Mack J. H.,and Miller , R. E. (2014) Phylogenetics and diversification of morning glories (tribe *Ipomoeeae*, Convolvulaceae) based on whole plastome sequences. *American Journal of Botany* 101(1): 92–103.

Hallier, H. (1893) Versuch einer natürlichen Gliederung der Convolvulaceen auf morphologischer und anatomischer Grundlage. *Botanische Jahrbücher für Systematik, Pflanzengeshichte und Pflanzengeographie* 16: 453–591.

(1894) Convolvulaceae Africanae. *Botanische Jahrbücher für Systematik, Pflanzengeshichte und Pflanzengeographie* 28: 28–54.

Hennig, W. (1966) *Phylogenetic Systematics [Internet]*. University of Illinois Press, Urbana [accessed 6 July 2020]. https://doi.org/10.1002/mmnd.19820290131

House, H. D. (1908) The North American species of the genus Ipomoea. *Annals of the New York Academy of Sciences* 18(1): 181–263.

McDonald, J. A. (1991) Origin and diversity of Mexican Convolvulaceae. *Anales del Instituto de Biología de la Universidad Nacional Autónoma de México, Serie Botánica* 62(1): 65–82.

Mickevich, M. F. and Platnick, N. I. (1989) On the information content of classifications. *Cladistics* 5(1): 33–47.

Miller, R. E., McDonald, J. A., and Manos, P. S. (2004) Systematics of *Ipomoea* subgenus *Quamoclit* (Convolvulaceae) based on ITS sequence data and a Bayesian phylogenetic analysis. *American Journal of Botany* 91(8): 1208–1218.

Miller, R.E., Rausher, M. D., and Manos, P. S. (1999) Phylogenetic systematics of *Ipomoea* (Convolvulaceae) based on ITS and Waxy sequences. *Systematic Botany* 24(2): 209–227.

Moonlight, P. W., Ardi, W. H., Padilla, L. A. et al. (2018) Dividing and conquering the fastest-growing genus: towards a natural sectional classification of the mega-diverse genus Begonia (Begoniaceae). *Taxon* 67(2): 267–323.

Muñoz-Rodríguez, P., Carruthers, T, Wood, J. R. I. et al. (2019) A taxonomic monograph of

Ipomoea integrated across phylogenetic scales. *Nature Plants* 5:1136–1144.

van Ooststroom, S. J. (1953) Convolvulaceae. In: *Flora Malesiana [Internet].* Vol. 4. Noordhoff-Kolff N.V., Djakarta, pp. 388–512. http://biodiversitylibrary.org/page/28682917

Patterson, C. (1982) Morphological characters and homology. In: K. A. Joysey and A. E. Friday (eds.) *Problems in Phylogenetic Reconstruction.* Academic Press, London, pp. 21–74.

Peralta, I. E. and Spooner, D.M. (2000) Classification of wild tomatoes: A review. *Kurtz* 28(1): 45–54.

Royal Botanic Gardens, Kew. (2019) World Checklist of Selected Plant Families. World Checklist of Selected Plant Families [Internet]. [accessed 1 August 2019]. http://wcsp.science.kew.org/

Sauquet, H., von Balthazar, M., Magallón, S. et al. (2017) The ancestral flower of angiosperms and its early diversification. *Nature Communications* 8(1): 16047.

Scotland, R. W. (2010) Deep homology: A view from systematics. *Bioessays* 32(5): 438–449.

(2011) What is parallelism? *Evolution & Development* 13(2): 214–227.

Simões, A. R. and Staples, G. (2017) Dissolution of Convolvulaceae tribe Merremieae and a new classification of the constituent genera. *Botanical Journal of the Linnean Society* 183 (4): 561–586.

Simões, A. R., Culham, A., and Carine, M. (2015) Resolving the unresolved tribe: A molecular phylogenetic framework for the Merremieae (Convolvulaceae). *Botanical Journal of the Linnean Society* 179(3): 374–387.

Sneath, P. H. and Sokal, R. R. (1973) *Numerical Taxonomy: The Principles and Practice of Numerical Classification.* 1st ed. W.H. Freeman & Co., San Francisco.

Stefanovic, S., Austin, D. F., and Olmstead, R. G. (2003) Classification of Convolvulaceae: A phylogenetic approach. *Systematic Botany* 28(4): 791–806.

Stefanovic, S., Krueger, L., and Olmstead, R. G. (2002) Monophyly of the Convolvulaceae and circumscription of their major lineages based on DNA sequences of multiple chloroplast loci. *American Journal of Botany* 89(9): 1510–1522.

Stevens, P. F. (1984) Metaphors and typology in the development of botanical systematics 1690–1960, or the art of putting new wine in old bottles. *Taxon* 33(2): 169–211.

(2020) Angiosperm Phylogeny Website, version 14. Angiosperm Phylogeny Website [Internet]. [accessed 1 September 2020]. www.mobot.org/MOBOT/research/APweb/

Verdcourt, B. (1957) Typification of the subdivisions of *Ipomoea* L. (Convolvulaceae) with particular regard to the East African species. *Taxon* 6(5): 150–152.

Wilkin, P. (1999) A morphological cladistic analysis of the *Ipomoeeae* (Convolvulaceae). *Kew Bulletin* 54(4): 853–876.

Wojciechowski, M. F. (2013) Towards a new classification of Leguminosae: Naming clades using non-Linnaean phylogenetic nomenclature. *South African Journal of Botany* 89: 85–93.

Wood, J. R. I., Williams, B. R. M., Mitchell, T. C. et al. (2015) A foundation monograph of *Convolvulus* L. (Convolvulaceae). *Phytokeys* 51: 1–282.

Wood, J. R. I., Muñoz-Rodríguez, P., Williams, B. R. M., and Scotland, R. W. (2020) A foundation monograph of *Ipomoea* (Convolvulaceae) in the New World. *Phytokeys* 143: 1–823.

6

Connecting Micro- and Macro-Evolutionary Research

Extant Cryptic Species as Systems to Understand Macro-Evolutionary Stasis

Torsten H. Struck and José Cerca

Introduction

Considerations on evolution focus predominantly on the degree to which organisms are capable of change, including how fast they can respond to changing environments, which processes are responsible for these changes, and how rates of change can vary and be traced through time. This, for example, is reflected by the major topics addressed by the different disciplines dealing with evolution. In evolutionary biology, major research topics revolve around the causes of adaptive radiations, ecological speciation, or the build-up of reproductive isolation. In macroevolution and palaeontology, rates of diversification and the impact of major events such as the asteroid impact at the K/T boundary have received considerable attention. In systematics, major goals include the delimitation of species by diagnostic characters and the detection of apomorphic characters defining higher taxonomic units, both of which are essentially about detecting differences between separate natural units. Even though these processes have shaped the way we view evolution, one should consider the limits of evolution in the generation of biological diversity, which are also at least implicit assumptions in the studies concerning these mentioned phenomena.

In evolutionary biology, investigating change occurs against the null hypothesis that species, by default, remain relatively constant or change occurs only at rates of genetic drift and random fixation as proposed by the neutral theory of evolution (Kimura 1968). In the fossil record, no or little change in individual characters or whole organisms is a predominant pattern in fossil time series over thousands or millions of generations

(Gould and Eldredge 1977). In systematics, an implicit assumption of any species concept is that species are coherently recognisable as persistent natural units over time and space for a considerable time. Similarly, the recognition of apomorphic characters for taxonomic units such as vertebrates, insects, fungi, or angiosperms requires that these characters are still detectable in extant species even though they evolved hundreds of million years ago. Hence, no or very little change of either characters or whole organisms is a dominant part of evolution; however, it obtains considerably less attention than other properties of evolution (Gingerich 2019).

This neglect is probably due to several reasons. While there is no shortage of suggested hypotheses explaining no or very little change, testing these is not straightforward (Hansen and Houle 2004). It is much easier to derive testable predictions for hypotheses of change and conceive experimental setups to reject these predictions. Additionally, in the era prior to the formulation of the evolutionary theory, the permanence of species was the prevailing assumption and species were regarded as constant, never changing entities. Hence, historically, proponents of the evolutionary theory had to make the case for change and variation. Even though evolutionary theory is now unanimously accepted as one of the fundamental theories of biology, these historic debates might still be unintentionally entrenched in the traditions of research on evolution across disciplines. Finally, even though it has long been recognised by biologists, the integration of long-term periods of no or little change into evolutionary theory are still a substantial challenge and unsolved research topic (e.g. Hansen and Houle 2004; Futuyma 2010; Lidgard and Love 2018). As a result, stasis or the absence of change causes vigorous discussions, as can be exemplified by the 'punctuated equilibrium vs. gradualism' debate (Gingerich 2019). For the same reasons, the detection and recognition of cryptic species caused an ongoing debate about the validity of the concept of cryptic species. The debate revolves around whether cryptic species actually constitute a natural phenomenon or are just an artificial taxonomic artefact.

6.1 Cryptic Species

According to Winker (2005), the English clergyman William Derham was the first to apply a concept resembling what is nowadays called cryptic species when studying the genus *Phylloscopus* (Aves) in 1718. The first formal introduction of a concept for identical or extremely similar species, which were reproductively isolated, was the term sibling species (Mayr 1963). With the advent of molecular taxonomy and, later on, barcoding initiatives, the detection of cryptic species substantially increased. Cryptic species are being increasingly reported across all environments and taxonomic groups from unicellular eukaryotes to gymnosperms and vertebrates (e.g. Bickford et al. 2007; Pfenninger and Schwenk 2007; Perez-Ponce de Leon and Poulin 2016; Hawksworth and Lücking 2017; Struck et al. 2018b), suggesting that cryptic species are a substantial part of biodiversity (Appeltans et al. 2012).

This has consequences for studies in ecology and conservation biology, which rely on estimates of species diversity, and these consequences have already been intensively outlined elsewhere (Bickford et al. 2007; Bernardo 2011; Pante et al. 2015; Fišer et al.

2018). Moreover, cryptic species have the potential to contribute to our understanding of how evolution proceeds and what factors affect different evolutionary processes such as convergence or stasis (Fišer et al. 2018; Struck et al. 2018a,b). However, this potential contribution to evolutionary biology and macroevolution has been barely explored and exploited thus far: potentially because cryptic species have primarily been regarded as a purely taxonomic topic. It is thus our goal here to provide an outlook on how cryptic species can be a surrogate to investigate the evolutionary phenomenon of stasis, given the recent progress in the theoretical understanding of the concept of cryptic species.

To be able to integrate cryptic species into evolutionary biological and macroevolutionary research it is important that an appropriate theoretical concept of cryptic species is applied across studies. This concept needs to bridge across studies and allow general conclusions to be drawn. Unfortunately, the term cryptic species is often applied with very different meanings, which have different implications. Bickford et al. (2007) provided the first comprehensive and most influential definition of cryptic species as 'two or more distinct species that are erroneously classified (and hidden) under one species name'. This definition is easy to understand and to apply to different study systems, as it only requires knowledge of the taxonomic history of the investigated species complex.

However, this definition offers several disadvantages. First, very similar or identical phenotypes, or at least morphotypes, are often implicitly or explicitly associated with cryptic species. While Bickford et al. (2007) also imply this in their text, their definition does not explicitly require morphological similarity. Accordingly, cryptic species have been described solely based on newly detected morphological differences (e.g. Tsoumani et al. 2013; Grishin 2014; Sasakawa 2016). In consequence, any practice resulting in the splitting of an already described species could result in a cryptic species. For example, a previous record of *Marphysa sanguinea* (Annelida) from Egyptian waters has been described as a new species *M. aegypti* (Elgetany et al. 2018). Both species are substantially different in morphology, but could, nonetheless, be called cryptic species under this definition. However, this case is more an example of sloppy practices in the taxonomic determination of new records (Hutchings and Kupriyanova 2018).

Second, as a consequence of its dependence on the taxonomic history the definition is also tightly linked to practices in different taxonomic schools favouring different rigour in splitting species (the splitter vs. lumper debate). While *M. aegypti* could be called cryptic species, the species pairs *Polygordius lacteus*/*P. neapolitanus* and *P. jouinae*/*P. triestinus* (Annelida) could not, as they were previously considered different species based on their different geographic distributions, even though their adult morphology is indistinguishable (Ramey-Balcı et al. 2018). Third, the reliance on taxonomic history also tightly links this definition and any similar to the applied species concepts in both the definition of the original and the new species (Nygren 2013; Pante et al. 2015; Sukumaran and Knowles 2017; Fišer et al. 2018). Thus, the application of different concepts to a group of species may artificially generate cryptic species (Heethoff 2018). Fourth, any new species, however similar it may be to a previously described species, but which has never been recorded before, can also not be called a cryptic species under this definition, as it was never 'erroneously classified'. For example, within the *Stygocapitella* species complex

(Annelida) two new species from Russia, which are morphologically identical to species from the US-Pacific coast, have recently been described (Cerca et al. 2020a). As these were described as new species right away given the molecular evidence, they cannot be considered as cryptic species based on the Bickford et al. (2007) definition.

Hence, this definition does not guarantee that the cryptic species have similar properties, which could be meaningful compared across different study systems. One outcome is that meta-analyses investigating the impact of cryptic species in biodiversity studies (e.g. Pfenninger and Schwenk 2007; Perez-Ponce de Leon and Poulin 2016; Poulin and Pérez-Ponce de León 2017) may essentially compare apples with oranges.

Even though the explicit definition of what constitutes a cryptic species is rarely stated in manuscripts, those which do still provide a definition relying heavily on the taxonomic history, and accompany this with additional criteria such as a prerequisite for molecular data, sympatric occurrence, reproductive isolation, reduced gene flow, or lack of morphological differences (Struck et al. 2018b). Given that these definitions depend not only on biological but also non-biological factors such as taxonomic tradition, several authors concluded that cryptic species are not a true natural phenomenon but only reflect our limitation in recognising species and their diagnostic characters. For example, Korshunova et al. (2017) stated 'the terms "cryptic"/"pseudocryptic" species [should be avoided] as a reference to a "natural phenomenon" because it is obscuring multilevel character diversity within a complicated taxonomy-dependent framework' and instead 'the term "cryptic species" only [be used] for a temporary formalisation of the problems with delineation of the species from the same geographic region, when those species demonstrate significant molecular phylogenetic differences, but are hardly distinguished morphologically, ethologically, etc.'. From a purely taxonomic point of view, restricting the problem of cryptic species to only the lack of diagnostic characters and hence to our limitations of recognition might be sensible and reasonable. It is most likely the case for any cryptic species or set of structured populations that if one searches hard enough, small yet consistent differences in ultrastructure, physiology, behaviour, or something else can eventually be found that could serve as taxonomically diagnostic features. However, we argue that this might also prevent the recognition and detection of properly defined cryptic species, which can reveal or contribute to interesting natural phenomena, which are interesting beyond taxonomy. In this way, taxonomy could provide model systems to address different evolutionary processes including stasis.

Recently, two reviews have independently highlighted two general properties of cryptic species that were usually applied in the practical procedures of delineating cryptic species (Fišer et al. 2018; Struck et al. 2018b). First, cryptic species are generally thought to be true species. Even though species concepts are rarely defined or given explicitly, the text often makes it obvious that the authors at least implicitly applied a species concept, be it a phylogenetic, evolutionary, or other one, and regarded the species within this framework as 'good' species. Second, these species are extremely similar or even identical to each other in morphology. Hence, these two properties were put forward as defining characters for cryptic species. Struck et al. (2018a, 2018b) provide a more explicit and formalised way to define cryptic species, and hereafter we focus on this definition.

The definition is separated into two steps, each reflecting one of these properties, and ultimately seeks to disentangle the assignment of the 'cryptic' status from the species delimitation. First, it has to be established that the species are truly separate, given any applied species concept. Essentially, this first step is not any different from any species delimitation process in taxonomy. Moreover, in practical terms most authors already usually do this as their first step in the study of cryptic species, though not explicitly in their manuscripts. However, the difference from customary taxonomic practices comes in the second step, which explicitly requires testing the 'cryptic' status. Previously, it was sufficient to show that the separated species had earlier been combined into one and this had essentially already been done with the delimitation process. Following Struck et al. (2018a; 2018b), one should explicitly show that the species are morphologically more similar to each other than one would expect given the time that has passed since their last common ancestor (or the level of genetic divergence as a proxy for time). Only species exhibiting a level of morphological similarity that is significantly higher than expected should be called cryptic species. Hence, in this definition the delimitation process and the assignment of cryptic status are clearly separated from each other. Disentangling both steps leads to a definition of cryptic species, which is independent of their taxonomic history. Its property-based nature (i.e. time since divergence, degree of morphological disparity) allows for easier comparison across studies, life histories, habitats, and so forth, as it provides a quantitative framework.

6.2 Four Processes: Recent Divergence, Parallelism, Convergence, and Stasis

Four processes have been considered to lead to morphologically similar species: recent divergence, parallelism, convergence, and stasis (Fišer et al. 2018; Struck et al. 2018b). Recent divergence assumes that a cryptic species complex results from recent speciation, and morphological similarity results from the lack of time to accumulate substantial morphological differences (Knowlton 1993; Reidenbach et al. 2012; Korshunova et al. 2017). Several examples have been provided for this process and it is often assumed to be the most common reason for the presence of cryptic species. Such cases occur because speciation is not necessarily accompanied by morphological change, as changes in physiological, immunological, reproductive, or behavioural traits may be more important and result in reproductive isolation (Bensch et al. 2004; Damm et al. 2010; Derycke et al. 2016). However, although a thorough quantification of the underlying causes for morphological similarity in cryptic species complexes is still lacking, it is unlikely that most cryptic species can be ascribed to recent divergence. Struck et al. (2018b), in their analyses of more than 600 papers from the literature, found that long-term stasis can be assumed as often as recent divergence. Similarly, Fišer et al. (2018) found that only five out of 43 investigated amphipod cryptic species complexes have a time of divergence <1 million years, while 24 others were older than 10 million years, including two diverging between 30 and 40 million years ago. Hence, while recent divergence certainly is the relevant process for several cryptic species, it seems not to be the only or at least most dominant explanation for their occurrence.

In the cases of convergence and parallelism, the same or very similar morphotypes evolve independently of each other. While for convergence the ancestral morphologies and their genomic backgrounds are dissimilar between the cryptic species, for parallelism they are the same (Swift et al. 2016; Fišer et al. 2018; Struck et al. 2018b). In practice, it might be difficult to discern between these two processes. Nonetheless, cases of independent evolution of the same morphology are known. In the *Mastigias* species complex (Scyphozoa) the 'lake' phenotypes evolved several times independently from the 'ocean' phenotype (Swift et al. 2016). In the Holarctic *Enallagma* species (Hexapoda), selective pressures from predators have led to the independent evolution of the same morphotype (Stoks et al. 2005). In species complexes of amphipods, independent evolution of cryptic species has occurred due to adaptation to freshwater-semiterrestrial habitats and subterranean ones including micro-niches (Villacorta et al. 2008; Trontelj et al. 2009; Trontelj et al. 2012; Yang et al. 2013). Cases of convergence resulting in cryptic species were also found in the deep sea (Vrijenhoek 2009). Fišer et al. (2018) suggested that as much 26 per cent of all cases in amphipod cryptic species might result from independent evolution as they do not differentiate between convergence and parallelism. However, this number is likely to be an overestimate as they considered all cases of non-sister species to be convergently evolved. However, some of these might be cases of stasis due to retention of a symplesiomorphic morphotype. For example, in the *Cavernacmella* species complex (Gastropoda) morphologically different species evolved within a complex of several cryptic species. One of the cryptic species was not the sister species to another cryptic species, but sister to the clade of morphologically divergent species. The similarity of this cryptic species to the other cryptic species in the complex did not result from convergence, but rather from the retention of a symplesiomorphic, ancestral morphotype and hence stasis. Similarly, cryptic species in the *Mastigias* species with the 'ocean' morphotype are often not sister to one another; however, they retain the ancestral morphotype. In another such case, clades of cryptic species in *Stygocapitella* are nested within clades of other cryptic species, which are morphologically slightly different (Cerca et al. 2020a,b). Here again, ancestral morphotypes are retained and did not evolve by convergence. It is important to note that cryptic species evolving under parallel and convergent evolution have undergone actual morphological change since at least two species independently arrive at the same or a very similar morphotype. The independent evolution of similar or identical morphotypes could possibly result from similar shifts in selective regimes (Fišer et al. 2018), as in the *Mastigias*, *Enallagma*, and some amphipod examples given earlier independent evolution seems to be linked to shifts in biotic or abiotic ecological parameters. However, explicit investigations of the causes and driving forces have not yet been carried out; instead, convergence has merely been suggested as the process. The independent evolution of phenotypically similar or identical species makes them very interesting study systems, but since our focus in this contribution is on long-term stasis, we will not explore this type of speciation any further here.

Stasis is the fourth process suggested as a possible explanation for the evolution of cryptic species. Cryptic species under stasis retain similar phenotypes for millions of years or generations. As mentioned previously, stasis may be frequent (Fišer et al. 2018; Struck et al. 2018b). For example, in the *Cavernacmella* complex (Gastropoda) the '*C. minima*' morphotype is present in five cryptic species and is likely unaltered for over 3 million years (Wada et al. 2013). Similar examples of stasis have been found in cichlids, amphipods, *Stygocapitella* (Annelida), *Diporiphora* (Squamata), *Mastigias* (Scyphozoa), and *Cletocamptus* (Crustacea) (Rocha-Olivares et al. 2001; Seehausen 2006; Smith et al. 2011; Swift et al. 2016; Struck et al. 2017; Fišer et al. 2018). The most extreme example of stasis to date is probably that of *Stygocapitella*. The cryptic species *S. pacifica*, *S. furcata*, and *S. australis* all possess the same morphotype, which has not changed for 140 million years (Cerca et al. 2020a,b). Other cryptic species in the complex retained unaltered morphotypes for ~65 and ~18 million years. Moreover, only minimal morphological differences can in fact be detected between the morphotypes of the different groups of cryptic species within the genus *Stygocapitella*. It can, hence, be said that they evolved very slowly for more than 240 million years and presently exhibit a morphological disparity which is about five to eight times lower than in closely related annelid groups (Struck et al. 2017; Cerca et al. 2020b). Hence, cryptic species such as those presented in these examples are ideal systems to investigate stasis using extant taxa. However, as in the case of independent evolution, stasis has been suggested only as a possible process, and the possible causes of long-term stasis have not been further explored.

6.3 Phenotypic Stasis in Macroevolution

To date, the causes of long-term stasis have been discussed almost exclusively in the context of macro-evolutionary and paleontological research. Stasis figured prominently in the debate on punctuated equilibrium. While certain aspects of the punctuated equilibrium are dismissed nowadays, such as the assumption that all change is linked to speciation events, stasis has remained part of the debate. Gingerich (2019) pointed out that stasis is a central component for both punctuated equilibrium and gradualism, suggesting that across timescales stasis is the predominant process and generally fits the data better than random walk or directional selection. Hence, stasis is not only the most prominent process in the fossil record but also in observations encompassing only a few generations. This is in agreement with the evidence that models of stasis are often the best fit to different datasets (e.g. Gould and Eldredge 1977; Lynch 1990; Hunt and Rabosky 2014). Additionally, renewed interest in stasis also stems from research in understanding the tempo and mode of speciation and drivers of morphological diversification in datasets encompassing large clades (Rabosky and Adams 2012). This includes consideration of 'evolvability' within and between clades, that is, the potential rates of phenotypic variation, and the potential production of radiations of lineages and occupation of different evolutionary niches. Generally, there is an expectation that groups with high evolvability are able to supersede groups with low evolvability, as they are able to respond more rapidly to environmental changes and to occupy a broader set of niches (Estes and Arnold 2007; Rabosky and Adams 2012). In the light of these expectations, stasis as seen in cryptic species and the paleontological record contradicts this theory because high evolvability and rapid responses to environmental change resulting in ecological speciation should be favoured (Hansen and Houle 2004; Rabosky and Adams 2012). Clades of cryptic species should have a lower 'adaptive zone' and hence be replaced by clades with the potential to evolve and adapt faster (Rabosky and Adams 2012). Accordingly, stasis is still considered one of the most important unresolved problems in evolutionary biology (e.g. Hansen and Houle 2004; Estes and Arnold 2007; Futuyma 2010; Hunt and Rabosky 2014; Lidgard and Love 2018).

However, addressing this problem is challenging (Hansen and Houle 2004). First, stasis may persist during thousands or millions of generations, and thus be inaccessible to the standard experimental toolkit of evolutionary biology. Furthermore, where stasis has been reported, species usually have long generation times (weeks to years). Hence, experiments can only reasonably be run for a few generations. Even though Gingerich (2019) showed that stasis operates similarly at all timescales, the effects of possible causes of stasis might be different on short timescales than on longer ones (see Hansen and Houle 2004 for details). Hence, short-term experiments may lack the power to inform on long-term stasis, as it is observed in the fossil record and for some cryptic species. Second, there is no shortage of possible causes explaining the occurrence of stasis. Moreover, many of the causes are not mutually exclusive and may act in synergy and complement each other (e.g. Hansen and Houle 2004; Estes and Arnold 2007). The most prominent of these causes include stabilising selection, lack of genetic diversity, repeated bottlenecks that decrease standing genetic variation, large populations, genetic and developmental constraints, ecological niche tracking, niche conservatism, ephemeral, stressful or fluctuating environments, high short-term variability coupled with long-term stability, phenotypic landscapes with limited differentiation, evolutionarily stable configurations, fluctuation around a long-term stable mean, or source populations impeding specialisation (metapopulation dynamics) (Maynard Smith 1983; Sheldon 1996; Eldredge et al. 2005; Bickford et al. 2007; Futuyma 2010; Haller and Hendry 2014; Voje 2016; Chomicki and Renner 2017). Many of these causes can be addressed only indirectly using the fossil record, or not at all when genetic or detailed ecological data are necessary. To disentangle them, detailed knowledge about their effects and the factors influencing them is needed (Hansen and Houle 2004; Futuyma 2010; Lidgard and Love 2018). Third, stasis probably operates differently at the various levels at which it is often considered, such as individual character traits and entire organisms. This needs to be taken into account when considering the different causes of stasis (Futuyma 2010; Lidgard and Love 2018). Research on stasis in macroevolution mostly concentrates on individual character traits rather than entire organisms. Moreover, it concentrates on individual quantitative character traits, while qualitative character traits are largely neglected. On the other hand, identical qualitative character traits rather than quantitative ones often result in high similarity values between cryptic species. The causes for these different character traits might be different if, for example, their developmental basis is different. A qualitative trait is probably more restricted in the number of potential character states it can exhibit than a quantitative one. The latter can in theory have infinitive potential states whose number depends solely on precision of measurement.

6.4 Suggested Causes of Stasis and the Potential of Cryptic Species

Cryptic species characterised by long-term stasis (e.g. Lee and Frost 2002; Lavoué et al. 2011; Smith et al. 2011; Van Bocxlaer and Hunt 2013; Wada et al. 2013; Davis et al. 2014; Swift et al. 2016; Cerca et al. 2020b; Leria et al. 2020) can complement the ongoing research on stasis in macroevolution. First, stasis in cryptic species does not only affect individual character traits but the entire organism. Hence, this allows causes of stasis to be addressed, which potentially could affect the entire organism. Second, in these cases, it is often relatively easy to assess quantitative as well as qualitative character traits, making it possible to investigate whether both are affected differently by stasis. Tools that allow the study of the 'whole morphotype', such as decomposition of variance across several characters and 3D morphometrics (e.g. Foote 1997; Struck et al. 2017), could be usefully deployed for this purpose. Third, since these species have been separated and reproductively isolated for thousands or millions of generations, they can serve as independent replicates of the investigation, and commonalities and differences may therefore be scrutinised. The commonalities provide the first indications of possible causes of long-term stasis. Such shared properties, for example, in genome architecture, conserved genomic regions, ecological parameters, or developmental patterns, can then be investigated in more detail. Fourth, as these species are extant taxa they make possible the integration of macroevolutionary patterns, from, for example, ecological and fossil data, with microevolutionary processes such as gene flow, population bottlenecks, migration rates, and so forth. Fifth, many of these species still possess substantial distribution ranges, which allows assessment of their ecology in much more detail than is possible with fossil data. Moreover, the migration rate between different habitats can be assessed as well. Sixth, the developmental processes of these cryptic species can be investigated to reveal how static these processes in cryptic species may be.

Considering that cryptic species might provide a bridge between microevolution and macroevolution for understanding stasis, we provide a proof-of-concept by means of a literature survey (Table 6.1). On 22 June 2018 we searched the ISI web-of-knowledge for '(cryptic sp* OR cryptic diversity) AND stasis'. From this search, we removed all theoretical, opinion, and review papers without new data, ending up with 64 manuscripts. For these, we scored the research discipline (biogeography, ecology, evolutionary biology, systematics, and paleontology), the higher taxon of the focal group, the general environment and if mentioned, the focal environmental habitat, the usage of molecular and/or morphological data, and whether the study identifies stasis. Finally, for papers describing stasis, we scored which underlying causes for stasis the study proposed, if any. One conclusion is that, at present, stasis is usually just suggested as a possible process for evolution of cryptic species but not further investigated. Of the 53 studies in this survey that suggested stasis as a potential process, 28 did not discuss any cause, that is, 53 per cent of the papers. Of the remaining 25 studies, several just mention general terms such as stabilising selection and constraints but do not provide specific mechanisms and, more importantly, do not test

Table 6.1 Literature survey of studies mentioning stasis and cryptic species, speciation, or diversity in the text. For details, see the text. y = yes, n = no, p = potentially, na = not applicable, nd = none discussed, T = terrestrial, F = freshwater, M = marine, P = parasitic, mol = molecular data only, mor = morphological data only, mm = molecular and morphological data, none = no molecular and morphological data

Reference	Taxon	Environment	Data	Stasis	Supposed cause
Biogeography					
Bauret et al. (2017)	Polypodiopsida	T, different	mol	p	nd
Dornburg et al. (2016)	Acanthopterygii	M	mm	y	Bottlenecks
Valtueña et al. (2016)	Magnoliophyta	T, different	mol	y	Bottlenecks Short and asexual life cycle
Liang et al. (2015)	Magnoliophyta	T, high altitude	mol	y	nd
Cooke et al. (2012)	Actinopterygii	F, river	mol	y	Non-visual mating signals High ecological opportunity Specific selective pressures
Seidel et al. (2009)	Crustacea	F, river spring	mol	n	na
McDaniel and Shaw (2003)	Bryophyta	T, wet temperate forest	mol	y	Constraint Stabilising selection
Ecology					
Gabaldón et al. (2017)	Rotifera	na	none	n	na
Wen et al. (2016)	Rotifera	F, lake	mol	y	nd
Gabaldón et al. (2015)	Rotifera	na	none	y	nd
Montero-Pau et al. (2011)	Rotifera	F, salt lake	none	n	na
Mant et al. (2005)	Magnoliaphyta	T	mm	p	Ephemeral divergence
Evolutionary biology					
Wellborn and Broughton (2008)	Crustacea	M, coastal lake	mol	y	Ecological constraints
Leavitt et al. (2007)	Squamata	T, desert	mol	y	nd

Table 6.1 (*cont.*)

Reference	Taxon	Environment	Data	Stasis	Supposed cause
Paleontology					
Si et al. (2018)	Foraminifera	M, deep sea	mor	y	nd
Rosso et al. (2016)	Gastropoda	na	mor	y	nd
Pachut and Anstey (2009)	Bryozoa	na	mor	y	Stabilising selection Constraints
Alizon et al. (2008)	Foraminifera	na	none	y	nd
Knappertsbusch (2000)	Coccolithophora	na	mor	y	nd
Systematics					
Mas-Peinado et al. (2018)	Hexapoda	T, xeric	mm	y	nd
Zuccarello et al. (2018)	Rhodophyta	M, subtidal	mm	p	Stabilising selection Constraints
Michaloudi et al. (2017)	Rotifera	M, plankton, sediment	mm	n	na
Mills et al. (2017)	Rotifera	na	mol	y	Niche conservatism
Swift et al. (2016)	Cnidaria	M, coastal lake, ocean	mm	y	Constraints Similar environments
Papakostas et al. (2016)	Rotifera	F, lake, river	mm	y	nd
Dool et al. (2016)	Hexapoda	na	mol	y	nd
Santamaria et al. (2016)	Crustacea	M, supralittoral	mm	y	Stabilising selection Contraints
Moussalli and Herbert (2016)	Gastropoda	na	mm	y	nd
Muggia et al. (2015)	Ascomycota	T, desert	mm	y	nd
Razo-Mendivil et al. (2015)	Platyhelminthes	P, fish intestines	mm	y	Non-morphological evolution (physiology)

Table 6.1 (*cont.*)

Reference	Taxon	Environment	Data	Stasis	Supposed cause
Dominguez et al. (2015)	Annelida	T	mm	n	na
Brandt et al. (2014)	Crustacea	M, deep-sea	mm	na	na
Lemer et al. (2014)	Bivalvia	M	mol	y	nd
Armenteros et al. (2014)	Nematoda	M	mm	y	Stabilising selection
Fontaneto (2014)	Rotifera	na	mol	n	na
Malekzadeh-Viayeh et al. (2014)	Rotifera	M, F, lagoons, ponds, pools	mm	n	na
Novo et al. (2013)	Annelida	T, soil	mol	y	nd
Mathers et al. (2013)	Crustacea	na	mol	y	Constraints Unchanging selection patterns
Bilandžija et al. (2013)	Bivalvia	T, cave	mm	y	nd
Weigand et al. (2013)	Gastropoda	T, different	mol	y	nd
Novo et al. (2012)	Annelida	T, soil	mol	y	Extreme environments Physiological evolution Stabilising selection
Vanschoenwinkel et al. (2012)	Crustacea	M, F	mol	y	Niche conservatism Simple body plan Extreme environments
Richards et al. (2012)	Crustacea	M	mol	y	Stabilising selection Lifestyle Niche conservatism
Smith et al. (2011)	Squamata	T, savannah, woodlands	mm	y	Neutral genetic drift Stabilising selection Constraints

Table 6.1 *(cont.)*

Reference	Taxon	Environment	Data	Stasis	Supposed cause
Lavoué et al. (2011)	Actinopterygii	F, river	mm	y	Stabilising selection Niche conservatism Contraints Behavioural and physiological evolution
Novo et al. (2010)	Annelida	T, soil	mm	y	Non-visual reproductive signals Stabilising selection Extreme environments
Sánchez Herrera et al. (2010)	Hexapoda	T, foothills	mm	y	nd
Xavier et al. (2010)	Porifera	M	mol	y	nd
Thum and Harrison (2008)	Crustacea	F, lake, pond	mol	y	Physiological, chromosomal or cytonuclear incompatibilities
Thum and Derry (2008)	Crustacea	F, lake	mol	y	nd
Fontaneto et al. (2007)	Rotifera	F, lake	mor	n	na
Johannesen et al. (2007)	Chelicerata	na	mol	n	na
Machordom and Macpherson (2004)	Crustacea	M	mm	y	Constraints Vicariance and neutral evolution
Penton et al. (2004)	Crustacea	F, lake, pond	mol	y	nd
Parra-Olea (2003)	Caudata	T, could forest	mol	y	nd
Dodson et al. (2003)	Crustacea	F, shallow lake	mor	y	nd
Witt et al. (2003)	Crustacea	F, lake, pond, river	mol	y	Biological interactions

Table 6.1 *(cont.)*

Reference	Taxon	Environment	Data	Stasis	Supposed cause
Gómez et al. (2002)	Crustacea	M, F, pond, lake, lagoon	mm	y	Sexual selection Divergence in non-morphological traits
Lee and Frost (2002)	Crustacea	M, F, marsh, estuary, lake	mm	y	Fluctuating environments Stabilising selection Constraints Ephemeral divergence Sexual selection
Koufopanou et al. (2001)	Ascomycota	T, different	mol	n	na
Rocha-Olivares et al. (2001)	Crustacea	M, brine seep	mol	y	nd
King and Hanner (1998)	Crustacea	F, pool	mol	y	nd
Rogers et al. (1995)	Nemertea	M, infaunal	mm	y	nd
Darda (1994)	Caudata	T, different	mm	y	nd

these assumptions. Hence, cryptic species do not yet contribute to the debate about stasis, despite their potential.

6.5 Stabilising Selection as a Suggested Cause

Similar to many studies in Table 6.1 listing stabilising selection and constraints as causes of stasis in cryptic species, Hansen and Houle (2004) also grouped the different causes of stasis into these two major categories. They suggested that the two categories may not be mutually exclusive and could act simultaneously. For instance, several papers in the survey suggest that both are potential causes for stasis in the system they studied (Lee and Frost 2002; McDaniel and Shaw 2003; Pachut and Anstey 2009; Lavoué et al. 2011; Smith et al. 2011; Santamaria et al. 2016; Zuccarello et al. 2018). In particular, stabilising selection is often somehow treated as a 'deus-ex-machina' and introduced without any further reference to possible causes. A requirement for stabilising selection to explain stasis is that the selective optimum varies only within a narrow range over long periods of time (Hansen and Houle 2004). One way this could come about is through a combination of niche conservatism and efficient dispersal. In such a scenario, species under stasis are effective niche trackers, allowing them to stay in the same selective, stable environment (Eldredge 1999), and indeed the idea of niche conservatism was suggested in five papers in the literature surveyed. For instance, stasis in Notostraca might be due to persistence in extreme aquatic habitats from the Devonian/Carboniferous until modern times (Vanschoenwinkel et al. 2012). In *Leucothoe ashleyae* and *L. kensleyi*, Richards et al. (2012) have suggested that stasis may be linked to the persistence of these species in their habitats. These two species have morphological adaptations for a filter feeding lifestyle within the canals of sponges, which have not changed for 580 million years. Therefore, using cryptic species, one may assess the degree of migration and the persistence of a stable population size through time because effective niche tracking predicts that the species will find the perfect niche relatively quickly. Recently, Cerca et al. (2020a) linked stasis in *Stygocapitella* to its ability to track new patches of favourable habitat (sediments on sandy beaches). Beaches, in their composition and properties, are most likely very similar to how they were hundreds of millions of years ago (Noodt 1974; Westheide 1977; Westheide and Rieger 1987; Giere 2009). However, the lack of fossil record for *Stygocapitella* means that changes through time can only be traced indirectly by using statistical approaches such as ancestral state reconstructions. Applying a similar set-up to a widely dispersed group of cryptic species under stasis with a well-established fossil record would be most informative. An additional advantage of cryptic species is the ability to test whether all potential niches in an area are occupied by the cryptic species or not. Effective niche tracking would also predict that a large proportion of these potential niches should be occupied.

Another explanation for stasis in macroevolution is that the selective optimum is not stable but fluctuates around a long-term stable mean (Voje 2016). This can mean that different populations track different optima, but on average and over long periods the fluctuation is stable (Lieberman and Dudgeon 1996). Alternatively, the environmental conditions in a habitat may be highly variable in the short term but form a stable mean in the long term (Sheldon 1996). For example, Lee and Frost (2002) have argued that for *Eurytemora affinis* stasis may be associated with wide fluctuations in salinity, temperature, and other variables, at the scale of hours, seasons, years, or geological periods, thus suggesting the optimisation of a generalist strategy and the evolution of physiological tolerance instead of morphological change. If populations adapt locally, one would expect slight differences would be detectable between cryptic species across populations, for example, in quantitative traits, but that at the same time high degrees of gene flow over longer periods would be detectable between the populations of a cryptic species. Moreover, the pattern should be the same for all cryptic species in that complex. If environmental change is predominant, then environments with short-term instability but long-term stability should harbor more cryptic species with long-term stasis than other environments. For *Stygocapitella*, Cerca et al. (2020b) suggested that stasis may result from the specific properties of the interstitial habitat, which is characterised by wide changes in the short term but seems to be a very stable environment at geological timescales (Noodt 1974; Westheide 1977; Westheide and Rieger 1987; Giere 2009).

In contrast, an alternative suggestion is that the niches of species under stasis are hyperstable (Williams 1992). Hence, such cryptic species should occur in environments that have more or less identical environmental conditions across their entire distribution range, and these should be the same across all cryptic species of a complex. On the other hand, closely related species that are not cryptic should show substantially more variation in the environmental parameters of the habitats they are occupying. A variation of this hypothesis is that it is not the entire range that is relevant but core habitats possessing the most favourable conditions (Holt and Gaines 1992; Kawecki 1995; Holt 1996). Hence, populations in these core habitats would be source populations, while the ones in the marginal, less favourable habitats would be sink populations. This would predict that a given species would have populations occupying habitats with stable and identical conditions, while other populations occur in habitats that are more variable. This hypothesis requires the occurrence of predominant migration and gene flow from the core to the marginal habitats and has also been named 'ephemeral divergence' (Futuyma 2010). In such a scenario, mariginal populations may respond to selection, but they may not persist long enough to be registered in the fossil record, or local adaptation may be suppressed by gene flow from a source population (Futuyma 1987; Eldredge et al. 2005). In our survey, Mant et al. (2005) suggest that pollinator-mediated selection may be strong enough to guide divergence in certain traits. However, over long timescales, accumulated genetic differences are lost due to gene flow across populations – in a 'merge-and-diverge oscillation scenario'.

Finally, directional selection of a trait or a suite of traits towards a new optimum in one factor might be prevented due to the selective forces towards the original state by several other factors (Hansen 1997). Adaptation, and hence change, might require in this case several changes also in other traits that are not directly selected. Hence, the burden for change is too high for it to occur. In this case, one would expect that both the phenotypic traits as well as their underlying genetic pathways would be highly correlated with each other across populations within a cryptic species and across cryptic species within a complex.

6.6 Constraint as a Suggested Cause for Stasis

The other category suggested for causes of stasis was constraint. Hansen and Houle (2004) define constraint as any mechanism that may limit or bias the response to selection, which they divided into variational and selective constraints. Variational constraints are limited in the expressed variability of characters, while selective ones 'derive from conflicting selection pressures, and are constraints only from the perspective of achieving specific adaptations, and not from the perspective of optimising the fitness of the organism as a whole'. Since cryptic species concern whole organisms we will not consider selective constraints further, but Hansen and Houle (2004) also point out that the distinction between the two is often a matter of perspective.

One cause of variational constraints can be pleiotropic effects (Hansen and Houle 2004), that is, the genes underlying one character trait might also be involved in the development or regulation of one or more others. Hence, changes concerning one trait might be limited because these would also cause changes in another trait that are either not possible functionally or negative selected. This could be tested in cryptic species in different ways. First, laboratory experiments could be conducted in which one trait was selected to produce change and other traits would be measured to see (1) whether phenotypic changes occur and are consistent across populations and cryptic species, and (2) if they are accompanied by similar changes in the genotype. This could be done for a suite of characters. Second, in complexes of cryptic species that also include morphologically different species, experiments could investigate if changes in one trait are always accompanied by changes in other traits and again whether the observed genotypes undergo correlated changes.

Another source of constraints stems from the fact that mutations enhancing fitness can be rare and hence the rate of evolution is limited (Futuyma 2010). This could be due to a small target size for genetic change and consequently the resulting phenotypic change (Houle 1998). However, this seems unlikely for cryptic species because it is not just single traits but entire organisms that are affected by stasis. Another reason could be epistasis (Hansen and Houle 2004). For example, a theoretically possible character state of trait might only be realisable on a certain genetic background, which itself may not be realisable for some reason. One factor that could limit the variation of the genetic background is pleiotropy, and another could be low levels of standing genetic variation in general (Futuyma 2010). A given reservoir of standing genetic variation can be maintained in a population due to balancing selection rather than by a mutation-selection sequence. Hence, if the degree of balancing selection is low, one would also expect the standing genetic variation to be low. Another reason for low levels of standing genetic variation could be recurring bottlenecks that regularly minimise effective population size and hence diminish the genetic variation in a population. Cryptic species complexes could be tested across all populations and species for low or negligible balancing selection or for indications of multiple bottlenecks in the genome. For two cryptic species complexes covered in our survey, the occurrence of bottlenecks was reported and linked to the absence of phenotypic variation (Dornburg et al. 2016; Valtueña et al. 2016). For the genus *Lepidonotothen*, it has been suggested that glacial cycles may force population bottlenecks, which strip variation from populations. While interglacial periods may offer opportunities for expansion, diversification, and colonisation of new habitat areas due to the retreat of glaciers, glacial periods may diminish this diversification and lead to bottlenecks in the refugial habitats (Dornburg et al. 2016).

Finally and paradoxically, stasis can also stem from excessive degrees of variability (Futuyma 2010), which could promote selection on several traits and potentially in opposing directions, with the result that over long periods of time the phenotype remains stable within certain boundaries. This is essentially similar to the scenario of fluctuations around a mean mentioned previously and could be tested by seeking to show whether both intraspecific phenotypic and genotypic variability of cryptic species are substantially higher than in closely related, non-cryptic species. Curiously, ecological, phylogenetic, developmental, and genetic constraints were the most frequently suggested causes of cryptic species under stasis in our survey (Table 6.1). However, as discussed previously, although invoked as possible explanations, constraints have not been explicitly tested in these studies.

Conclusions

Our discussion shows that many possible causes of stasis could be investigated using cryptic species and that they could complement studies based on the fossil record. The advantage of cryptic species outlined in this chapter are that genetic, phenotypic, and ecological data can be obtained relatively easily for most cryptic species and that for some, even explicit experiments to test predictions are possible. Moreover, in cryptic species more or less the whole organism is affected by stasis, which indicates that stasis is much more effective in these species than in species where only a single or a few character traits are affected.

Acknowledgements

The authors gratefully acknowledge the comments by Simon Mayo that improved the flow of the manuscript substantially, and discussion about stasis with Kjetill Voje and Jostein Starrfelt (both University of Oslo). We also thank Alexandre Monro for organising the symposium on cryptic species, where parts of this paper were presented for the first time. This is NHM Evolutionary Genomics Lab contribution No. #24.

References

Alizon, S., Kucera, M., and Jansen, V. A. A. (2008) Competition between cryptic species explains variations in rates of lineage evolution. *Proceedings of the National Academy of Sciences* 105: 12382–12386. doi: 10.1073/pnas.0805039105

Appeltans, W., Angel, Martin V., Artois, T. et al. (2012) The magnitude of global marine species diversity. *Current Biology* 22: 2189–2202. doi: 10.1016/j.cub.2012.09.036

Armenteros, M., Ruiz-Abierno, A., and Decraemer, W. (2014) Taxonomy of Stilbonematinae (Nematoda: Desmodoridae): Description of two new and three known species and phylogenetic relationships within the family. *Zoological Journal of the Linnean Society* 171: 1–21. doi: 10.1111/zoj.12126

Bauret, L., Rouhan, G., Hirai, R. Y. et al. (2017) Molecular data, based on an exhaustive species sampling of the fern genus *Rumohra* (Dryopteridaceae), reveal a biogeographical history mostly shaped by dispersal and several cryptic species in the widely distributed *Rumohra adiantiformis*. *Botanical Journal of the Linnean Society* 185: 463–481. doi: 10.1093/botlinnean/box072

Bensch, S., Pérez-Tris, J., Waldenström, J., and Hellgren, O. (2004) Linkage between nuclear and mitochondrial DNA sequences in avian Malaria parasites: Multiple cases of cryptic speciation? *Evolution* 58: 1617–1621. doi: 10.1111/j.0014-3820.2004.tb01742.x

Bernardo, J. (2011) A critical appraisal of the meaning and diagnosability of cryptic evolutionary diversity, and its implications for conservation in the face of climate change. In: T. R. Hodkinson, M. B. Jones, S. Waldren, and J. A. N. Parnell (eds.) *Climate Change, Ecology and Systematics.* Cambridge University Press, Cambridge, pp. 380–438.

Bickford, D., Lohman, D. J., Sodhi, N. S. et al. (2007) Cryptic species as a window on diversity and conservation. *Trends in Ecology & Evolution* 22: 148–155. doi: 10.1016/j.tree.2006.11.004

Bilandžija, H., Morton, B., Podnar, M., and Ćetković, H. (2013) Evolutionary history of relict *Congeria* (Bivalvia: Dreissenidae): Unearthing the subterranean biodiversity of the Dinaric Karst. *Frontiers in Zoology* 10: 5. doi: 10.1186/1742-9994-10-5

Brandt, A., Brix, S., Held, C., and Kihara, T. C. (2014) Molecular differentiation in sympatry despite morphological stasis: Deep-sea *Atlantoserolis* Wägele, 1994 and *Glabroserolis* Menzies, 1962 from the south-west Atlantic (Crustacea: Isopoda: Serolidae). *Zoological Journal of the Linnean Society* 172: 318–359. doi: 10.1111/zoj12178

Cerca, J., Meyer, C., Purschke, G., and Struck, T. H. (2020a) Delimitation of cryptic species drastically reduces the geographical ranges of marine interstitial ghost-worms (*Stygocapitella*; Annelida, Sedentaria). *Molecular Phylogenetics and Evolution* 143: 106663. doi: 10.1016/j.ympev.2019.106663

(2020b) Deceleration of morphological evolution in a cryptic species complex and its link to paleontological stasis. *Evolution* 74: 116–131. doi: 10.1111/evo.13884

Chomicki, G. and Renner, S. S. (2017) Partner abundance controls mutualism stability and the pace of morphological change over geologic time. *Proceedings of the National Academy of Sciences* 114: 3951–3956. doi: 10.1073/pnas.1616837114

Cooke, G. M., Chao, N. L., and Beheregaray, L. B. (2012) Five cryptic species in the Amazonian catfish *Centromochlus existimatus* identified based on biogeographic predictions and genetic data. *PLoS ONE* 7: e48800. doi: 10.1371/journal.pone.0048800

Damm, S., Schierwater, B., and Hadrys, H. (2010) An integrative approach to species discovery in odonates: From character-based DNA barcoding to ecology. *Molecular Ecology* 19: 3881–3893. doi: 10.1111/j.1365-294X.2010.04720.x

Darda, D. M. (1994) Allozyme variation and morphological evolution among Mexican salamanders of the genus *Chiropterotriton* (Caudata: Plethodontidae). *Herpetologica* 50: 164–187.

Davis, C. C., Schaefer, H., Xi, Z. et al. (2014) Long-term morphological stasis maintained by a plant–pollinator mutualism. *Proceedings of the National Academy of Sciences* 111: 5914–5919. doi: 10.1073/pnas.1403157111

Derycke, S., De Meester, N., Rigaux, A. et al. (2016) Coexisting cryptic species of the *Litoditis marina* complex (Nematoda) show differential resource use and have distinct microbiomes with high intraspecific variability. *Molecular Ecology* 25: 2093–2110 doi: 10.1111/mec.13597

Dodson, S. I., Grishanin, A. K., Gross, K., and Wyngaard, G. A. (2003) Morphological analysis of some cryptic species in the *Acanthocyclops vernalis* species complex from North America. *Hydrobiologia* 500: 131–143. doi: 10.1023/A:1024678018090

Dominguez, J., Aira, M., Breinholt, J. W. et al. (2015) Underground evolution: New roots for the old tree of lumbricid earthworms. *Molecular Phylogenetics and Evolution* 83: 7–19. doi: 10.1016/j.ympev.2014.10.024

Dool, S. E., Puechmaille, S. J., Foley, N. M. et al. (2016) Nuclear introns outperform mitochondrial DNA in inter-specific phylogenetic reconstruction: Lessons from horseshoe bats (Rhinolophidae: Chiroptera). *Molecular Phylogenetics and Evolution* 97: 196–212. doi: 10.1016/j.ympev.2016.01.003

Dornburg, A., Federman, S., Eytan, R. I. , and Near, T. J. (2016) Cryptic species diversity in sub-Antarctic islands: A case study of *Lepidonotothen*. *Molecular Phylogenetics and Evolution* 104: 32–43. doi: 10.1016/j.ympev.2016.07.013

Eldredge, N. (1999) *The Pattern of Evolution*. Freeman, New York, 250 pp.

Eldredge, N., Thompson, J. N., Brakefield, P. M. et al. (2005) The dynamics of evolutionary stasis. *Paleobiology* 31: 133–145. doi: 10.1666/0094-8373(2005)031[0133:TDOES]2.0.CO;2

Elgetany, A. H., El-Ghobashy, A. E., Ghoneim, A. M., and Struck, T. H. (2018) Description of a new species of the genus Marphysa (Eunicidae), *Marphysa aegypti* sp.n., based on molecular and morphological evidence. *Invertebrate Zoology* 15: 71–84. doi: 10.15298/invertzool.15.1.05

Estes, S. and Arnold, S. J. (2007) Resolving the paradox of stasis: Models with stabilizing selection explain evolutionary divergence on all timescales. *The American Naturalist* 169: 227–244. doi: 10.1086/510633

Fišer, C., Robinson, C. T. , and Malard, F. (2018) Cryptic species as a window into the paradigm shift of the species concept. *Molecular Ecology* 27: 613–635. doi: 10.1111/mec.14486

Fontaneto, D. (2014) Molecular phylogenies as a tool to understand diversity in rotifers. *International Review of Hydrobiology* 99: 178–187. doi: 10.1002/iroh.201301719

Fontaneto, D., Giordani, I., Melone, G., and Serra, M. (2007) Disentangling the morphological stasis in two rotifer species of the *Brachionus plicatilis* species complex. *Hydrobiologia* 583: 297–307. doi: 10.1007/s10750-007-0573-1

Foote, M. (1997) The evolution of morphological diversity. *Annual Review of Ecology and Systematics* 28: 129–152. doi: 10.1146/annurev.ecolsys.28.1.129

Futuyma, D. J. (1987) On the role of species in anagenesis. *The American Naturalist* 130: 465–473. doi: 10.1086/284724

(2010) Evolutionary constraint and ecological consequences. *Evolution* 64: 1865–1884. doi: 10.1111/j.1558-5646.2010.00960.x

Gabaldón, C., Fontaneto, D., Carmona, M. J., Montero-Pau, J., and Serra, M. (2017) Ecological differentiation in cryptic rotifer species: What we can learn from the *Brachionus plicatilis* complex. *Hydrobiologia* 796: 7–18. doi: 10.1007/s10750-016-2723-9

Gabaldón, C., Serra, M., Carmona, M. J., and Montero-Pau, J. (2015) Life-history traits, abiotic environment and coexistence: The case of two cryptic rotifer species. *Journal of Experimental Marine Biology and Ecology* 465: 142–152. doi: 10.1016/j.jembe.2015.01.016

Giere, O. (2009) *Meiobenthology: The Microscopic Motile Fauna of Aquatic Sediments*. Springer-Verlag, Berlin, Heidelberg, 527 pp.

Gingerich, P. D. (2019) *Rates of Evolution: A Quantative Synthesis*. Cambridge University Press, New York, 381 pp.

Gómez, A., Serra, M., Carvalho, G. R., and Lunt, D. H. (2002) Speciation in ancient cryptic species complexes: Evidence from the molecular phylogeny of *Brachionus plicatilis* (Rotifera). *Evolution* 56: 1431–1444. doi: 10.1111/j.0014-3820.2002.tb01455.x

Gould, S. J. and Eldredge, N. (1977) Punctuated equilibria: The tempo and mode of evolution reconsidered. *Paleobiology* 3: 115–151. doi: 10.1017/S0094837300005224

Grishin, N. V. (2014) Two new species of *Clito* from South America and a revision of the *Clito littera* group (Lepidoptera: Hesperiidae: Pyrginae). *Zootaxa* 3861: 231–248. doi: 10.11646/zootaxa.3861.3.2

Haller, B. C. and Hendry, A. P. (2014) Solving the paradox of stasis: Squashed stabilizing selection and the limits of detection. *Evolution* 68: 483–500. doi: 10.1111/evo.12275

Hansen, T. F. (1997) Stabilizing selection and the comparative analysis of adaptation. *Evolution* 51: 1341–1351. doi: 10.1111/j.1558-5646.1997.tb01457.x

Hansen, T. F. and Houle, D. (2004) Evolvability, stabilizing selection, and the problem of stasis. In: M. Pigliucci and K. Preston (eds.) *Phenotypic Integration: Studying the Ecology and Evolution of complex Phenotypes*. Oxford University Press, Oxford, pp. 130–154.

Hawksworth, D. L. and Lücking, R. (2017) Fungal Diversity Revisi ted: 2.2 to 3.8 Million Species. Microbiology *Spectrum* 5. doi: 10.1128/microbiolspec.FUNK-0052-2016

Heethoff, M. (2018) Cryptic species: Conceptual or terminological chaos? A response to Struck et al. *Trends in Ecology & Evolution* 33: 310. doi: 10.1016/j.tree.2018.02.006

Holt, R. D. (1996) Demographic constraints in evolution: Towards unifying the evolutionary theories of senesence and niche conversatism. *Evolutionary Ecology* 10: 1–11. doi: 10.1007/BF01239342

Holt, R. D. and Gaines, M. S. (1992) Analysis of adaptation in heterogeneous landscapes: Implications for the evolution of fundamental niches. *Evolutionary Ecology* 6: 433–447. doi: 10.1007/BF02270702

Houle, D. (1998) How should we explain variation in the genetic variance of traits? *Genetica* 102/103: 241–253. doi: 10.1023/A:1017034925212

Hunt, G. and Rabosky, D. L. (2014) Phenotypic evolution in fossil species: Pattern and process. *Annual Review of Earth and Planetary Sciences* 42: 421–441. doi: 10.1146/annurev-earth-040809-152524

Hutchings, P. and Kupriyanova, E. (2018) Cosmopolitan polychaetes – fact or fiction? Personal and historical perspectives. *Invertebrate Systematics* 32: 1–9. doi: 10.1071/IS17035

Johannesen, J., Lubin, Y., Smith, D. R., Bilde, T., and Schneider, J. M. (2007) The age and evolution of sociality in *Stegodyphus* spiders: A molecular phylogenetic perspective. *Proceedings of the Royal Society B: Biological Sciences* 274: 231–237. doi: 10.1098/rspb.2006.3699

Kawecki, T. J. (1995) Deomgraphy of source-sink populations and the evolution of ecological niches. *Evolutionary Ecology* 9: 38–44. doi: 10.1007/BF01237695

Kimura, M. (1968) Evolutionary rate at the molecular level. *Nature* 217: 624–626. doi: 10.1038/217624a0

King, J. L. and Hanner, R. (1998) Cryptic species in a 'living fossil' lineage: Taxonomic and phylogenetic relationships within the genus *Lepidurus* (Crustacea: Notostraca) in North America. *Molecular Phylogenetics and Evolution* 10: 23–36. doi: 10.1006/mpev.1997.0470

Knappertsbusch, M. (2000) Morphologic evolution of the coccolithophorid *Calcidiscus leptoporus* from the Early Miocene to Recent. *Journal of Paleontology* 74: 712–730. doi: 10.1017/S0022336000032820

Knowlton, N. (1993) Sibling species in the sea. *Annual Review of Ecology and Systematics* 24: 189–216. doi: 10.1146/annurev.es.24.110193.001201

Korshunova, T., Martynov, A., Bakken, T., and Picton, B. (2017) External diversity is restrained by internal conservatism: New nudibranch mollusc contributes to the cryptic species problem. *Zoologica Scripta* 46: 683–692. doi: 10.1111/zsc.12253

Koufopanou, V., Burt, A., Szaro, T., and Taylor, J. W. (2001) Gene genealogies, cryptic species, and molecular evolution in the human pathogen *Coccidioides immitis* and relatives (Ascomycota, Onygenales). *Molecular Biology and Evolution* 18: 1246–1258. doi: 10.1093/oxfordjournals.molbev.a003910

Lavoué, S., Miya, M., Arnegard, M. E. et al. (2011) Remarkable morphological stasis in an extant vertebrate despite tens of millions of years of divergence. *Proceedings of the Royal Society B: Biological Sciences* 278: 1003–1008. doi: 10.1098/rspb.2010.1639

Leavitt, D. H., Bezy, R. L., Crandall, K. A., and Sites, J. W., Jr. (2007) Multi-locus DNA sequence data reveal a history of deep cryptic vicariance and habitat-driven convergence in the desert night lizard *Xantusia vigilis* species complex (Squamata: Xantusiidae). *Molecular Ecology* 16:

4455–4481. doi: 10.1111/j.1365-294X.2007.03496.x

Lee, C. E. and Frost, B. W. (2002) Morphological stasis in the *Eurytemora affinis* species complex (Copepoda: Temoridae). *Hydrobiologia* 480: 111–128. doi: 10.1023/A:1021293203512

Lemer, S., Buge, B., Bemis, A. et al. (2014) First molecular phylogeny of the circumtropical bivalve family Pinnidae (Mollusca, Bivalvia): Evidence for high levels of cryptic species diversity. *Molecular Phylogenetics and Evolution* 75: 11–23. doi: 10.1016/j.ympev.2014.02.008

Leria, L., Vila-Farré, M., Álvarez-Presas, M. et al. (2020) Cryptic species delineation in freshwater planarians of the genus *Dugesia* (Platyhelminthes, Tricladida): Extreme intraindividual genetic diversity, morphological stasis, and karyological variability. *Molecular Phylogenetics and Evolution* 143: 106496. doi: 10.1016/j.ympev.2019.05.010

Liang, Q., Hu, X., Wu, G., and Liu, J. (2015) Cryptic and repeated 'allopolyploid' speciation within *Allium przewalskianum* Regel. (Alliaceae) from the Qinghai-Tibet Plateau. *Organisms Diversity & Evolution* 15: 265–276. doi: 10.1007/s13127-014-0196-0

Lidgard, S. and Love, A. C. (2018) Rethinking living fossils. *Bioscience* 68: 760–770. doi: 10.1093/biosci/biy084

Lieberman, B. S. and Dudgeon, S. (1996) An evaluation of stabilizing selection as a mechanism for stasis. *Palaeogeography, Palaeoclimatology, Palaeoecology* 127: 229–238. doi: 10.1016/S0031-0182(96)00097-1

Lynch, M. (1990) The rate of morphological evolution in mammals from the standpoint of the neutral expectation. *American Naturalist* 136: 727–741. doi: 10.1086/285128

Machordom, A. and Macpherson, E. (2004) Rapid radiation and cryptic speciation in squat lobsters of the genus *Munida* (Crustacea, Decapoda) and related genera in the South West Pacific: Molecular and morphological evidence. *Molecular Phylogenetics and Evolution* 33: 259–279. doi: 10.1016/j.ympev.2004.06.001

Malekzadeh-Viayeh, R., Pak-Tarmani, R., Rostamkhani, N., and Fontaneto, D. (2014) Diversity of the rotifer *Brachionus plicatilis* species complex (Rotifera: Monogononta) in Iran through integrative taxonomy. *Zoological Journal of the Linnean Society* 170: 233–244. doi: 10.1111/zoj.12106

Mant, J., Peakall, R., and Weston, P. H. (2005) Specific pollinator attraction and the diversification of sexually deceptive *Chiloglottis* (Orchidaceae). *Plant Systematics and Evolution* 253: 185–200. doi: 10.1007/s00606-005-0308-6

Mas-Peinado, P., Buckley, D., Ruiz, J. L., and Garcia-Paris, M. (2018) Recurrent diversification patterns and taxonomic complexity in morphologically conservative ancient lineages of *Pimelia* (Coleoptera: Tenebrionidae). *Systematic Entomology* 43: 522–548. doi: 10.1111/syen.12291

Mathers, T. C., Hammond, R. L., Jenner, R. A., Hänfling, B., and Gómez, A. (2013) Multiple global radiations in tadpole shrimps challenge the concept of 'living fossils'. *PeerJ* 1: e62. doi: 10.7717/peerj.62

Maynard Smith, J. (1983) The genetics of stasis and punctuation. *Annual Review of Genetics* 17: 11–25. doi: 10.1146/annurev.ge.17.120183.000303

Mayr, E. (1963) *Animal Species and Evolution.* Belknap Press of Harvard University, Cambridge, MA, 797 pp.

McDaniel, S. F. and Shaw, A. J. (2003) Phylogeographic structure and cryptic speciation in the trans-Antarctic moss *Pyrrhobryum mnioides*. *Evolution* 57: 205–215. doi: 10.1111/j.0014-3820.2003.tb00256.x

Michaloudi, E., Mills, S., Papakostas, S. et al. (2017) Morphological and taxonomic demarcation of *Brachionus asplanchnoidis* Charin within the *Brachionus plicatilis* cryptic species complex (Rotifera, Monogononta). *Hydrobiologia* 796: 19–37. doi: 10.1007/s10750-016-2924-2

Mills, S., Alcántara-Rodríguez, J. A., Ciros-Pérez, J. et al. (2017) Fifteen species in one: Deciphering the *Brachionus plicatilis* species complex (Rotifera, Monogononta) through DNA taxonomy. *Hydrobiologia* 796: 39–58. doi: 10.1007/s10750-016-2725-7

Montero-Pau, J., Ramos-Rodríguez, E., Serra, M., and Gómez, A. (2011) Long-term coexistence of rotifer cryptic species. *PLoS ONE* 6: e21530. doi: 10.1371/journal.pone.0021530

Moussalli, A. and Herbert, D. G. (2016) Deep molecular divergence and exceptional morphological stasis in dwarf cannibal snails *Nata* sensu lato Watson, 1934 (Rhytididae) of southern Africa. *Molecular Phylogenetics and Evolution* 95: 100–115. doi: 10.1016/j.ympev.2015.11.003

Muggia, L., Kocourková, J., and Knudsen, K. (2015) Disentangling the complex of *Lichenothelia* species from rock communities in the desert. *Mycologia*, 107: 1233-1253. doi: 10.3852/15-021

Noodt, W. (1974) Anpassungen an interstitielle Bedingungen: Ein Faktor in der Evolution höherer Taxa der Crustacea. *Faunistische-ökologische Mitteilungen* 4: 445–452.

Novo, M., Almodóvar, A., Fernández, R., Trigo, D., and Díaz Cosín, D. J. (2010) Cryptic speciation of hormogastrid earthworms revealed by mitochondrial and nuclear data. *Molecular Phylogenetics and Evolution* 56: 507–512. doi: 10.1016/j.ympev.2010.04.010

Novo, M., Almodóvar, A. N. A., Fernández, R. et al. (2012) Appearances can be deceptive: Different diversification patterns within a group of Mediterranean earthworms (Oligochaeta, Hormogastridae). *Molecular Ecology* 21: 3776–3793. doi: 10.1111/j.1365-294X.2012.05648.x

Novo, M., Riesgo, A., Fernández-Guerra, A., and Giribet, G. (2013) Pheromone evolution, reproductive genes, and comparative transcriptomics in Mediterranean earthworms (annelida, oligochaeta, hormogastridae). *Molecular Biology and Evolution* 30: 1614–1629. doi: 10.1093/molbev/mst074

Nygren, A. (2013) Cryptic polychaete diversity: A review. *Zoologica Scripta* 43: 172–183. doi: 10.1111/zsc.12044

Pachut, J. F. and Anstey, R. L. (2009) Inferring evolutionary modes in a fossil lineage (Bryozoa: *Peronopora*) from the Middle and Late Ordovician. *Paleobiology* 35: 209–230. doi: 10.1666/07055.1

Pante, E., Puillandre, N., Viricel, A. et al. (2015) Species are hypotheses: Avoid connectivity assessments based on pillars of sand. *Molecular Ecology* 24: 525–544. doi: 10.1111/mec.13048

Papakostas, S., Michaloudi, E., Proios, K. et al. (2016) Integrative taxonomy recognizes evolutionary units despite widespread mitonuclear discordance: Evidence from a rotifer cryptic species complex. *Systematic Biology* 65: 508–524. doi: 10.1093/sysbio/syw016

Parra-Olea, G. (2003) Phylogenetic relationships of the genus *Chiropterotriton* (Caudata: Plethodontidae) based on 16S ribosomal mtDNA. *Canadian Journal of Zoology* 81: 2048–2060. doi: 10.1139/z03-155

Penton, E. H., Hebert, P. D. N. and Crease, T. J. (2004) Mitochondrial DNA variation in North American populations of *Daphnia obtusa*: Continentalism or cryptic endemism? *Molecular Ecology* 13: 97–107. doi: 10.1046/j.1365-294X.2003.02024.x

Perez-Ponce de Leon, G. and Poulin, R. (2016) Taxonomic distribution of cryptic diversity among metazoans: Not so homogeneous after all. *Biology Letters* 12: 20160371. doi: 10.1098/rsbl.2016.0371

Pfenninger, M. and Schwenk, K. (2007) Cryptic animal species are homogeneously distributed among taxa and biogeographical regions. *BMC Evolutionary Biology* 7: 121. doi: 10.1186/1471-2148-7-121

Poulin, R. and Pérez-Ponce de León, G. (2017) Global analysis reveals that cryptic diversity is linked with habitat but not mode of life. *Journal of Evolutionary Biology* 30: 641–649. doi: 10.1111/jeb.13034

Rabosky, D. L. and Adams, D. C. (2012) Rates of moprhological evolution are correlated with species richness in salamanders. *Evolution* 66: 1807–1818. doi: 10.1111/j.1558-5646.2011.01557.x

Ramey-Balcı, P., Fiege, D., and Struck, T. H. (2018) Molecular phylogeny, morphology, and distribution of *Polygordius* (Polychaeta: Polygordiidae) in the Atlantic and Mediterranean. *Molecular Phylogenetics and Evolution* 127: 919–930. doi: 10.1016/j.ympev.2018.06.039

Razo-Mendivil, U., Rosas-Valdez, R., Rubio-Godoy, M., and Pérez-Ponce de León, G. (2015) The use of mitochondrial and nuclear sequences in prospecting for cryptic species in *Tabascotrema verai* (Digenea: Cryptogonimidae), a parasite of *Petenia splendida* (Cichlidae) in Middle America. *Parasitology International* 64: 173–181. doi: 10.1016/j.parint.2014.12.002

Reidenbach, K. R., Neafsey, D. E., Costantini, C. et al. (2012) Patterns of genomic differentiation between ecologically differentiated M and S forms of *Anopheles gambiae* in West and Central Africa. *Genome Biology and Evolution* 4: 1202–1212. doi: 10.1093/gbe/evs095

Richards, V. P., Stanhope, M. J., and Shivji, M. S. (2012) Island endemism, morphological stasis, and possible cryptic speciation in two coral reef, commensal *Leucothoid* amphipod species throughout Florida and the Caribbean. *Biodiversity and Conservation* 21: 343–361. doi: 10.1007/s10531-011-0186-x

Rocha-Olivares, A., Fleeger, J. W., and Foltz, D. W. (2001) Decoupling of molecular and morphological evolution in deep lineages of a meiobenthic harpacticoid copepod. *Molecular Biology and Evolution* 18: 1088–1102. doi: 10.1093/oxfordjournals.molbev.a003880

Rogers, A. D., Thorpe, J. P., and Gibson, R. (1995) Genetic evidence for the occurrence of a cryptic species with the littoral nemerteans *Lineus ruber* and *L. viridis* (Nemertea: Anopla). *Marine Biology* 122: 305–316. doi: 10.1007/BF00348944

Rosso, A., Sanfilippo, R., Di Geronimo, I., and Bonfiglio, L. (2016) Pleistocene occurrence of recently discovered cryptic vermetid species from the Mediterranean. *Bollettino della Società Paleontologica Italiana* 55: 105–109.

Sánchez Herrera, M., Realpe, E., and Salazar, C. (2010) A neotropical polymorphic damselfly shows poor congruence between genetic and traditional morphological characters in Odonata. *Molecular Phylogenetics and Evolution* 57: 912–917. doi: 10.1016/j.ympev.2010.08.016

Santamaria, C. A., Mateos, M., DeWitt, T. J., and Hurtado, L. A. (2016) Constrained body shape among highly genetically divergent allopatric lineages of the supralittoral isopod *Ligia occidentalis* (Oniscidea). *Ecology and Evolution* 6: 1537–1554. doi: 10.1002/ece3.1984

Sasakawa, K. (2016) Two new species of the ground beetle subgenus *Sadonebria* Ledoux & Roux, 2005 (Coleoptera, Carabidae, Nebria) from Japan and first description of larvae of the subgenus. *ZooKeys* 578: 97–113. doi: 10.3897/zookeys.578.7424

Seehausen, O. (2006) African cichlid fish: A model system in adaptive radiation research. *Proceedings of the Royal Society B: Biological Sciences* 273: 1987–1998. doi: 10.1098/rspb.2006.3539

Seidel, R. A., Lang, B. K., and Berg, D. J. (2009) Phylogeographic analysis reveals multiple cryptic species of amphipods (Crustacea: Amphipoda) in Chihuahuan Desert springs. *Biological Conservation* 142: 2303–2313. doi: 10.1016/j.biocon.2009.05.003

Sheldon, P. R. (1996) Plus ça change: A model for stasis and evolution in different environments. *Palaeogeography, Palaeoclimatology, Palaeoecology* 127: 209–227. doi: 10.1016/S0031-0182(96)00096-X

Si, W., Berggren, W. A., and Aubry, M.-P. (2018) Mosaic evolution in the middle Miocene planktonic foraminifera *Fohsella* lineage. *Paleobiology* 44: 263–272. doi: 10.1017/pab.2017.23

Smith, K. L., Harmon, L. J., Shoo, L. P., and Melville, J. (2011) Evidence of constrained phenotypic evolution in a cryptic species complex of agamid lizards. *Evolution* 65: 976–992. doi: 10.1111/j.1558-5646.2010.01211.x

Stoks, R., Nystrom, J. L., May, M. L. McPeek, M. A., and Benkman, C. (2005) Parallel evolution in ecological and reproductive traits to produce cryptic damselfly species across the Holarctic. *Evolution* 59: 1976–1988. doi: 10.1554/05-192.1

Struck, T. H., Koczula, J., Stateczny, D. et al. (2017) Two new species in the annelid genus *Stygocapitella* (Orbiniida, Parergodrilidae) with comments on their biogeography. *Zootaxa* 4286: 301–332. doi: 10.11646/zootaxa.4286.3.1

Struck, T. H., Feder, J. L., Bendiksby, M. et al. (2018a) Cryptic species: More than terminological chaos: A reply to Heethoff. *Trends in Ecology & Evolution* 33: 310–312. doi: 10.1016/j.tree.2018.02.008

(2018b) Finding Evolutionary Processes Hidden in Cryptic Species. *Trends in Ecology & Evolution*. doi: 10.1016/j.tree.2017.11.007

Sukumaran, J. and Knowles, L. L. (2017) Multispecies coalescent delimits structure, not species. *Proceedings of the National Academy of Sciences* 114: 1607–1612. doi: 10.1073/pnas.1607921114

Swift, H. F., Gómez Daglio, L., and Dawson, M. N. (2016) Three routes to crypsis: Stasis, convergence, and parallelism in the *Mastigias* species complex (Scyphozoa, Rhizostomeae). *Molecular Phylogenetics and Evolution* 99: 103–115. doi: 10.1016/j.ympev.2016.02.013

Thum, R. A. and Derry, A. M. (2008) Taxonomic implications for diaptomid copepods based on contrasting patterns of mitochondrial DNA sequence divergences in four morphospecies. *Hydrobiologia* 614: 197. doi: 10.1007/s10750-008-9506-x

Thum, R. A. and Harrison, R. G. (2008) Deep genetic divergences among morphologically similar and parapatric *Skistodiaptomus* (Copepoda: Calanoida: Diaptomidae) challenge the hypothesis of Pleistocene speciation. *Biological Journal of the Linnean Society* 96: 150–165. doi: 10.1111/j.1095-8312.2008.01105.x

Trontelj, P., Blejec, A., and Fišer, C. (2012) Ecomorphological convergence of cave communities. *Evolution* 66: 3852–3865. doi: 10.1111/j.1558-5646.201201734.x

Trontelj, P., Douady, C. J., Fišer, C. et al. (2009) A molecular test for cryptic diversity in ground water: How large are the ranges of macro-stygobionts? *Freshwater Biology* 54: 727–744. doi: 10.1111/j.1365-2427.2007.01877.x

Tsoumani, M., Apostolidis, A. P., and Leonardos, I.D. (2013) Biogeography of *Rutilus* species of the southern Balkan Peninsula as inferred by multivariate analysis of morphological data. *Journal of Zoology* 289: 204–212. doi: 10.1111/j.1469-7998.2012.00979.x

Valtueña, F. J., López, J., Álvarez, J., Rodríguez-Riaño, T., and Ortega-Olivencia, A. (2016) *Scrophularia arguta*, a widespread annual plant in the Canary Islands: A single recent colonization event or a more complex phylogeographic pattern? *Ecology and Evolution* 6: 4258–4273. doi: 10.1002/ece3.2109

Van Bocxlaer, B. and Hunt, G. (2013) Morphological stasis in an ongoing gastropod radiation from Lake Malawi. *Proceedings of the National Academy of Sciences* 110: 13892–13897. doi: 10.1073/pnas.1308588110

Vanschoenwinkel, B., Pinceel, T., Vanhove, M. P. M. et al. (2012) Toward a global phylogeny of the 'living fossil' crustacean order of the notostraca. *PLoS ONE* 7: e34998. doi: 10.1371/journal.pone.0034998

Villacorta, C., Jaume, D., Oromí, P., and Juan, C. (2008) Under the volcano: Phylogeography and evolution of the cave-dwelling *Palmorchestia hypogaea* (Amphipoda, Crustacea) at La Palma (Canary Islands). *BMC Biology* 6: 7. doi: 10.1186/1741-7007-6-7

Voje, K. L. (2016) Tempo does not correlate with mode in the fossil record. *Evolution* 70: 2678–2689. doi: 10.1111/evo.13090

Vrijenhoek, R. C. (2009) Cryptic species, phenotypic plasticity, and complex life histories: Assessing deep-sea faunal diversity with molecular markers. *Deep Sea Research Part II: Topical Studies in Oceanography* 56: 1713–1723. doi: 10.1016/j.dsr2.2009.05.016

Wada, S., Kameda, Y., and Chiba, S. (2013) Long-term stasis and short-term divergence in the phenotypes of microsnails on oceanic islands. *Molecular Ecology* 22: 4801–4810. doi: 10.1111/mec.12427

Weigand, A. M., Jochum, A., Slapnik, R. et al. (2013) Evolution of microgastropods (Ellobioidea, Carychiidae): Integrating taxonomic, phylogenetic and evolutionary hypotheses. *BMC Evolutionary Biology* 13: 18. doi: 10.1186/1471-2148-13-18

Wellborn, G. A. and Broughton, R. E. (2008) Diversification on an ecologically constrained adaptive landscape. *Molecular Ecology* 17: 2927–2936. doi: 10.1111/j.1365-294X.2008.03805.x

Wen, X., Xi, Y., Zhang, G., Xue, Y., and Xiang, X. (2016) Coexistence of cryptic *Brachionus calyciflorus* (Rotifera) species: Roles of environmental variables. *Journal of Plankton Research* 38: 478–489. doi: 10.1093/plankt/fbw006

Westheide, W. (1977) The geographical distribution of interstitial polychaetes. *Mikrofauna Meeresboden* 61: 287–302.

Westheide, W. and Rieger, R. M. (1987) Systematics of the amphiatlantic *Microphthalmus listensis* species-group (Polychaeta: Hesionidae): Facts and concepts for reconstruction of phylogeny and speciation. *Zeitschrift für zoologische Systematik and Evolutionsforschung* 25: 12–39. doi: 10.1111/j.1439-0469.1987.tb00911.x

Williams, Gary C. (1992) *Natural Selection: Domains, Levels, and Challenges.* Oxford University Press, Oxford, 222 pp.

Winker, K. (2005) Sibling species were first recognized by William Derham (1718). *Auk* 122: 706–707. doi: 10.1642/0004-8038(2005)122[0706:SSWFRB]2.0.CO;2

Witt, J. D., Blinn, D. W., and Hebert, P. D. (2003) The recent evolutionary origin of the phenotypically novel amphipod *Hyalella montezuma* offers an ecological explanation for morphological stasis in a closely allied species complex. *Molecular Ecology* 12: 405–413. doi: 10.1046/j.1365-294x.2003.01728.x

Xavier, J. R., Rachello-Dolmen, P. G., Parra-Velandia, F. et al. (2010) Molecular evidence of cryptic speciation in the 'cosmopolitan' excavating sponge *Cliona celata* (Porifera, Clionaidae). *Molecular Phylogenetics and Evolution* 56: 13–20. doi: 10.1016/j.ympev.2010.03.030

Yang, L., Hou, Z., and Li, S. (2013) Marine incursion into East Asia: A forgotten driving force of biodiversity marine incursion into East Asia. *Proceedings of the Royal Society B: Biological Sciences* 280: 1–8. doi: 10.1098/rspb.2012.2892

Zuccarello, G. C., West, J. A., and Kamiya, M. (2018) Non-monophyly of *Bostrychia simpliciuscula* (Ceramiales, Rhodophyta): Multiple species with very similar morphologies, a revised taxonomy of cryptic species. *Phycological Research* 66: 100–107. doi: 10.1111/pre.12207

7

Coexisting Cryptic Species as a Model System in Integrative Taxonomy

Cene Fišer and Klemen Koselj

Introduction

The term 'cryptic' species refers to pairs of species with no clear morphological differences that could have been delimited from one another only with molecular tools (Knowlton 1993; Bickford et al. 2007; Fišer et al. 2018). Morphological crypsis is a common phenomenon present in all animal phyla (Pfenninger and Schwenk 2007; Pérez-Ponce de León and Poulin 2016). In some lineages, cryptic species represent a small fraction of undiscovered species, whereas in others they account for more than half of species richness (Adams et al. 2014; Eme et al. 2017; Lukić et al. 2019). Cryptic species are thus integral to basic and applied research, from macroecology (Gill et al. 2016), conservation (Delić et al. 2017) to ecotoxicology (Zettler et al. 2013). As such, they need to be described and named and this requires taxonomic expertise.

Taxonomy consists of two steps: species delimitation and naming (Pante et al. 2015; Renner 2016). The former step is an analysis of divergent traits, whereas the latter formally completes the taxonomic work. The analysis is the essential part of taxonomy. Delimitation methods are theoretically rooted in evolutionary theory, namely the process of speciation (Figure 7.1) (Hey 2006; De Queiroz 2007). Since each speciation event takes place under different circumstances and under different conditions (Schluter 2009; Butlin et al. 2012), it is impossible to generalise about what level of trait divergence justifies the recognition of two or more species. Instead, each species should be delimited individually where possible, with a critical assessment of the ecological and geographic circumstances of its divergence. Many taxonomists agree that delimitation procedures should consider divergence in as many types of trait as possible: so-called integrative taxonomy (Padial et al. 2010; Yeates et al. 2011). This approach is able to capture different aspects of inter-population divergence, link them with the process of speciation (Figures 7.1 and 7.2) and provide a justification for the final taxonomic decision (Fišer et al. 2018).

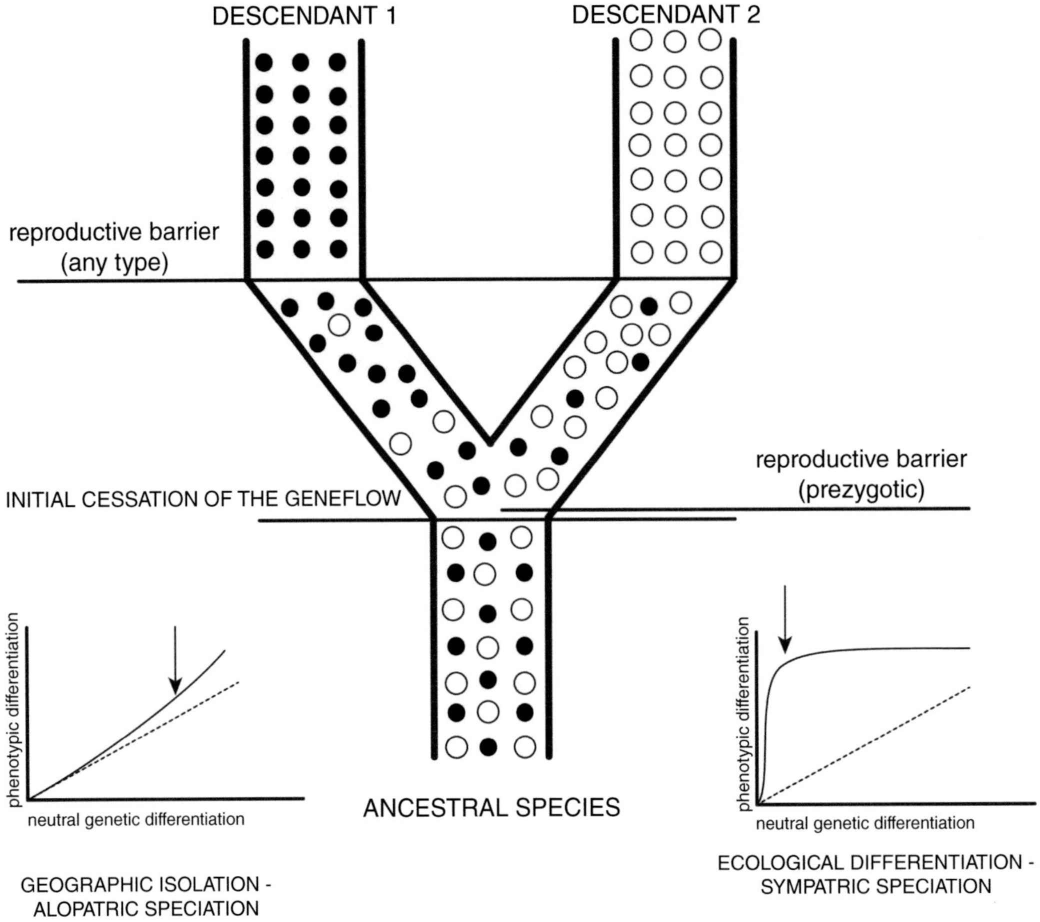

Figure 7.1 Speciation is a complex process that unfolds in multiple, context-specific ways. An initial cessation of gene flow between the two subsets of ancestral populations is followed by divergent evolution of the emerging descendant species. While neutral genetic differentiation presumably increases linearly with time, divergence in morphology, physiology, behaviour, life histories, or reproductive barrier evolves unpredictably, and at different pace. Species delimitation in the transitional period, when the measurable traits still diverge, is a matter of dispute. In this period, hybridisation is still possible. The left and right inset plots illustrate differences in two scenarios, allopatric and sympatric speciation, initiated by geographic and ecological factors, respectively. Geographic isolation physically limits dispersion and gene flow. In theory, the emerging species may diverge solely due to accumulation of mutations. The divergence is slow and reproductive barriers appear with delay. In reality, both environments likely differ, and selection may accelerate the processes of divergence. By contrast, disruptive selection may lead to ecological differentiation among individuals that initially still interbreed. A reproductive barrier may emerge as a by-product of phenotypic differentiation at the onset of the speciation, as in sensory drive speciation.

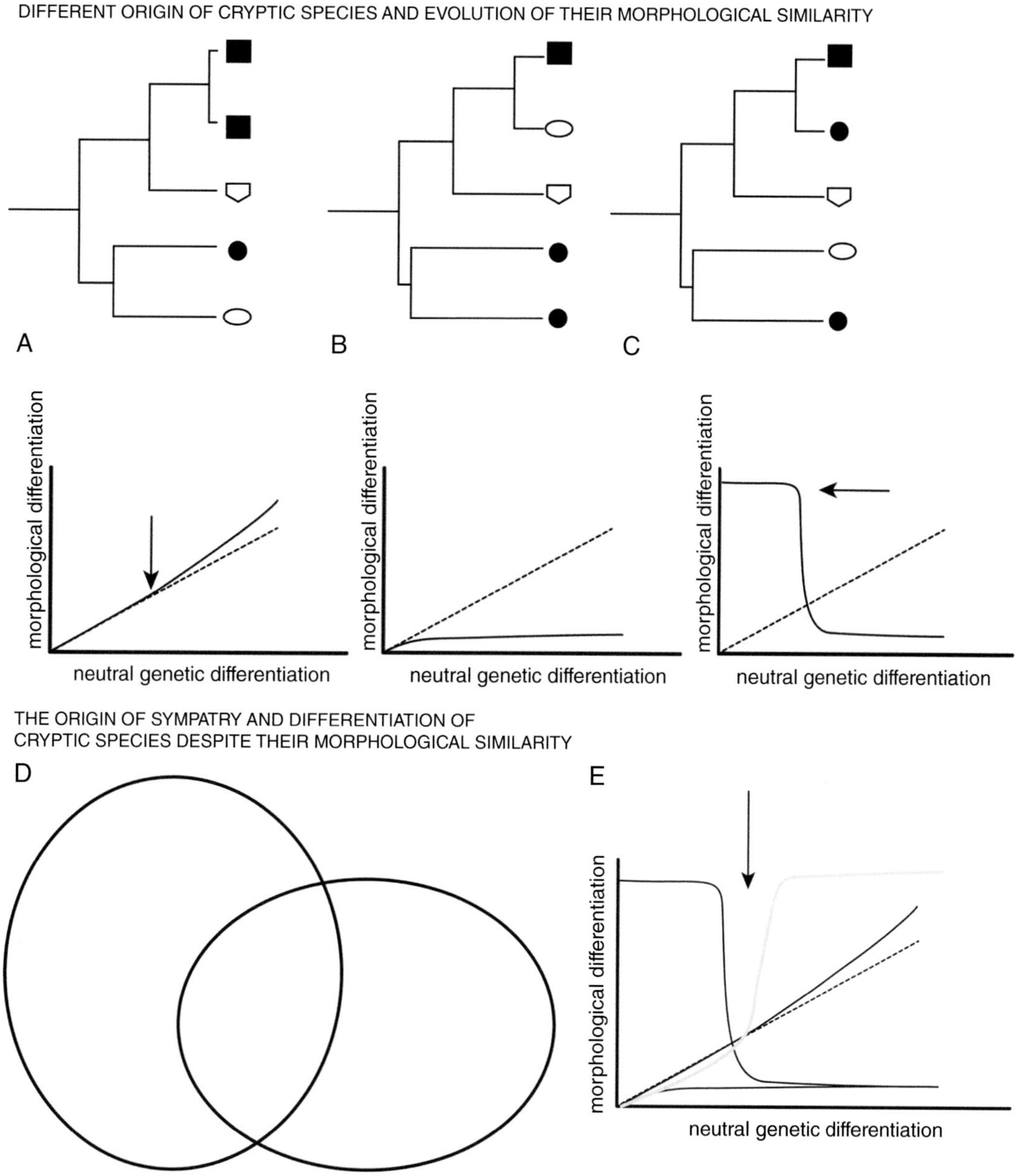

Figure 7.2 The origins of cryptic species and emergence of hidden differences in sympatry. The cryptic species, labelled with identical symbols on trees, are a complex phenomenon. Dashed lines indicate the expected neutral evolution as inferred from neutral genetic markers. Morphological similarity (black lines) may reflect recent speciation prior to morphological differentiation (A, arrow shows the onset of morphological differentiation). Alternatively, it can be subject to stabilising selection among sister pairs of cryptic species living in similar environments (B), or the species attain their morphological similarity independently due to convergent evolution (C). If cryptic species overlap their ranges (D), they need to establish a reproductive barrier and diverge in their ecological traits (grey line on a plot E, arrow shows the onset of sympatry), regardless of their origin (E). The detection of these differences can be hard, because they can be inaccessible to human senses.

Detection and delimitation of cryptic species has greatly accelerated in recent decades with the availability of molecular data and development of species delimitation methods. Despite their power and the progress made in the field (Pons et al. 2006; Fujita et al. 2012; Leache et al. 2014; Tang et al. 2014; Fontaneto et al. 2015; Kapli et al. 2017), molecular methods are based on many assumptions (Luo et al. 2018; Campillo, Barley, and Thomson 2019; Chambers and Hillis 2019), which are in practice often untested. If these are violated, the results may be misleading. Dependent on a single type of data, the taxonomy of cryptic species is the Achilles heel of the Linnean taxonomy. The question is whether molecular taxonomy can be strengthened within the integrative taxonomic framework. Many taxonomists who re-analysed cryptic species delimited by molecular data claimed to have found morphological differences among them (Marsteller et al. 2009; Lajus et al. 2015; Karanovic et al. 2016). In the absence of demonstrable morphological differences, however, morphological re-analyses can be a laborious undertaking with questionable prospects for success (Dinca et al. 2011; Jugovic et al. 2012). Moreover, post-delimitation analyses of morphology are not independent of molecular delimitations.

In this chapter, we therefore ask whether cryptic species themselves can tell us how and by what means they discriminate each other. We propose that co-occurring cryptic species offer an opportunity to pinpoint taxonomically informative traits that could form the basis of integrative taxonomy.

Cryptic species relatively frequently co-occur (Marsteller et al. 2009; Henry et al. 2013; Fišer et al. 2018) and are subject to two biological processes with potentially negative impacts on individual fitness and population growth, namely interbreeding and interspecific competition. It is reasonable to assume that co-occurring species evolved mechanisms protecting them from either process. In the next sections, we first briefly introduce speciation and the origin of cryptic species. We then review how co-occurring species have overcome the challenges that interbreeding and interspecific competition represent, and show that understanding these two processes has a common denominator in sensory ecology. We conclude the chapter by discussing the incorporation of this knowledge into taxonomic practice.

7.1 Speciation and the Origin of Cryptic Species

Speciation is a process in which two or more descendant species arise from a single ancestral species (Figure 7.1) or by the hybridisation of several. In a narrow sense, speciation involves cessation of gene flow between the two subsets of ancestral populations accompanied by divergence in morphology, ecology, and behaviour. Speciation is a context-specific process. Individual traits may diverge in a different order and at a different pace. Here we review some models of speciation and link them to speciation where morphological differentiation substantially lags behind or completely fails.

Speciation is simpler to understand when external factors are involved. The incipient species can be isolated either spatially in case of allopatric speciation or temporally in case of allochronic speciation (Friesen et al. 2007), where reproduction in each lineage has become seasonally separated. The newly isolated populations diverge either by selection or genetic drift and once they have accumulated a sufficient number of differences remain reproductively isolated, even if they later come into contact (Figure 7.1).

Speciation can also take place in sympatry, when disruptive selection favours extreme over intermediate phenotypes and initiates divergence. If disruptive selection is associated with assortative mating, interbreeding does not break the divergence and speciation unfolds (Figure 7.1). Mathematical models have shown that divergence can only take place if selection is frequency dependent, for example through competition for some resource (Dieckmann and Doebeli 1999; Doebeli and Dieckmann 2000). Otherwise, recombination breaks up the linkage disequilibrium between ecologically relevant characters and characters that determine assortative mating. This holds on condition that natural selection mediates divergence.

Natural selection can be assisted or opposed by sexual selection. At least theoretically, sexual selection could replace natural selection, although compelling evidence from natural populations is lacking (Servedio and Boughman 2017). Sexual selection occurs when the frequency of a trait changes as a result of differential mating success of trait carriers (Servedio and Boughman 2017). A trait determining mating preferences of the choosing sex has to diverge in concordance with divergence of preferred traits in the opposite sex. Both divergences have to run in parallel. According to theory (van Doorn et al. 2004), this requires frequency dependent selection within both sexes, otherwise, recombination again breaks linkage disequilibrium between mating preference and preferred trait loci. For example, females have to compete for males and males have to engage in competition for resources, for example, territories.

Cryptic species by definition evolved without morphological change. Evolution of morphological differentiation fails in numerous circumstances. Cryptic species can be either young species, that is, species that have diverged only recently, and have not accumulated many differences (Figure 7.2A), or they have evolved in similar selective environments through parallelism, stasis, regression, or convergence (Figure 7.2B,C; Fišer et al. 2018; Struck et al. 2018). We highlight two case studies of the evolution of morphologically extremely similar species.

The first case study is the discovery of a recent divergence of the cichlid fish *Pundamilia* in Lake Victoria (Seehausen et al. 2008). *Pundamilia nyererei* and *P. pundamilia* are rather dull coloured fishes lacking obvious differences except during the time of mating, when males develop nuptial coloration. Owing to particles in the lake water, the spectrum of light changes with depth. Blue components are filtered out and in deeper water the spectrum is shifted towards red light. *Pundamilia nyererei,* which lives at greater depths than *P. pundamilia,* has a different composition of visual pigments in the retina. Their alleles make individuals of *P. nyererei* and *P. pundamilia,* respectively, more sensitive to the red and blue parts of the spectrum. Hence, divergent sensitivity of vision has led to two lineages occupying different depths, where each lineage is more successful in prey detection. Red nuptial coloration is more pronounced in deep water whereas blue males make a stronger impression near the surface. Owing to correspondingly different spectral sensitivity, females of *P. nyererei* prefer red-coloured males, whereas females of *P. pundamilia* search for blue-coloured males. This has driven the evolution of nuptial colour in males of each species. Sensory adaptation to life in different habitats has resulted in the evolution of divergent mating signals and corresponding mating preferences that sustained assortative mating and eventually led to speciation. The evolution of signals to match receiver sensitivity, which in turn becomes adapted to the environmental conditions, is termed sensory drive (Endler and Basolo 1998; Boughman 2002; Bradbury and Vehrencamp 2011). Despite the low gene flow retained between the two lineages, they deserve species status.

The second case study is a stunning convergent origin of cryptic species via evolution of Müllerian mimicry in neotropical passion vine butterflies *Heliconius* (Mallet and Gilbert 1995). The species in this genus are non-palatable and signal this to predators by colour patterns on their wings. Curiously, within each species there are several races that differ strongly in wing pattern and are sometimes distributed over relatively small areas. Local predators learn only the local wing pattern and avoid eating butterflies that sport it. In any one area, two convergent mimetic species occur as a pair of sympatric races showing the local wing pattern. For example, the races of *H. erato* and *H. melpomene* are mutual mimics and have adapted their wing patterns to resemble each other (Hoyal Cuthill et al. 2019). In each mimic pair, males of one species mate with teneral females before they completely eclose from pupae, whereas males of the other species mate with adult females. Hence, males of the former species search for pupae on larval plants and males of its co-mimetic species search for adult females, thus presumably preventing mate recognition mistakes by males (Gilbert 1991). However, *H. melpomene* has a sister species *H. cydno* that has completely different wing patterns (its races are co-mimetic with those of *H. sapho*), and these two species mate assortatively (Naisbit et al. 2001). Very rarely, hybridisation occurs and hybrid males are non-sterile, but in this case they have a mixed wing pattern, which females of neither species prefer and does not prevent predators from eating them. Therefore, hybrids have lower fitness than pure breeds – a form of post-zygotic reproductive isolation (Naisbit et al. 2001).

These studies are instructive in two ways. First, they clearly show very different origins of morphological crypsis. *Pundamilia* species are young species that may diverge further in future, whereas *Heliconius* butterflies have attained their morphological similarity through convergence. In both cases species status was not questioned despite their morphological similarity and in the cichlid case even some low-level hybridisation. The authors' conclusions were based on deep insights into the biology of the species rather than just observation of genetic divergence. They demonstrate the important role played by species identity cues or signals, and their perception by potential mates, in maintaining reproductive isolation. We develop this further in the next section.

7.2 How Do Co-occurring Cryptic Species Maintain Reproductive Barriers?

Reproductive barriers were long ago proposed as a fundamental criterion for species boundaries (Mayr 1942). Although many contemporary researchers employ different and often less restrictive criteria for delimiting species (Mayden 1997; De Queiroz 2005, 2007), the concept of a reproductive barrier has retained its appeal as strong evidence for a hypothesis of species status (Figure 7.1). When two species regularly co-occur, some degree of reproductive isolation is mandatory or the two species would coalesce. Reproductive barriers emerge through diverse pre- and postzygotic mechanisms (Mayr 1942). Here we limit ourselves to the pre-mating mechanisms that cause speciation and mate recognition.

Premating isolation mechanisms can manifest themselves in diverse ways. One is physical separation along spatial or ecological gradients and another is different timing of mating. These mechanisms are based on ecological properties of interacting species and are categorised here as mechanisms of coexistence in sympatric cryptic species, discussed in the next section. Alternatively, premating isolation may evolve through differential mate recognition. The cues and signals of sensory systems have a fundamental role in mate recognition. Cues can be any parameters that encode information about the environment. Their function is not communication (cues can be of non-biological origin, e.g. celestial cues, or of biological origin, e.g. species identity cues; Bradbury and Vehrenkamp 2011). Signals, in contrast, have an adaptive function in providing communication between sender and receiver from which each party derives a net benefit (Bradbury and Vehrenkamp 2011). Therefore, mechanisms of signal production (sender) and signal perception (receiver) have coevolved. A cue for species identity could, for example, be the size of an animal, whereas male nuptial coloration in cichlids is a signal. In the context of mate and species recognition we will discuss only signals, because they are used more commonly than cues. However, we are aware that both play a role.

Signals are of diverse kinds (visual, auditory, vibroreceptive, chemoreceptive, tactile, and electroreceptive) and usually provoke a behavioural response from the receiver, for example, either mate acceptance or rejection. As we have illustrated in Section 7.1, differentiation of signals can result in speciation (Smadja and Butlin 2009; Wilkins et al. 2013).

From a taxonomic perspective, signals are species traits that can be quantified and may have diagnostic value. Behavioural response to signals in appropriately designed experiments is a test of their biological meaning. Relatively few studies explicitly show that co-occurring cryptic species recognise each other and do not mate in nature. Horseshoe bats are able to discriminate congeneric species of an entire community using echolocation calls, even if these overlap in frequency (Schuchmann and Siemers 2010). Signals used in species recognition that could be translated into taxonomic traits are in most cases not known (Dinca et al. 2013; Lagrue et al. 2014).

Animal senses differ from ours in type, range of sensitivity, and dominance of particular modalities (Stevens 2013). Visual signalling that uses morphology or colour patterns is only one of many communication channels used for mate recognition. In humans it is usually the dominant modality, whereas in other species it may be absent, less important, or they may use different wavelengths, polarisation, and other information that we do not perceive. If co-occurring cryptic species recognise each other, we should learn what stimuli they respond to and ask if these stimuli could be useful in taxonomy.

Below we use the terms ‘hidden signals/traits’ for signals or traits other than those that are accessible to humans (i.e. size, shape, visible spectrum of colours, and audible sounds).

We first briefly present some examples of traits beyond our perception. For each of these hidden traits, we evaluate the evidence that they are involved in species recognition and can represent reproductive barriers. Finally, we investigate whether any of these mechanisms has been reported in cryptic species.

7.2.1 Light

Shape and colour are central to visual communication in many species, including ourselves. We detect light within the wavelength range of 400–800 nm. Many taxa, however, can detect information delivered by ultraviolet wavelengths (300–400 nm, UV) and some taxa infrared wavelengths (>800 nm). Our inability to detect UV is an exception rather than the norm among animals, meaning that our perception of colours can differ substantially from other species (Tovée 1995; Lim et al. 2007).

The UV patterns reflected by butterfly wings are an illustrative example. Members of at least ten butterfly families reflect UV (Kemp and Rutowski 2011) and more UV-reflecting species are continually being uncovered (Stella Rindoš et al. 2018). UV reflectance is mediated by pigments and/or scale structures (Finkbeiner et al. 2017; Stella Rindoš et al. 2018). It has a genetic basis and is inherited (Obara et al. 2010), but butterfly age and environmental conditions add to the observed range of variation (Stella Pecháček et al. 2018; Stella Rindoš et al. 2018). Butterflies possess additional photo-receptors with the highest sensitivity in the ultraviolet part of spectrum (Arikawa 2017). Behavioural experiments found evidence that UV reflectance transmits biologically relevant inter- and intraspecific signals used for predator–prey interaction, aposematism, mimicry, and mate recognition (Kemp and Rutowski 2011; Finkbeiner et al. 2017). The degree of UV reflectance often differs between males and females (Kemp 2007; Stella Pecháček et al. 2018). Variation in UV patterns within a species mediates recognition and quality of mates, female choice, and correlates with male reproductive success (Brunton and Majerus 1995; Kemp 2007; Obara et al. 2010; Kemp and Rutowski 2011; Rutowski and Rajyaguru 2013). Several authors have noted that butterfly UV-reflecting patterns are more conspicuous than those visible to the human eye (Imafuku 2008; Pecháček et al. 2019), and some authors have suggested that UV-reflecting patterns are relevant for butterfly taxonomy (Knuttel and Fiedler 2000; Imafuku 2008).

Silberglied and Taylor (1973) were the first to suggest that differences in UV reflectance contribute to reproductive barriers between closely related species. Subsequent studies provided some evidence that variation in UV reflectance could be explained by reproductive character displacement and reinforcement. A comparative study of *Agrodiaetus* that integrated phylogeny and species distributions showed that pairs of sister species differed in wing coloration when sympatric but not when allopatric. If contrasting coloration indeed impeded interbreeding, differences in sympatry would indicate the role of coloration in reproductive character displacement and even reinforcement (Lukhtanov et al. 2005; Lukhtanov 2010). Similar patterns were documented among lycaenid and pierid butterflies: sympatric and co-occurring species pairs have contrasting UV patterns, whereas allopatric species pairs, or species that do not co-occur, have similar UV wing patterns (Meyer-Rochow 1991; Imafuku 2008; Pecháček et al. 2019). Some of these species are morphologically very similar (Meyer-Rochow 1991; Imafuku 2008). This evidence is still preliminary and differences in UV patterns among cryptic species remain to be studied. Notably, cryptic species were found in UV reflecting families (Dinca et al. 2011; Voda et al. 2015a), and it seems reasonable to explore whether divergence among these species evolved in the UV part of the spectrum.

7.2.2 Sound and Vibration

The sounds we hear are air-borne oscillations in pressure that span frequencies between 20 and 20,000 Hz; however, other species also detect ultra- (>20,000 Hz) or infrasound (<20 Hz) frequencies. Additionally, several animal groups detect sounds in water and vibrations transmitted through solid substrates. In this section, we overview communication by mechanical waves in both air and solids.

Vocal communication is widespread among vertebrates and invertebrates. Sound is produced by diverse mechanisms and carries information about territoriality, reproductive status, disturbance, danger, or aggression (Claridge 1985; Wilkins et al. 2013). In some animal groups, audition is more important for species recognition than vision. Probably the best-known examples are calls of morphologically similar birds of the family Sylvidae, so different as to be routinely used in species' monitoring. Cryptic species in the sylvid genus *Phylloscopus* were reported as early as 1718 (Struck et al. 2018). Communication with songs has been recognised as a premating isolation mechanism in many vertebrate and invertebrate species (Hobel and Gerhardt 2003; Onda-Sumi 2005; Kirschel et al. 2009). Species in sympatry augment differences in songs through reproductive character displacement (Hobel and Gerhardt 2003). The songs of insects and most other taxa are heritable (Butlin and Hewitt 1986). Primates, pinnipeds, bats, whales, songbirds, parrots, and hummingbirds are vocal learners that inherit only the basic means of song production but learn singing and songs from conspecific and occasionally other models, either during a limited developmental phase or even over an entire lifetime (Catchpole and Slater 2008; Bradbury and Verhrenkamp 2011). Differences in vocal communication among cryptic species are well documented among singing cicadas, crickets, anurans, birds, and primates (Gogala and Trilar 2004; Izzo and Gray 2004; Braune et al. 2008; Catchpole and Slater 2008; Funk et al 2012; Hertach et al. 2016).

Many insects use different modes of vibrational communication including tremulation, drumming, vibration, near-field air pulsing, or near-field vibration (Henry 1994; Cocroft and Rodriguez 2005). These signals can only be recorded with special equipment. Animals produce vibratory signals using various body parts and detect them using scolopidial and chordotonal organs in legs (Čokl and Virant-Doberlet 2003). Such 'private songs' are energetically less expensive and thus more appropriate means of communication for small-bodied animals but only over a short range. They are often species-specific and important for mate localisation, recognition, and pair formation (Čokl and Virant-Doberlet 2003; Cocroft and Rodriguez 2005; Henry et al. 2013). In cryptic species, speciation and songs were extensively studied in a complex of morphologically similar lacewings. Lacewings of the complex *Chrysoperla carnea* group are a complex of morphologically and genetically similar species with distinct, species-specific duetting songs produced between male and female. Songs are produced by hitting the substrate with the abdomen. Both sexes exercise mate choice equally and emit sexually monomorphic vibratory songs (Henry et al. 2013). While postzygotic (e.g. molecular) reproductive barriers are weak, songs act as a strong reproductive barrier. Many species can forcibly hybridise, however, and the hybrids have unfavourable intermediate song phenotypes and consequently cannot persist in nature due to low mating success (Henry and Wells 2010). Sympatric species clearly partition acoustic signal space, whereas signals among allopatric species do not differ (Henry et al. 2013). Moreover, similar vibratory signals among geographically isolated lineages have evolved several times independently (Henry et al. 2014). Although clear evidence for reproductive character displacement is lacking (Henry et al. 2013), these studies clearly demonstrate that 'private songs' contribute to reproductive barriers among morphologically cryptic species.

7.2.3 Chemicals

Chemoreception is perhaps the most widespread sense in animals (Smadja and Butlin 2009; Stevens 2013). Albeit underappreciated by humans, chemicals communicate information about food, shelter, predators, and mates (Hay 2009), and defend organisms from predators or parasites (Bakus et al. 1986; Lindquist 2002). Chemical signalling is by far the commonest means for communicating species and sex identity (Bradbury and Vehrenkamp 2011). Studying this form of communication is a challenge for at least four reasons: (1) chemical signals may be effective at low concentrations and are hard to detect; (2) they are often complex blends, effective in specific ratios of signalling molecules; (3) communication may include specific cascades of chemical signals; and (4) signalling chemicals are quickly degradable in order to prevent mismatch between old and new signals (Smadja and Butlin 2009; Hay 2010). The chemical identity of most of these odours remains unknown, despite recent progress in the field (Hardege and Terschak 2010).

Pheromone-mediated signals elicit strong behavioural responses. Copepods are small aquatic crustaceans that leave a hydrodynamic wake specific for their swimming style. Females of some pelagic families secrete pheromones into this hydrodynamic envelope, leaving a gradually attenuating chemical trail. On reception, chemoreceptive males accelerate their swimming and accurately follow this path (Yen and Lasley 2010). In crabs, pheromones stimulate mate guarding and copulatory behaviour. Males of *Erimacus isenbeckii* guard and attempt to copulate with bath sponges soaked with water conditioned by a receptive female (Asai et al. 2000). Males of *Carcinus maenas* form homosexual precopula pairs when treated with female urine containing pheromone blends (Hardege et al. 2002). Their behavioural response to chemicals is typically sex-specific (Berry and Breithaupt 2010), in which males respond more accurately to female signals than females to male (Barbosa et al. 2006).

The chemical ecology of cryptic species has been little studied. The moth *Eriocrania semipurpurella* comprises three cryptic species. Females of this genus produce a pheromone blend of short-chain alcohols and ketones that is used for male attraction over a distance. Behavioural essays showed that these three species produce species-specific responses to relative ratios of two major pheromone components, suggesting an assortative mating (Lassance et al. 2019). Sister species of tobacco budworm *Heliothis virescens* and *H.*

subflexa have overlapping ranges. In the pheromone blend the females secrete, most components are produced in both species, but others present at lower concentrations differ. *Heliothis virescens* produces two aldehydes and *H. subflexa* several acetates that have been shown to attract conspecific and repel heterospecific males (Vickers 2002). Male sac-winged bats mix pheromone secretions from glands and urine in special wing sacs. The brew also contains different bacterial floras and is released into the faces of roosting females during courtship. The blend simultaneously contains information about the species (*Saccopteryx bilineata* or *S. leptura*), age, body size, and physiological condition of scenting males (Caspers et al. 2009).

7.2.4 Other Modalities and Multimodal Signalling

Animals also use other, less known modalities, such as electro- and magnetoreception. These senses may be particularly important in environments where other sources of information are limited (Soares and Niemiller 2013). Although rare, there are reports of cryptic species that use these modalities. Weakly electric fish of family Mormyridae have electric organs for the generation of electric fields. Electrolocation – orientation by distortions in self-generated electric fields – has itself an additional communicative role. In these fish, the electrical organ discharge (EOD) pattern is further modulated for use in communication during social interactions, pair formation, and mating. EOD is species- and sex-specific and some species emit what can only be described as electro-songs (Bradbury and Vehrenkamp 2011). A study of mormyrid *Campylomormyrus numenius* provided evidence that this complex comprised three sympatric, molecularly well-defined cryptic species, which in adulthood differed in their EOD patterns. These differences indirectly suggest that these species are reproductively isolated (Feulner et al. 2006).

So far, we have considered the role of individual modalities. Premating communication, however, often integrates two or more modalities. Multimodal courtship signals of spiders from the genus *Schizocoza* are combinations of chemical, visual, and vibrational communication (Uetz and Roberts 2002; Roberts and Uetz 2005; Moskalik and Uetz 2011; Uetz et al. 2013). A between-species comparison of *Schizocoza* female responses showed that all species used vibrational communication that was supplemented to a various extent by visual displays and chemicals (Uetz and Roberts 2002).

7.3 Do Co-occurring Cryptic Species Compete for Environmental Resources?

Individuals occupying the same habitat compete for space and resources. Whenever individuals of two species come in contact, a between-species interaction may affect their demography. Whether a species pair has the capacity to coexist or not is determined by the relative strengths of intra- and interspecific competition (Chesson 2000). If intraspecific competition is substantially lower or equal to interspecific competition, species pairs will form a so-called stable or unstable coexistence, respectively (Chesson 2000). In stable coexistence, species exploit different ecological resources and, despite sharing the same space, do not interfere with each other's demography. In the case of unstable coexistence, species are ecologically equivalent. Neither species in the co-occurring pair monopolises the limiting resources, and so their demographies depend on stochastic fluctuations, eventually leading to exclusion of one of the species (Chesson 2000; Holt 2006; Scheffer and van Nes 2006). In the remaining intermediary cases, the stronger competitor outcompetes the weaker one, unless environmentally mediated trade-offs compensate their different competitive strengths. Factors that may contribute to co-occurrence include differential dispersal, local divergence, presence of other competitors, differential predation or parasitism, or environmental fluctuations (Chesson 2000; Leibold and McPeek 2006; Siepielski and McPeek 2010; Li and Chesson 2016; Leibold et al. 2018). Identification of stable coexistence is possible only on the basis of experimental evidence that the population of one species grows in the presence of the presumed competitor, even though the former may be rare in the system (Siepielski and McPeek 2010; Shinen and Navarrete 2014). Coexistence can be established by character displacement. This can be shown experimentally, where population growth with a sympatric or allopatric competitor is studied (Germain et al. 2018).

Co-occurring cryptic species are interesting theoretically. In the absence of obvious morphological differences, it seems reasonable to assume they are ecologically equivalent and that they establish an unstable coexistence (Voda et al. 2015a,b). Other authors consider that equivalence of ecological niches cannot be inferred from morphological similarity and so co-occurring cryptic species can have a stable coexistence, at least by means of trade-offs. This puzzle has been solved only partially, mainly because the studied species have only rarely been manipulated experimentally (De Meester et al. 2011; Cothran et al. 2015). Additionally, only a few studies have attempted to evaluate interspecific relationships by studying patterns of co-occurrence in space (Voda et al. 2015a,b; Fišer et al. 2015). While stable and unstable coexistence cannot be inferred from distributional data alone, strong competition decreases frequency of co-occurrence with respect to overall occurrence records and can be reliably detected (Veech 2013, 2014). With these caveats in mind, we overview some emerging patterns.

To date, studies have examined different dimensions of ecological niches, such as spatial segregation, climatic envelope, differences in trophic niches, phenologies, or life histories, as well as trade-offs, such as differential predation. All showed some level of ecological niche differentiation. One series of studies pointed to spatial segregation in which different cryptic species were spatially segregated at a microscale. There is a strong confounding effect between space and ecological conditions. Co-occurrence of cryptic species has so far been attributed to spatially distributed abiotic conditions in specified habitats (Hebert et al. 2004; Wellborn and Cothran 2004; Eisenring et al. 2016), climatic envelope (Fišer et al. 2015; Macher et al. 2016), or host body parts (Kaliszewska et al. 2005; Condon et al. 2008).

A second series of studies examined density-dependent temporal adjustments of life histories. The evidence is mixed. Some studies found no differences in phenologies of co-occurring species (Dionne et al. 2011), whereas others did (Montero-Pau and Serra 2011; Montero-Pau et al. 2011; Scriven et al. 2016). Fig wasps adjust the sex ratio of their offspring to balance local mate competition and inbreeding. Zhang et al. (2004) demonstrated theoretically that density-dependent adjustment of sex ratio can enable the coexistence of ecologically equivalent species that pollinate the same host. This is in line with empirical data on cryptic fig wasp species (Molbo et al. 2003).

The components of an ecological niche are not entirely independent of each other. Both spatio-ecological segregation and temporal variation of population density may lead to different diets. This is in agreement with field observations of feeding (Scriven et al. 2016) as well as analyses of the feeding apparatus of sympatric cryptic species (Fišer et al. 2015).

None of these studies explicitly tested whether the differences detected in the field truly translate to meaningful ecological differences, that is, if the degree of niche differentiation is sufficient to underlie stable coexistence. For example, it is not always possible to distinguish spatial segregation as an outcome of past ecological differentiation or the habitat choice made by marginally differentiated species when each seeks advantage during competition in a spatially heterogeneous habitat. Several authors have cautioned that detected differences in ecology may represent a basis for coexistence, which may be controlled by differential predation (Wellborn and Cothran 2004, 2007) or environmental heterogeneity (De Meester et al. 2011), possibly in combination with limited dispersal (Fišer et al. 2015). Only a few studies found no difference in ecological niches among coexisting species (Cothran et al. 2013, 2015; Gabaldón et al. 2013).

Thus the hypothesis of ecological equivalence has apparently received little evidential support. Instead, many species show some level of differentiation along at least one dimension of their ecological niche. If this is indeed the case, it could be supposed that individuals of a species can perceive the limits of their ecological niches. Detection of the extent of their ecological space must depend on sensory capability and is manifested by habitat choice mechanisms (Jones 2001; Morris 2003; Kubisch et al. 2014).

7.4 Interacting Mechanisms of Premating Isolation and Coexistence Pinpoint the Importance of Sensory Ecology

From the previous discussion, it follows that species identity is signalled through species-specific sensory channels and detected with adapted sensory systems. Likewise, reduction of interspecific competition depends on differential habitat choice mechanisms and foraging strategies (Siemers and Schnitzler 2004; Koselj et al. 2011; Stevens 2013), that is, through sensory ecology. Different species perceive different properties of the same environment by using different sensory systems that are adapted to selectively gather information of ecological importance for the species, that is, its 'Umwelt' (von Uexküll 1956). Sensory ecology is thus the common denominator of two seemingly unrelated problems, species self-recognition and ecological niche separation. In other words, differentiation of the 'Umwelt' translates into reproductive barriers and ecological niche differentiation.

This is illustrated by a series of studies on bats. Bats evolved high-frequency echolocation with which they orient themselves spatially, and find and catch prey. Echolocation calls also contain information about the caller's species, sex, age, individual signature, and current activity (Jones and Siemers 2010). Besides echolocation calls, bats emit vocalisations that are used in communication, that is, social calls and songs (Smotherman et al. 2016). Cryptic diversity among European bats is well documented (Mayer et al. 2007). *Pipistrellus pipistrellus* and *P. pygmaeus* exemplify the main tenet of our paper, because they have diverged in characters hidden to unaided human perception (both echolocation calls and social calls), yet can hardly be distinguished morphologically (Jones and Parijs 1993, Kalko and Schnitzler 1993; Barlow and Jones 1997; Häussler et al. 1999; Russo and Jones 2000). This is no surprise, because echolocating bats live in the dark and rely more strongly on audition than vision. Both species are common in Europe and are sympatric (Mayer and Helversen 2001). After their acoustic differences were recognised, they were considered as subspecies or even phonic types of a single species. Then data started to accumulate showing that breeding groups contained only individuals of one type, that is, they breed assortatively (Jones and van Parijs 1993; Park et al. 1996) and that they differed in foraging habitat use and resource partitioning (reviewed by Jones 1997). This led to a cryptic species hypothesis that was confirmed conclusively when clear genetic divergence was discovered (Barratt et al. 1997).

It is important to note that premating reproductive barriers controlled by different signalling systems can emerge as a by-product of a habitat shift, that is, ecological speciation through sensory drive. Emerging species, in several studies labelled as races, retain their morphological similarity. Cases of such habitat-driven evolution of reproductive barriers have been reported for organisms with different sensory systems (e.g. vision in the *Pundamilia* cichlids described previously). The apple maggot fly, *Rhagoletis pomonella,* mates on host fruits of hawthorn or apple. Individuals attracted by volatiles of ripening apples are repelled by volatiles of hawthorn fruits and vice versa, leading to host segregation and premating reproductive isolation (Smadja and Butlin 2009; Powell et al. 2012). Analogous cases were documented from vibratory signalling systems that enable private communication on a local scale (Čokl and Virant-Doberlet 2003; Cocroft and Rodriguez 2005) and are transmitted through the substrate, typically through a plant stem or leaf. Background noise, such as wind, rain, or other disturbances interfere with the transmission of signals and attenuate them over a short distance. Structural characteristics of the plant, along with the receiver's sensory system, result in a particular optimal signal frequency (Cocroft and Rodriguez 2005), and so a shift in host plant inevitably modifies communication signals and on some occasions leads to signal divergence. Studies of the treehopper *Enchenopa binotata* have shown that differences in signals are important for species recognition (Rodríguez et al. 2004). Moreover, host shift altered the selection acting on private songs. In response to modified selection, the songs changed (Rodríguez et al. 2006). An additional large-scale study confirmed that song differences between populations should be ascribed to host plant divergence rather than their geographic isolation (Cocroft et al. 2010).

7.5 Sensory Ecology and Taxonomy

7.5.1 What Diversity Do We Detect?

Previous debates about cryptic species have questioned what degree of morphological similarity is needed to qualify species as cryptic (Jugovic et al. 2012; Lajus et al. 2015; Karanovic et al. 2016; Struck et al. 2018). Authors who found small yet stable and statistically significant differences argued that recognition of species as cryptic might be the result of methodological error rather than a natural phenomenon. A central argument for this claim was that morphological crypsis could be due to either insufficient investigation or inadequate analytic tools (Heethoff 2018).

Unlike previous researchers, we find this discussion unconvincing (see also Fišer et al. 2018). We nevertheless agree that cryptic species do reflect methodological error. Cases from previous sections clearly illustrate that human sensory perception captures only a fraction of the overall available information, implying that on many occasions variation that cannot be detected by the human eye passes unnoticed. Cryptic species are based on diverse and less easily perceived traits and signals, reminding us just how much we are channelled anthropocentrically in our exploration of life forms

Taxonomy would benefit if we took into consideration our current knowledge of species' biology. Evolution over eons appears to have diversified life forms beyond our imagination. If we could design experiments or use technological means to uncover the information accessible to animals through their senses, we could more effectively study and describe their species. Identification of the traits that play an essential role in the lives of individuals and study of their variation will pinpoint key diagnostic traits, whether they be UV colour patterns or electric fields. Integrative taxonomy should thus employ new techniques that can capture signals beyond our perception (Mielewczik et al. 2012). Crucially, it is morphologically similar species that seem to be the door to exploring this unknown and diverse world that we have so far barely accessed.

7.5.2 Sensory Ecology in Era of Molecular Taxonomy

A quest for bringing sensory ecology into taxonomy these days may seem paradoxical. Molecular tools are developing quickly; they are becoming ever more powerful and capture ever larger amounts of data (Elbrecht and Leese 2015; Morinière et al. 2017). These data are easy to store and retrieve through public databases, for example, GenBank or BOLD. Analytic methods are developing with increased computer capacity. Molecular tools hold the promise of improved biomonitoring and rapid biodiversity assessments, for example, research using DNA barcodes and eDNA (Leese et al. 2016). The development of new data collection and analytical methods are major changes, raising new and completely different issues (Deiner et al. 2016; Olds et al. 2016) but may be poorly compatible with Linnaean taxonomy. These new research programmes often rely on molecular taxonomic units rather than formally recognised species, giving the impression that taxonomy is no longer either sustainable or necessary. Yet, the reduction of taxonomy to clusters of sequences with MOTU numbers would represent a huge step back and impede basic research on biodiversity that aims to integrate information beyond an individual's identity.

We hold that the description of cryptic species using molecular diagnoses is often legitimate (Delić et al. 2017; Fišer et al. 2018) but should represent the first rather than the last step in the taxonomy of cryptic species. The ecological speciation of cryptic species illustrates why we need to develop classical taxonomy further. Ecological speciation can be fast and can emerge despite the initial persistent geneflow (Nosil 2012). Commonly used genetic markers such as CO1 overlook such ecologically differentiated species (Herrel et al. 2008). If the ecological differences do not translate into obvious morphological variation, as in *Rhagoletis* discussed earlier, such cases of speciation remain unnoticed. It is clear that the extent of ongoing speciation cannot be quantified, let alone studied, without additional biological information. Taxonomy has been accumulating intellectual capital over previous centuries, and this knowledge should be enriched by available technological progress rather than impoverished by excessively routine approaches.

7.5.3 A Shift in Taxonomic Practice?

Consideration of information hidden from our unaided senses inevitably leads to changed taxonomic practice. Detection of such traits requires special equipment, and different methodological approaches are needed to test how animals recognise their own species or ecological niche. Exploring novel sources of information would require standardisation of laboratory protocols and agreement on methods of quantification of information to assure comparative analyses (Knuttel and Fiedler 2000). Such data could be processed with current analytic methods. Focusing on sensory ecology, however, also offers an opportunity for an upgrade in species delimitations.

Current taxonomic practice searches for gaps in morphological, ecological, and genetic variation, and aims to distinguish variation between populations from variation between species. Morphological and genetic variation is extracted from preserved material and analysed using more or less explicit models, which identify the shift from reticulate to cladogenic evolution (Sites and Marshall 2003, 2004; Pons et al. 2006; Flot et al. 2010; Zhang et al. 2013; Fontaneto et al. 2015). However advanced, this approach relies on models and assumptions, and is, in the broadest perspective, of a correlative nature. Likewise, ecological variation is often inferred from models calculated from field data and pre-defined taxa (Raxworthy et al. 2007), thereby blurring the relationship between ecology and geographic distribution (Warren et al. 2014).

By contrast, sensory ecologists design experiments and behavioural tests, using playbacks of potential signals (Cocroft and Rodriguez 2005; Henry et al. 2013) for recording and analysis and equipment for establishing natural conditions during the tests, such as flight tunnels (Powell et al. 2012). This allows direct evaluation of the biological significance of the studied signals and traits (Figure 7.3). The experimental approach is beneficial in two aspects. First, it introduces hypothesis testing into taxonomy, thereby providing a stronger argument than currently predominating correlational evidence. Second, it discovers the biological function of individual traits. If the trait/signal was experimentally studied in several taxa, conclusions about its function can be used for studies of other related species. If taxonomic revisions relied on such biologically meaningful traits, the differences among the species could be better justified even in a correlative analysis. All these aspects are a quality upgrade of taxonomic practice, which would intensify the links among taxonomic expertise, sensory biology, ecology, evolutionary biology, and nature conservation (Doctrow 2018; Fišer et al. 2018).

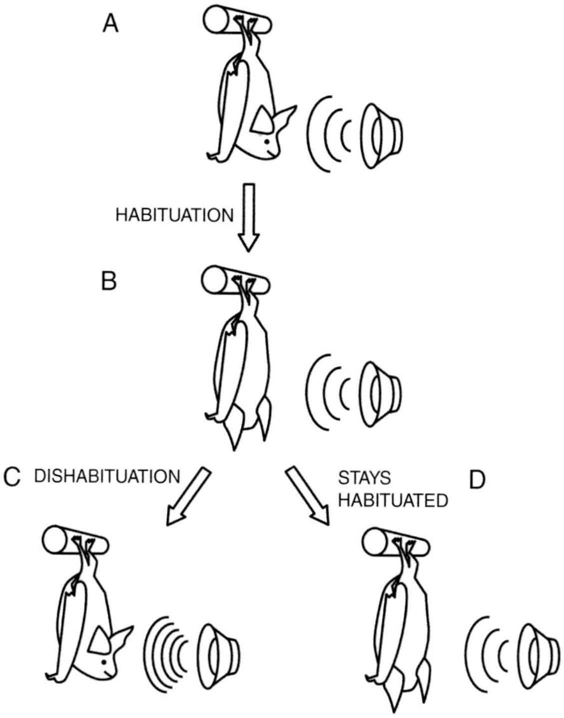

Figure 7.3 An introduction to sensory ecology toolbox for taxonomists, based on a study that tested if bats can discriminate species based on echolocation calls (Schuchmann and Siemers 2010), but can be applied to other species and other stimuli. The test is based on a habituation-dishabituation paradigm. Its main advantage is that it does not require training of animals. (A) An animal is presented with a playback of a stimulus. The term playback is used irrespective of the modality. (B) The animal is exposed to stimulus until it habituates to it, i.e. stops showing any arousal or interest in the stimulus. Subsequently, playback stimulus is abruptly replaced with a testing stimulus, which belongs to either the same or a different species (denoted by a different wavelength of stimulus). If the animal reacts to the change by showing arousal or interest, actively scanning towards the test stimulus (i.e. dishabituation), the reaction is recorded as a discrimination (C). If the animal stays habituated, it is considered that it did not recognise the change (D). Care needs to be taken that the change in stimuli is not accompanied by any artefact and that a sufficiently large library of stimuli is used. The statistical analysis compares the probability of reaction to the same or different species stimuli.

Conclusions

We show the diverse ways in which species extract relevant information from their environment, what ecological and reproductive importance this information has, and how differentiation in sensory ecology may lead to speciation. We argue that morphological variation and the information it represents inherently is only one of many information sources that the organisms themselves rely on. Studying species through morphological variation alone embodies a bias rooted in our own perception of the living world, and morphologically cryptic species are a result of this bias as expressed in conventional taxonomic practice.

Molecular methods have shaken taxonomy, and their implementation into routine practice is a noteworthy milestone. Taxonomy, however, needs to move forward and open itself also towards other fields of biology. We believe that studying the sensory ecology of species is a promising, albeit technically demanding and time-consuming research field. In our opinion, the results will be rewarding. They will provide insights into the functional and ecological significance of between-individual differences and fill a deep gap in basic biology that has long hampered the interpretation of biological data. It can contribute significantly to estimating the extent of global biodiversity with greater accuracy. This is particularly pressing at a time of massive anthropogenic biodiversity loss. We can only protect species that we can recognise.

Acknowledgements

We thank Alexandre Monro for his invitation to contribute this chapter. CF was supported by the Slovenian Research Agency through programme P1-0184.

References

Adams, M., Raadik T. A., Burridge, C. P., and Georges, A. (2014) Global biodiversity assessment and hyper-cryptic species complexes: More than one species of elephant in the room? *Systematic Biology* 63: 518–533. doi: 10.1093/sysbio/syu017

Arikawa, K. (2017) The eyes and vision of butterflies. *The Journal of Physiology* 16: 5457–5464. doi: 10.1113/JP273917

Asai, N., Fusetani, N., Matsunaga, S., and Sasaki, J. (2000) Sex pheromones of the hair crab *Erimacrus isenbeckii*. Part 1: Isolation and Structures of Novel Ceramides. *Tetrahedron* 56: 9895–9899. doi: 10.1021/np010177m

Bakus, G. J., Targett, N. M., and Schulte, B. (1986) Chemical ecology of marine organisms: An overview. *Journal of Chemical Ecology* 12: 951–987.

Barbosa, D., Font, E., Desfilis, E., and Carretero, M. A. (2006) Chemically mediated species recognition in closely related *Podarcis* wall lizards. *Journal of Chemical Ecology* 32: 1587–1598. doi: 10.1007/s10886-006-9072-5

Barlow, K. E. and Jones, G. (1997) Function of pipistrelle social calls: Field data and a playback experiment. *Animal Behaviour* 53: 991–999. doi: 10.1006/anbe.1996.0398

Barratt, E. M., Deaville, R., Burland, T. M. et al. (1997) DNA answers the call of pipistrelle bat species. *Nature* 387: 138-139.

Berry, F. C. and Breithaupt, T. (2010) To signal or not to signal? Chemical communication by urine-borne signals mirrors sexual conflict in crayfish. *BMC Biology* 8: 1–11. doi: 10.1186/1741-7007-8-25

Bickford, D., Lohman, D. J., Sodhi, N. S. et al. (2007) Cryptic species as a window on diversity and conservation. *Trends in Ecology & Evolution* 22: 148–155. doi: 10.1016/j.tree.2006.11.004

Boughman, J. W. (2002) How sensory drive can promote speciation. *Trends in Ecology & Evolution* 17: 571–577.

Bradbury, J. W. and Vehrenkamp, S. L. (2011) *Principals of Animal Communication*. 2nd ed. Sinauer, Sunderland, MA, 697 pp.

Braune, P., Schmidt, S., and Zimmermann, E. (2008) Acoustic divergence in the communication of cryptic species of nocturnal primates (*Microcebus* ssp.). *BMC Biology* 6: 19. doi: 10.1186/1741-7007-6-19

Brunton, C. F. A. and Majerus, M. E. N. (1995) Ultraviolet colours in butterflies: Intra- or inter-specific communication? *Proceedings of the Royal Society B* 260: 199–204. doi: 10.1098/rspb.1995.0080

Butlin, R., Debelle, A., Kerth, C. et al. (2012) What do we need to know about speciation? *Trends in Ecology and Evolution* 27: 27–39. doi: 10.1016/j.tree.2011.09.002

Butlin, R. K. and Hewitt, G. M. (1986) Heritability estimates for characters under sexual selection in the grasshopper, *Chorthippus brunneus*. *Animal Behaviour* 34: 1256–1261. doi: 10.1016/S0003-3472(86)80185-3

Campillo, L. C., Barley, A. J., and Thomson, R. C. (2020) Model-based species delimitation: are coalescent species reproductively isolated? *Systematic Biology* 69: 708–721. doi: 10.1093/sysbio/syz072

Caspers, B. A., Schroeder, F. C., Franke, S. et al. (2009) Odour-based species recognition in two sympatric species of sac-winged bats (*Saccopteryx bilineata, S. leptura*): Combining chemical analyses, behavioural observations and odour preference tests. *Behavioral Ecology and Sociobiology* 63: 741–749.

Catchpole, C. K. and Slater, P. J. B. (2008) *Bird Song: Biological Themes and Variations*. 2nd ed. Cambridge University Press, Cambridge, 335 pp.

Chambers, A. E. and Hillis, D. M. (2019) The multispecies coalescent over-splits species in the case of geographically widespread taxa. *Systematic Biology* 69: 184–193. doi: 10.1093/sysbio/syz042

Chesson, P. (2000) Mechanisms of maintenance of species diversity. *Annual Review of Ecology and Systematics* 31: 343–366. doi: 10.1146/annurev.ecolsys.31.1.343

Claridge, M. (1985) Acoustic signals in the homoptera: Behavior, taxonomy, and evolution. *Annual Review of Entomology* 30: 297–317. doi: 10.1146/annurev.ento.30.1.297

Cocroft, R. B. and Rodríguez, R. L. (2005) The behavioral ecology of insect vibrational communication. *BioScience* 55: 323–334.

Cocroft, R. B., Rodríguez, R. L., and Hunt, R. E. (2010) Host shifts and signal divergence: Mating signals covary with host use in a complex of specialized plant-feeding insects. *Biological Journal of the Linnean Society* 99: 60–72. doi: 10.1111/j.1095-8312.2009.01345.x

Čokl, A. and Virant-Doberlet, M. (2003) Communication with substrate-borne signals in small plant-dwelling insects. *Annual Review of Entomology* 48: 29–50. doi: 10.1146/annurev.ento.48.091801.1

Condon, M., Adams, D. C., Bann, D. et al. (2008) Uncovering tropical diversity: Six sympatric cryptic species of Blepharoneura (Diptera: Tephritidae) in flowers of *Gurania spinulosa* (Cucurbitaceae) in eastern Ecuador. *Biological Journal of the Linnean Society* 93: 779–797. doi: 10.1111/j.1095-8312.2007.00943.x

Cothran, R. D., Noyes, P., and Relyea, R. A. (2015) An empirical test of stable species coexistence in an amphipod species complex. *Oecologia* 178: 819–831. doi: 10.1007/s00442-015-3262-1

Cothran, R. D., Henderson, K. A., Schmidenberg, D. et al. (2013) Phenotypically similar but ecologically distinct: Differences in competitive ability and predation risk among amphipods. *Oikos* 122: 1429–1440. doi: 10.1111/j.1600-0706.2013.00294.x

Deiner, K., Fronhofer, E. A., Mächler, E., Walser, J. C., and Altermatt, F. (2016) Environmental DNA reveals that rivers are conveyer belts of biodiversity information. *Nature Communications* 7: 12544. doi: 10.1038/ncomms12544

Delić, T., Trontelj, P., Rendoš, M., and Fišer, C. (2017) The importance of naming cryptic species and the conservation of endemic subterranean amphipods. *Scientific Reports* 7: 3391. doi: 10.1038/s41598-017-02938-z

De Meester, N. et al. (2011) Salinity effects on the coexistence of cryptic species: A case study on marine nematodes. *Marine Biology* 158: 2717–2726. doi: 10.1007/s00227-011-1769-5

De Queiroz, K. (2005) Different species problems and their resolution. *BioEssays* 27: 1263–1269. doi: 10.1002/bies.20325

(2007) Species concepts and species delimitation. *Systematic Biology* 56: 879–886. doi: 10.1080/10635150701701083

Dieckmann, U. and Doebeli, M. (1999) On the origin of species by sympatric speciation. *Nature* 400: 354–357.

Dinca, V., Lukhtanov, V. A., Talavera, G., and Vila, R. (2011) Unexpected layers of cryptic diversity in wood white. *Nature Communications* 2: 324. doi: 10.1038/ncomms1329

Dinca, V., Wiklund, C., Lukhtanov, V. A. et al. (2013) Reproductive isolation and patterns of genetic differentiation in a cryptic butterfly species complex. *Journal of Evolutionary Biology* 26: 2095–2106. doi: 10.1111/jeb.12211

Dionne, K., Vergilino, R., Dufresne, F., Charles, F., and Nozais, C. (2011) No evidence for temporal variation in a cryptic species community of freshwater amphipods of the *Hyalella azteca* species complex. *Diversity* 3: 390–404. doi: 10.3390/d3030390

Doctrow, B. (2018) QnAs with Mark Hay. *Proceedings of the National Academy of Sciences of the United States of America* 115: 6521–6522. doi: 10.1073/pnas.1808472115

Doebeli, M. and Dieckmann, U. (2000) Evolutionary branching and sympatric speciation caused by different types of ecological interactions. *American Naturalist* 156: S77–S101.

van Doorn, G. S., Dieckmann, U., and Weissing F.J. (2004) Sympatric speciation by sexual selection: A critical reevaluation. *American Naturalist* 163: 709–725.

Eisenring, M., Altermatt, F., Westram, A. M. et al. (2016) Habitat requirements and ecological niche of two cryptic amphipod species at landscape and local scales. *Ecoshphere* 7: e01319. doi: e01319.10.1002/ecs2.1319

Elbrecht, V. and Leese, F. (2015) Can DNA-based ecosystem assessments quantify species abundance? Testing primer bias and biomass–sequence relationships with an innovative metabarcoding protocol. *Plos One* 10: e0130324. doi: 10.1371/journal.pone.0130324

Eme, D., Zagmajster, M., Delić, T. et al. (2017) Do cryptic species matter in macroecology? Sequencing European groundwater crustaceans yields smaller ranges but does not challenge biodiversity determinants. *Ecography* 41: 1–13. doi: 10.1111/ecog.02683

Endler, J. A. and Basolo, A. L. (1998) Sensory ecology, receiver biases and sexual selection. *Trends in Ecology and Evolution* 13: 415-420.

Feulner, P. G. D., Kirschbaum, F., Schugardt, C. et al. (2006) Electrophysiological and molecular genetic evidence for sympatrically occurring cryptic species in African weakly electric fishes (Teleostei: Mormyridae: *Campylomormyrus*). *Molecular Phylogenetics and Evolution* 39: 198–208. doi: 10.1016/j.ympev.2005.09.008

Finkbeiner, S. D., Fishman, D. A., Osorio, D. et al. (2017) Ultraviolet and yellow reflectance but not fluorescence is important for visual discrimination of conspecifics by *Heliconius erato*. *Journal of Experimental Biology* 220: 1267–1276. doi: 10.1242/jeb.153593

Fišer, C., Robinson, C. T., and Malard, F. (2018) Cryptic species as a window into the paradigm shift of the species concept. *Molecular Ecology* 27: 613–635. doi: 10.1111/ijlh.12426

Fišer, Ž., Altermatt, F., Zakšek, V., Knapič, T., and Fišer, C. (2015) Morphologically cryptic Amphipod species are "ecological clones" at regional but not at local scale: A case study of four Niphargus species. *PLoS ONE* 10: e0134384. doi: 10.1371/journal.pone.0134384

Flot, J.-F., Couloux, A., and Tillier, S. (2010) Haplowebs as a graphical tool for delimiting species: A revival of Doyle's "field for recombination" approach and its application to the coral genus *Pocillopora* in Clipperton. *BMC Evolutionary Biology* 10: 372. doi: 10.1186/1471-2148-10-372

Fontaneto, D., Flot, J.-F., and Tang, C. Q. (2015) Guidelines for DNA taxonomy, with a focus on the meiofauna. *Marine Biodiversity* 45: 533–451. doi: 10.1007/s12526-015-0319-7

Friesen, V. L., Smith, A. L., Gómez-Díaz, E. et al. (2007) Sympatric speciation by allochrony in a seabird. *Proceedings of the National Academy of Sciences of the United States of America* 104: 18589–18594. doi: 10.1073/pnas.0700446104

Fujita, M. K., Leaché, A. D., Burbrink, F. T., McGuire, J. A., and Moritz, C. (2012) Coalescent-based species delimitation in an integrative taxonomy. *Trends in Ecology & Evolution* 27: 480–488. doi: 10.1016/j.tree.2012.04.012

Funk, C. W., Caminer, M., and Ron, S. R. (2012) High levels of cryptic species diversity uncovered in Amazonian frogs. *Proceedings of the Royal Society B* 279: 1806–1814. doi: 10.1098/rspb.2011.1653

Gabaldón, C., Montero-Pau, J., Serra, M., and Carmona, M.J. (2013) Morphological similarity and ecological overlap in two rotifer species. *PLoS ONE* 8: e57087. doi: 10.1371/journal.pone.0057087

Germain, R. M., Williams, J. L., Schluter, D., and Angert, A. L. (2018) Moving character displacement beyond characters using contemporary coexistence theory. *Trends in Ecology and Evolution* 33: 74-84 doi: 10.1016/j.tree.2017.11.002

Gilbert, L. E. (1991) Biodiversity of a Central American *Heliconius* community: pattern, process, and problems. In: P. W. Price, T. M. Lewinsohn, T. W. Fernandes, and W. W. Benson (eds.) *Plant-Animal Interactions: Evolutionary Ecology in Tropical and Temperate Regions.* John Wiley & Sons, New York, pp. 403–427.

Gill, B. A., Kondratieff, B. C., Casner, K. L. et al. (2016) Cryptic species diversity reveals biogeographic support for the "mountain passes are higher in the tropics" hypothesis. *Proceedings of The Royal Society B* 283: 7–12. doi: 10.1098/rspb.2016.0553

Gillespie, R. G., Bennett, G. M., De Meester, L. et al. (2020) Comparing adaptive radiations across space, time, and taxa. *Journal of Heredity* 111:1–20. doi: 10.1093/jhered/esz064

Gogala, M. and Trilar, T. (2004) Bioacoustic investigations and taxonomic considerations on the Cicadetta montana species complex (Homoptera : Cicadoidea : Tibicinidae). *Anais de Academia Brasileira de Ciencas* 76: 316–324.

Hardege, J. D., Jennings, A., Hayden, D. et al. (2002) Novel behavioural assay and partial purification of a female-derived sex pheromone in *Carcinus maenas. Marine Ecology Progress Series* 244: 179–189. doi: 10.3354/meps244179

Hardege, J. D. and Terschak, J. A. (2010) Identification of crustacean sex pheromones. In: T. Breithaupt and M. Thiel (eds.) *Chemical Communication in Crustaceans.* Springer, New York, pp. 373–392.

Häussler, U. et al. 1999. External characteristics discriminating species of European pipistrelles, *Pipistrellus pipistrellus* (Schreber, 1774) and *P. pygmaeus* (Leach, 1825). *Myotis* 37: 27-40.

Hay, M. E. (2009) Marine chemical ecology: Chemical signals and cues structure marine populations, communities, and ecosystems. *Annual Review of Marine Science* 1: 193–214. doi: 10.1146/annurev.marine.010908.163708

(2010) Crustaceans as Powerful Models in Aquatic Chemical Ecology. In: T. Breithaupt and M. Thiel (eds.) *Chemical Communication in Crustaceans.* Springer, New York, pp. 41–62. doi: 10.1007/978-0-387-77101-4

Hebert, P. D. N., Penton, E. H., Burns, J. M., Janzen, D. H., and Hallwachs, W. (2004) Ten species in one: DNA barcoding reveals cryptic species in the neotropical skipper butterfly *Astraptes fulgerator. Proceedings of the National Academy of Sciences of the United States of America* 101: 14812–14817. doi: 10.1073/pnas.0406166101

Heethoff, M. (2018) Cryptic species – conceptual or terminological chaos? A response to Struck et al. *Trends in Ecology & Evolution* 33: 310. doi: 10.1016/j.tree.2018.02.006

Henry, C. S. (1994) Singing and cryptic speciation in insects. *Trends in Ecology & Evolution* 9: 388–392.

Henry, C. S. and Wells, M. M. (2010) Acoustic niche partitioning in two cryptic sibling species of *Chrysoperla* green lacewings that must duet before mating. *Animal Behaviour* 80: 991–1003. doi: 10.1016/j.anbehav.2010.08.021

Henry, C. S., Brooks, S. J., Duelli, P. et al. (2013) Obligatory duetting behaviour in the *Chrysoperla carnea* -group of cryptic species (Neuroptera : Chrysopidae): Its role in shaping evolutionary history. *Biological Reviews* 88: 787–808. doi: 10.1111/brv.12027

(2014) A new cryptic species of the *Chrysoperla carnea* group (Neuroptera : Chrysopidae) from western Asia: Parallel speciation without ecological adaptation. *Systematic Entomology* 39: 380–393. doi: 10.1111/syen.12061

Herrel, A., Huyghe, K., Vanhooydonck, B. et al. (2008) Rapid large-scale evolutionary divergence in morphology and performance associated with exploitation of a different dietary resource. *Proceedings of the National Academy of Sciences of the United States of America* 105: 4792–4795. doi: 10.1073/pnas.0711998105

Hertach, T., Puissant, S., Gogala, M. et al. (2016) Complex within a complex : Integrative taxonomy reveals hidden diversity in *Cicadetta brevipennis* (Hemiptera : Cicadidae) and unexpected relationships with a song divergent relative. *PLoS ONE* 11: e0165562. doi: 10.1371/journal.pone.0165562

Hey, J. (2006) On the failure of modern species concepts. *Trends in Ecology and Evolution* 21: 447–450. doi: 10.1016/j.tree.2006.05.011

Hobel, G. and Gerhardt, C. H. (2003) Reproductive character displacement in the acoustic communication system of green tree frogs (*Hyla cinerea*). *Evolution* 57: 894–904.

Holt, R. D. (2006) Emergent neutrality. *Trends in Ecology and Evolution* 21: 531–533. doi: 10.1016/j.tree.2006.08.003

Hoyal Cuthill, J. W., Guttenberg, N., Ledger, S., Crowther, R., and Huertas, B. (2019) Deep learning on butterfly phenotypes tests evolution's oldest mathematical model. *Science Advances* 5: eaaw4967. doi: 10.1126/sciadv.aaw4967

Imafuku, M. (2008) Variation in UV light reflected from the wings of *Favonius* and *Quercusia* butterflies. *Entomological Science* 11: 75–80. doi: 10.1111/j.1479-8298.2007.00247.x

Izzo, A. S. and Gray, D. A. (2004) Cricket song in sympatry: Species specificity of song without reproductive character displacement in *Gryllus rubens*. *Annals of the Entomological Society of America* 97: 831–837. doi: 10.1603/0013-8746(2004)097

Jones, G. (1997) Acoustic signals and speciation: The roles of natural and sexual selection in the evolution of cryptic species. *Advances in the Study of Behavior* 26: 317–354.

Jones, G. and van Parijs, S. M. (1993) Bimodal echolocation in pipistrelle bats: Are cryptic species present? *Proceedings of the Royal Society B* 251: 119–125. doi: 10.1098/rspb.1993.0017

Jones, G. and Siemers, B. M. (2011) The communicative potential of bat echolocation pulses. *Journal of Comparative Physiology A* 197: 447–457. doi: 10.1007/s00359-010-0565-x

Jones, J. (2001) Habitat selection studies in avian ecology: A critical review. *The Auk* 118: 557–562.

Jugovic, J., Jalžić, B., Prevorčnik, S., and Sket, B. (2012) Cave shrimps *Troglocaris* s. str. (Dormitzer, 1853), taxonomic revision and description of new taxa after phylogenetic and morphometric studies. *Zootaxa* 3421: 1–31.

Kaliszewska, Z. A., Seger, J., Rowntree, V. J. et al. (2005) Population histories of right whales (Cetacea: Eubalaena) inferred from mitochondrial sequence diversities and divergences of their whale lice (Amphipoda: Cyamus). *Molecular Ecology* 14: 3439–3456. doi: 10.1111/j.1365-294X.2005.02664

Kalko, E. K. and Schnitzler, H. U. (1993) Plasticity in echolocation signals of European pipistrelle bats in search flight: Implications for habitat use and prey detection. *Behavioral Ecology and Sociobiology* 33(6): 415–428. https://link.springer.com/article/10.1007/BF00170257

Kapli, P., Lutteropp, S., Zhang, J. et al. (2017) Multi-rate Poisson tree processes for single-locus species delimitation under maximum likelihood and Markov chain Monte Carlo. *Bioinformatics* 33: 1630–1638. doi: 10.1093/bioinformatics/btx025

Karanovic, T., Djurakic, M., and Eberhard, S. M. (2016) Cryptic species or inadequate taxonomy? Implementation of 2D geometric morphometrics based on integumental organs as landmarks for delimitation and description of copepod taxa. *Systematic Biology* 65: 304–327. doi: 10.1093/sysbio/syv088

Kemp, D. J. (2007) Female butterflies prefer males bearing bright iridescent ornamentation. *Proceedings of the Royal Society B* 274: 1043–1047. doi: 10.1098/rspb.2006.0043

Kemp, D. J. and Rutowski, R. L. (2011) The role of coloration in mate choice and sexual interactions in butterflies. In: J. H. Brockmann and T. Roper (eds.) *Advances in the Study of Behavior, Volume 43*. Elsevier Inc., New York, pp. 55–92. doi: 10.1016/B978-0-12-380896-7.00002-2

Kirschel, A. N. G., Blumstein, D. T., and Smith T. B. (2009) Character displacement of song and morphology in African tinkerbirds. *Proceedings of the National Academy of Sciences* 106: 8256–8261.

Knowlton, N. (1993) Sibling species in the sea. *Annual Review of Ecology and Systematics* 24: 189–216.

Knuttel, H. and Fiedler, K. (2000) On the use of ultraviolet photography and ultraviolet wing patterns in butterfly morphology and taxonomy. *Journal of the Lepidopterists' Society* 54: 137–144.

Koselj, K., Schnitzler, H.-U., and Siemers, B. M. (2011) Horseshoe bats make adaptive prey-selection decisions, informed by echo cues. *Proceedings of the Royal Society B* 278, 3034–3041.

Kubisch, A., Holt, R. D., Poethke, H.-J., and Fronhofer, E.A. (2014) Where am I and why? Synthesizing range biology and the eco-evolutionary dynamics of dispersal. *Oikos* 123: 5–22. doi: 10.1111/j.1600-0706.2013.00706.x

Lagrue, C., Wattier, R., Galipaud, M. et al. (2014) Confrontation of cryptic diversity and mate discrimination within *Gammarus pulex* and *Gammarus fossarum* species complexes. *Freshwater Biology* 59: 2555–2570. doi: 10.1111/fwb.12453

Lajus, D., Sukhikh, N., and Alekseev, V. (2015) Cryptic or pseudocryptic: Can morphological methods inform copepod taxonomy? An analysis of publications and a case study of the *Eurytemora affinis* species complex. *Ecology and Evolution* 5: 2374–2385. doi: 10.1002/ece3.1521

Lassance, J.-M., Svensson, G. P., Kozlov, M. V., Francke, W., and Löfstedt, C. (2019) Pheromones and barcoding delimit boundaries between cryptic species in the primitive moth genus *Eriocrania* (Lepidoptera: Eriocraniidae). *Journal of Chemical Ecology* 45: 429—439. doi: 10.1007/s10886-019-01076-2

Leache, A. D., Fujita, M. K., Minin, V. N., and Bouckaert, R. R. (2014) Species delimitation using genome-wide SNP Data. *Systematic Biology* 63: 534–542. doi: 10.1093/sysbio/syu018

Leese, F., Altermatt, F., Bouchez, A. et al. (2016) DNAqua-Net: Developing new genetic tools for bioassessment and monitoring of aquatic ecosystems in Europe. *Research Ideas and Outcomes* 2: e11321. doi: 10.3897/rio.2.e11321

Leibold, M. A. and McPeek, M. A. (2006) Coexistence of the niche and neutral perspectives in community ecology. *Ecology* 87: 1399–410. doi: 10.1890/0012-9658(2006)87[1399:cotnan]2.0.co;2

Leibold, M. A., Urban, M. C., De Meester, L., Klausmeier, C. A., and Vanoverbeke, J.

(2018) Regional neutrality evolves through local adaptive niche evolution. *Proceedings of the National Academy of Sciences* 116: 2612–2617. doi: 10.1073/pnas.1808615116

Li, L. and Chesson, P. (2016) The effects of dynamical rates on species coexistence in a variable environment: The Paradox of the Plankton revisited. *The American Naturalist* 188: E46–E58. doi: 10.1086/687111

Lim, M. L. M., Land, M. F., and Li, D. (2007) Sex-specific UV and fluorescence signals in jumping spiders. *Science* 315: 481. doi: 10.1126/science.1134254

Lindquist, N. (2002) Chemical defense of early life stages of benthic marine invertebrates. *Journal of Chemical Ecology* 28: 1987–2000. doi: 10.1023/A:1020745810968

Lukhtanov, V. A. (2010) Dobzhansky's rule and reinforcement of pre-zygotic reproductive isolation in zones of secondary contact. *Zhurnal Obshchei Biologii* 71: 372–385. doi: 10.1134/s2079086411010051

Lukhtanov, V. A., Kandul, N. P., Plotkin, J. B. et al. (2005) Reinforcement of pre-zygotic isolation and karyotype evolution in *Agrodiaetus* butterflies. *Nature* 436: 385–389. doi: 10.1038/nature03704

Lukić M., Delić T., Pavlek M., Deharveng, L., and Zagmajster, M. (2019) Distribution pattern and radiation of the European subterranean genus Verhoeffiella (Collembola , Entomobryidae). *Zoologica Scripta*: 1–15. doi: 10.1111/zsc.12392

Luo, A., Ling, C., Ho, S. Y., and Zhu, C. (2018) Comparison of methods for molecular species delimitation across a range of speciation scenarios. *Systematic Biology* 67: 830–846. doi: 10.1093/sysbio/syy011/4866060

Maan, M. E., Hofker, K. D., van Alphen, J. J. M., and Seehausen, O. (2006) Sensory drive in cichlid speciation. *American Naturalist* 167: 947–954.

Macher, J. N., Weiss, M., Beermann, A. J., and Leese, F. (2016) Cryptic diversity and population structure at small scales: the freshwater snail *Ancylus* (Planorbidae, Pulmonata) in the Montseny mountain range. *Annales de Limnologie (International Journal of Limnology)* 52: 387–399. doi: 10.1051/limn/2016026

Mallet, J. and Gilbert, L. E. (1995) Why are there so many mimicry rings? Correlations between habitat, behaviour and mimicry in *Heliconius* butterflies. *Biological Journal of the Linnean Society* 55: 159–180.

Marsteller, S., Adams, D. C., Collyer, M. L., and Condon, M. (2009) Six cryptic species on a single species of host plant: Morphometric evidence for possible reproductive character displacement. *Ecological Entomology* 34: 66–73. doi: 10.1111/j.1365-2311.2008.01047.x

Mayden, R. L. (1997) A hierarchy of species concepts: The denouement in the saga of the species problem. In: M.F. Claridge, H. A. Dawah, and M. R. Wilson (eds.) *Species: The Units of Biodiversity*. Chapman and Hall, London, pp. 381–423.

Mayer, F. and Von Helversen, O. (2001) Sympatric distribution of two cryptic bat species across Europe. *Biological Journal of the Linnean Society* 74: 365–374. doi: 10.1006/bijl.2001.0586

Mayer, F., Dietz, C., and Kiefer, A. (2007) Molecular species identification boosts bat diversity. *Frontiers in Zoology* 4: 4. doi: 10.1186/1742-9994-4-4

Mayr, E. (1942) *Systematics and the Origin of Species from the Viewpoint of Zoologist.* Harvard University Press, Cambridge, MA, 337 pp.

Meyer-Rochow, V. B. (1991) Differences in ultraviolet wing patterns in the New Zealand lycaenid butterflies *Lycaena salustius, L. rauparaha*, and *L. feredayi* as a likely isolating mechanism. *Journal of the Royal Society of New Zealand* 21: 169–177. doi: 10.1080/03036758.1991.10431405

Mielewczik, M., Leibisch, F., Walter, A., and Greven, H. (2012) Near-Infrared (NIR) reflectance in insects: Phenetic studies of 181 species. *Entomologie Heute* 24: 183–215.

Molbo, D., Machado, C. A., Sevenster, J. G., Keller, L., and Herre, E. A. (2003). Cryptic species of fig-pollinating wasps: Implications for the evolution of the fig-wasp mutualism, sex allocation, and precision of adaptation. *Proceedings of the National Academy of Sciences of the United States of America* 100: 5867–5872.

Montero-Pau, J. and Serra, M. (2011) Life-cycle switching and coexistence of species with no niche differentiation. *PloS ONE* 6: e20314. doi: 10.1371/journal.pone.0020314

Montero-Pau, J., Ramos-Rodríguez, E., Serra, M., and Gómez, A. (2011) Long-term coexistence of rotifer cryptic species. *PloS ONE* 6: e21530. doi: 10.1371/journal.pone.0021530

Morinière, J., Hendrich, L., Balke, M. et al. (2017) A DNA barcode library for Germany's mayflies, stoneflies and caddisflies (Ephemeroptera, Plecoptera and Trichoptera). *Molecular Ecology Resources* 17: 1293–1307. doi: 10.1111/1755-0998.12683

Morris, D. W. (2003) Toward an ecological synthesis: A case for habitat selection. *Oecologia* 136: 1–13. doi: 10.1007/s00442-003-1241-4

Moskalik, B. and Uetz, G. W. (2011) Experience with chemotactile cues indicating female feeding history impacts male courtship investment in the wolf spider *Schizocosa ocreata*. *Behavioral Ecology and Sociobiology* 65: 2175–2181. doi: 10.1007/s00265-011-1225-z

Naisbit, R. E., Jiggins, C. D., and Mallet, J. (2001) Disruptive sexual selection against hybrids contributes to speciation between *Heliconius cydno* and Heliconius melpomene. *Proceedings of the Royal Society B* 268: 1849–1854. doi: 10.1098/rspb.2001.1753

Nosil, P. (2012) *Ecological Speciation*. Oxford Series in Ecology and Evolution, Oxford, 300pp.

Obara, Y., Watanabe, K., and Satoh, T. (2010) UV reflectance of inter-subspecific hybrid females obtained by crossing cabbage butterflies from Japan (*Pieris rapae crucivora*) with those from New Zealand (*P . rapae rapae*). *Entomological Science* 13: 156–158. doi: 10.1111/j.1479-8298.2010.00364.x

Olds, B. P., Jerde, C. L., Renshaw, M. A. et al. (2016) Estimating species richness using environmental DNA. *Ecology and Evolution* 6: 1–13. doi: 10.1002/ece3.2186

Onda-Sumi, E. (2005) Difference in calling song of three field crickets of the genus Teleogryllus: The role in premating isolation. *Animal Behaviour* 69: 881–889. doi: 10.1016/j.anbehav.2004.05.015

Padial, J. M., Miralles, A., De la Riva, I., and Vences, M. (2010) The integrative future of taxonomy. *Frontiers in Zoology* 7: 16. doi: 10.1186/1742-9994-7-16

Pante, E., Schoelinck, C., and Puillandre, N. (2015) From integrative taxonomy to species description: One step beyond. *Systematic Biology* 64: 152–160. doi: 10.1093/sysbio/syu083

Park, K. J., Altringham, J. D., and Jones, G. (1996) Assortative roosting in the two phonic types of *Pipistrellus pipistrellus* during the mating season. *Proceedings of the Royal Society B* 263: 1495–1499. doi: 10.1098/rspb.1996.0218

Pecháček, P., Stella, D., and Kleisner, K. (2019) A morphometric analysis of environmental dependences between ultraviolet patches and wing venation patterns in *Gonepteryx* butterflies (Lepidoptera, Pieridae). *Evolutionary Ecology* 33: 89–110. doi: 10.1007/s10682-019-09969-0

Pérez-Ponce de León, G. and Poulin, R. (2016) Taxonomic distribution of cryptic diversity among metazoans: Not so homogeneous after all. *Biology Letters* 12: 20160371. doi: 10.1098/rsbl.2016.0371

Pfenninger, M. and Schwenk, K. (2007) Cryptic animal species are homogeneously distributed among taxa and biogeographical regions. *BMC Evolutionary Biology* 7: 121. doi: 10.1186/1471-2148-7-121

Pons, J., Barraclough, T., Gomez-Zurita, J. et al. (2006) Sequence-based species delimitation for the DNA taxonomy of undescribed insects. *Systematic Biology* 55: 595–609. doi: 10.1080/10635150600852011

Powell, T. H. Q., Cha, D. H., Linn, C. E., and Feder, J. L. (2012) On the scent of standing variation for speciation: Behavioral evidence for native sympatric host races of *Rhagoletis pomonella* (Diptera: Tephritidae) in the southern United States. *Evolution* 66: 2739–2756. doi: 10.1111/j.1558-5646.2012.01625.x

Raxworthy, C., Ingram, C., Rabibisoa, N., and Pearson, R. (2007) Applications of ecological niche modeling for species delimitation: A review and empirical evaluation using day geckos (*Phelsuma*) from Madagascar. *Systematic Biology* 56: 907–923. doi: 10.1080/10635150701775111

Renner, S. S. (2016) A return to Linnaeus's focus on diagnosis, not description: The use of DNA characters in the formal naming of species. *Systematic Biology* 65: 1086–1095. doi: 10.1093/sysbio/syw032

Roberts, R. E. W. and Uetz, G. W. (2005) Information content of female chemical signals in the wolf spider, *Schizocosa ocreata*: Male discrimination of reproductive state and receptivity. *Animal Behaviour* 70: 217–223. doi: 10.1016/j.anbehav.2004.09.026

Rodríguez, R. L., Ramaswamy, K., and Cocroft, R. B. (2006) Evidence that female preferences have shaped male signal evolution in a clade of specialized plant-feeding insects. *Proceedings of the Royal Society B* 273: 2585–2593. doi: 10.1098/rspb.2006.3635

Rodríguez, R. L., Sullivan, L. E., and Cocroft, R. B. (2004) Vibrational communication and reproductive isolation in the *Enchenopa binotata* species complex of treehoppers (Hemiptera: Membracidae). *Evolution* 58: 571–578. doi: 10.1111/j.0014-3820.2004.tb01679.x

Russo, D. and Jones, G. (2000) The two cryptic species of *Pipistrellus pipistrellus* (Chiroptera: Vespertilionidae) occur in Italy: Evidence from echolocation and social calls. *Mammalia* 64: 187–197. doi: 10.1515/mamm.2000.64.2.187

Rutowski, R. L. and Rajyaguru, P. K. (2013) Male-specific iridescent coloration in the pipevine swallowtail (*Battus philenor*) is used in mate choice by females but not sexual discrimination by males. *Journal of Insect Behavior* 26: 200–211. doi: 10.1007/s10905-012-9348-2

Scheffer, M. and van Nes, E. H. (2006) Self-organized similarity, the evolutionary emergence of groups of similar species. *Proceedings of the National Academy of Sciences of the United States of America* 103: 6230–6235. doi: 10.1073/pnas.0508024103

Schuchmann, M. and Siemers, B. M. (2010) Behavioral evidence for community-wide species discrimination from echolocation calls in bats. *American Naturalist* 176: 72–82. doi: 10.1086/652993

Schluter, D. (2009) Evidence for ecological speciation and its alternative. *Science* 323: 737–741. doi: 10.1126/science.1160006

Scriven, J. J., Whitehorn, P. R., Goulson, D., and Tinsley, M. C. (2016) Niche partitioning in a sympatric cryptic species complex. *Ecology and Evolution* 6: 1328–1339. doi: 10.1002/ece3.1965

Seehausen, O., Terai, Y., Magalhaes, I. S. et al. (2008) Speciation through sensory drive in cichlid fish. *Nature* 455: 620–626. doi: 10.1038/nature07285

Servedio, M. R. and Boughman, J. W. (2017) The role of sexual selection in local adaptation and speciation. *Annual Review of Ecology and Systematics* 48: 85–109. doi: 10.1146/annurev-ecolsys-110316-022905

Shinen, J. L. and Navarrete, S. A. (2014) Lottery coexistence on rocky shores: Weak niche differentiation or equal competitors engaged in neutral dynamics? *The American Naturalist* 183: 342–362. doi: 10.1086/674898

Siemers, B. M. and Schnitzler, H.-U. (2004) Echolocation signals reflect niche differentiation in five sympatric congeneric bat species. *Nature* 429: 657–661.

Siepielski, A. M. and McPeek, M. A. (2010) On the evidence for species coexistence: A critique of the coexistence program. *Ecology* 91: 3153–3164. doi: 10.1890/10-0154.1

Silberglied, R. and Taylor, O. (1973) Ultraviolet differences between the sulphur butterflies, *Colia eurytheme* and *C. philodice*, and a possible isolating mechanism. *Nature* 241: 406–408. doi: 10.1038/246421a0

Sites, J. W. and Marshall, J. C. (2003) Delimiting species: A Renaissance issue in systematic biology. *Trends in Ecology and Evolution* 18: 462–470. doi: 10.1016/S0169-5347(03)00184-8

(2004) Operational criteria for delimiting species. *Annual Review of Ecology, Evolution, and Systematics* 35: 199–227. doi: 10.1146/annurev.ecolsys.35.112202.130128

Smadja, C. and Butlin, R. K. (2009) On the scent of speciation: The chemosensory system and its role in premating isolation. *Heredity* 102: 77–97. doi: 10.1038/hdy.2008.55

Smotherman, M., Knörnschild, M., Smarsh, G., and Bohn, K. (2016) The origins and diversity of bat songs. *Journal of Comparative Physiology A* 202: 535–554. doi: 10.1007/s00359-016-1105-0

Soares, D. and Niemiller, M. L. (2013) Sensory adaptations of fishes to subterranean environments. *BioScience* 63: 274–283. doi: 10.1525/bio.2013.63.4.7

Stella, D., Pecháček, P., Meyer-Rochow, V. B., and Kleisner, K. (2018) UV reflectance is associated with environmental conditions in Palaearctic *Pieris napi* (Lepidoptera: Pieridae). *Insect Science* 25: 508–518. doi: 10.1111/1744-7917.12429

Stella, D., Rindoš, M., Kleisner, K., and Pechá, P. (2018) Distribution of ultraviolet ornaments in *Colias* butterflies (lepidoptera : pieridae). *Environmetal Entomology* 47: 1344–1354. doi: 10.1093/ee/nvy111

Stevens, M. (2013) *Sensory Ecology, Behavior, and Evolution.* Oxford University Press, Glasgow, 260 pp.

Struck, T. H., Feder, J. L., Bendiksby, M. et al. (2018) Finding evolutionary processes hidden in cryptic species. *Trends in Ecology and Evolution* 33: 153–163. doi: 10.1016/j.tree.2017.11.007

Tang, C. Q., Humphreys, A., Fontaneto, D., and Barraclough, T. G. (2014) Effects of phylogenetic reconstruction method on the robustness of species delimitation using single locus data. *Methods in Ecology and Evolution* 5: 1086–1094. doi: 10.1111/2041-210X.12246

Tovée, M. J. (1995) Ultra-violet photoreceptors in the animal kingdom: Their distribution and function. *Trends in Ecology and Evolution* 10: 455–460. doi: 10.1016/S0169-5347(00)89179-X

Uetz, G. W. and Roberts, J. A. (2002) Multisensory cues and multimodal communication in spiders: Insights from video/audio playback studies. *Brain, Behavior and Evolution* 59: 222–230.

Uetz, G. W. Roberts, J. A., Clark, D. L., Gibson, J. S., and Gordon, S. D. (2013) Multimodal signals increase active space of communication by wolf spiders in a complex litter environment. *Behavioral Ecology and Sociobiology* 67: 1471–1482. doi: 10.1007/s00265-013-1557-y

von Uexküll, J. J. (1956) *Streifzüge durch die Umwelten von Tieren und Menschen/ Bedeutungslehre.* Reprint. Rowohlt, Hamburg, 182 pp.

Veech, J. A. (2013) A probabilistic model for analysing species co-occurrence. *Global Ecology and Biogeography* 22: 252–260. doi: 10.1111/j.1466-8238.2012.00789.x

(2014) The pairwise approach to analysing species co-occurrence. *Journal of Biogeography* 41: 1029–1035. doi: 10.1111/jbi.12318

Vickers, N. J. (2002) Defining a synthetic pheromone blend attractive to male *Heliothis subflexa* under wind tunnel conditions. *Journal of Chemical Ecology* 28: 1255–1267.

Voda, R., Dapporto, L., Dinca, V., and Vila, R. (2015a) Why do cryptic species tend not to co-occur? A case study on two cryptic pairs of butterflies. *PLoS ONE* 10: e0117802. doi: 10.1371/journal.pone.0117802

(2015b) Cryptic matters: Overlooked species generate most butterfly beta-diversity. *Ecography* 38: 405–409. doi: 10.1111/ecog.00762

Warren, D. L., Cardillo, M., Rosauer, D. F., and Bolnick, D. I. (2014) Mistaking geography for biology: Inferring processes from species distributions. *Trends in Ecology and*

Evolution 29: 572–580. doi: 10.1016/j.tree.2014.08.003

Wellborn, G. A. and Cothran, R. D. (2004) Phenotypic similarity and differentiation among sympatric cryptic species in a freshwater amphipod species complex. *Freshwater Biology* 49: 1–13.

(2007) Niche diversity in crustacean cryptic species: Complementarity in spatial distribution and predation risk. *Oecologia* 154: 175–183. doi: 10.1007/s00442-007-0816-x

Wilkins, M. R., Seddon, N., and Safran, R. J. (2013) Evolutionary divergence in acoustic signals: Causes and consequences. *Trends in Ecology & Evolution* 28: 156–166. doi: 10.1016/j.tree.2012.10.002

Yeates, D. K., Seago, A., Nelson, L. et al. (2011) Integrative taxonomy, or iterative taxonomy? *Systematic Entomology* 36: 209–217. doi: 10.1111/j.1365-3113.2010.00558.x

Yen, J. and Lasley, R. (2010) Chemical communication between copepods: Finding the mate in a fluid environment. In: T. Breithaupt and M. Thiel (eds.) *Chemical Communication in Crustaceans*. Springer, New York, pp. 177–198. doi: 10.1007/978-0-387-77101-4

Zettler, M. L., Proffitt, C. E., Darr, A. et al. (2013) On the myths of indicator species: Issues and further consideration in the use of static concepts for ecological applications. *PLoS ONE* 8: e78219. doi: 10.1371/journal.pone.0078219

Zhang, D. Y., Lin, K., and Hanski, I. (2004) Coexistence of cryptic species. *Ecology Letters* 7: 165–169. doi: 10.1111/j.1461-0248.2004.00569.x

Zhang, J., Kapli, P., Pavlidis, P., and Stamatakis, A. (2013) A general species delimitation method with applications to phylogenetic placements. *Bioinformatics* 29: 2869–2876. doi: 10.1093/bioinformatics/btt499

8

The Implications of Coalescent Conspecific Genetic Samples in Plants

Matt Lavin and R. Toby Pennington

Introduction

Defining cryptic species or cryptic speciation can be precise. For example, Struck et al. (2017, 2018) and Heethoff (2018) argue about the necessity of quantifying the degree of phenotypic divergence among potentially cryptic species. Large genetic differences, mainly in the form of coalesced (i.e., monophyletic) samples, combined with little if any phenotypic distinctions of those samples, mark unequivocal cases of cryptic species. Such cases potentially provide insights into underlying mechanisms responsible for cryptic speciation (Struck et al. 2017, 2018). In this context, we focus not just on cryptic species but whether conspecific genetic samples coalesce, or are monophyletic, during a phylogenetic analysis. This focus allows us to regard cryptic species as one form of overlooked species. Overlooked species are those that appear after genetic information reveals distinct clades within what was previously recognised as a single taxonomic species. This finding sends a taxonomist back to re-study field or museum specimens in search of covarying ecological, geographical, and phenotypic distinctions. Regardless of the degree of any such covariation discovered, the genetic result alone may suffice to recognise formally a previously overlooked species.

Our general perspective here encompasses the many diverse case studies that involve genetic evidence revealing underestimated levels of species diversity (e.g., Fišer et al. 2018; Struck and Cerca 2019). In the end, issues related to both cryptic and overlooked species dovetail with any inaccurate taxonomy wrought by geographically provincial and other forms of limited sampling, as suggested by Muñoz-Rodríguez et al. (2019).

Coalescent conspecific genetic samples are relatively uncommon in plants. Of the 11 presentations and the one poster paper presented in the symposium "Cryptic taxa: artefact of classification or evolutionary phenomena" at the 2019 Systematic Association Biannual Meeting Bristol, one (8 per cent) addressed cryptic plant species. Struck et al. (2017) analysed 606 randomly selected publications involving cryptic species or speciation and reported that plant studies represented 7.5 per cent of them. Hollingsworth et al. (2016) detail the applications of plant barcodes and devote only a small subsection to plant studies where new genetic information prompted re-examination of field or museum specimens followed by the formal taxonomic recognition of new species.

Genetic analysis revealing relatively small proportions of cryptic or overlooked species in plants compared to animals likely relates to the findings of Fazekas et al. (2009), who showed that phylogenies derived from two or more chloroplast DNA barcode loci often resolve multiple conspecific accessions as paraphyletic. In this context of species paraphyly, plant species may be less likely to be formally re-delimited taxonomically. That is, plant taxonomists are less likely to detect and formally recognise cryptic species in the absence of intraspecific monophyletic groups. For analogous studies of animals, in contrast, analysis of just a portion of the mitochondrial gene cytochrome c oxidase subunit 1 (*cox1/CO1*) in animals often resolves multiple conspecific accessions as monophyletic, which has resulted in the more frequent detection of cryptic animal species (Fazekas et al. 2009).

While these results of discovery of monophyletic species in animals but often not in plants may reflect faster rates of substitution for standard barcoding regions in animals (e.g., *cox1/CO1*) than in plants (e.g., *matK*, *rbcL*, and nuclear ribosomal internal transcribed spacers), we argue that other factors are at play. For example, Fazekas et al. (2009) argued that the addition of nuclear loci in plant studies, each with generally larger effective population sizes compared to plastid loci, would unlikely increase the proportion of monophyletic plant species. Phylogenomic studies in plants that include multiple conspecific accessions support Fasekas et al. (2009). For example, conspecific accessions not resolved as monophyletic with nuclear ribosomal ITS were likewise not necessarily resolved as monophyletic with genomic data in the plant genera *Ceiba* (Malvaceae; Pezzini 2019), *Ipomoea* (Convolvulaceae; Muñoz-Rodríguez et al. 2019), and *Inga* (Leguminosae; Dexter et al. 2017; Pennington et al. unpublished data; Figure 8.1). A phylogenomic analysis of *Eucalyptus* (Jones et al. 2016) included 161 taxa (species or subspecies), each represented by three to four genetic samples. Jones et al. resolved slight majority of these 161 taxa as not monophyletic. Vargas et al. (2017) detected high levels of reticulate evolution in the Andean genus *Diplostephium* (Asteraceae). Each of the few *Diplostephium* species represented by conspecific accessions, *Diplostephium pulchrum*, *D. rhomboidale*, *D. schultzii*, and *D.* sp. nov. "JUN," did not resolve, or at least consistently so, as monophyletic among four different data sets (i.e., complete nuclear ribosomal cistron, complete chloroplast genome, partial mitochondrial genome, and nuclear-ddRAD).

Fazekas et al. (2009), Hollingsworth et al. (2011), and Naciri and Linder (2015) suggest that species paraphyly, or lack of coalescence of multiple conspecific plant genetic samples, should be expected. Fazekas et al. (2009) suggest gene trees resolving paraphyletic plant species may be quite common because of gene exchange caused by hybridisation and polyploidy, retention of ancestral polymorphisms (incomplete lineage sorting), imperfect species definitions and taxonomy (e.g., cryptic or overlooked species), and high incidences of asexual reproduction. Large effective population size (Ne) combined with recent speciation events, for example, increases the probability of incomplete lineage sorting because the speed of sorting of ancestral genetic polymorphism is slower in large populations (Naciri and Linder 2015). Because the standard plant barcodes are the plastid markers

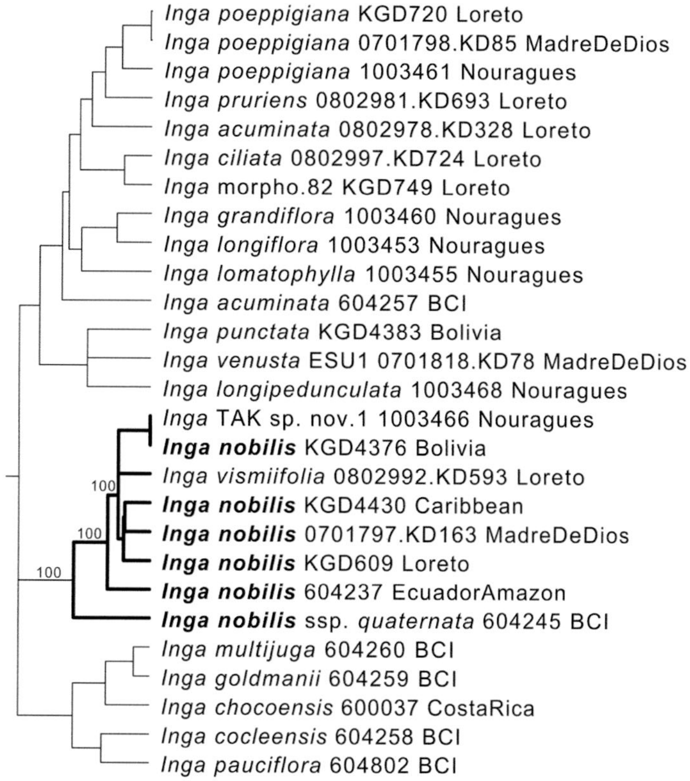

Figure 8.1 A small portion of the *Inga* phylogeny adapted from Dexter et al. (2017) and focusing on conspecific accessions of a widespread neotropical wet forest tree, *Inga nobilis* (bold font), which do not coalesce with respect to accessions of at least two other species of *Inga*, which are nested within the clade of *Inga nobilis* accessions. This result, common to most species of *Inga*, continues to be resolved by the analysis of an aligned DNA sequence data set of 1.3 M bps representing 818 nuclear loci (Pennington et al. unpublished data). Numbers adjacent to branches report relevant Bayesian posterior probabilities. Accession information follows each Latin binomial, as taken from Dexter et al. (2017).

matK, *psbA*, and *rbcL* (Hollingsworth et al., 2011), there is a further issue that limited seed dispersal compared to pollen dispersal will limit the degree to which these loci will track species boundaries (Hollingsworth et al. 2016).

Naciri and Linder (2015) detail seven processes explaining the expectation of paraphyly or non-coalescence of conspecific plant samples for both plastid and nuclear loci. These essentially relate to introgressive hybridisation and large effective population sizes at the genomic level and at the structured to unstructured population level. For example, Naciri and Linder (2015) indicate that a tree species with Ne = 1 M and a generation time of 10 years will require 50 Ma for monophyly to be achieved at all loci. In this context, Pennington and Lavin (2016), citing empirical data from rain forest ecological monitoring plots (e.g., ter Steege et al. 2013), point out that many tropical rain forest trees have population sizes exceeding this. Botanists should therefore expect to delimit species for which conspecific genetic samples resolve as uncoalesced or paraphyletic, as essentially advocated by Pennington and Lavin (2016), Freudenstein et al. (2017), Muñoz-Rodríguez et al. (2019), and Wood et al. (2020), for example.

Coalescent genetic samples are not required to delimit new species. Species-level molecular phylogenies resolving multiple conspecific accessions as monophyletic are not a necessary requirement for delimiting species, cryptic or otherwise, because ages of species or modes of speciation vary (e.g., Rieseberg and Brouillet 1994; Crisp and Chandler 1996; Gallardo 2017). In addition, species concepts accommodate non-coalescence of conspecific genetic samples. For example, the universal species concept (de Queiroz 2007) requires a combination of ecological, genetic, geographical, and phenotypic evidence for separately evolving metapopulations. Elephant species in Africa (Gilbert 2010; Rohland et al. 2010; Groves 2016) and Ponderosa pine tree species in northwestern North America (Latta and Mitton 1999; Willyard et al. 2017) exemplify this. Both cases involve large-bodied overlooked species exploited by humans for centuries, and genetic evidence suggests that supposed ecotypic variation in phenotype actually represented species differences. Forest elephants with longer generation times contrast to savanna elephants with shorter generation times. Tall statured west-of-the-divide Ponderosa pine contrast to short statured east-of-the-divide Ponderosa pine. Species delimitation followed even with the detection of introgressive hybridisation limited to contact zones, which resulted in the non-coalescence of conspecific genetic samples. The delimitation of separate species in each case is consistent with the perspective of separately evolving metapopulations. The universal species concept (de Queiroz 2007) is tantamount to generalised species definitions advocated by, for example, Cornetti et al. (2015), Naciri and Linder (2015), and, in our opinion, Freudenstein et al. (2017), where evidence for delimiting species should involve an integration of ecological, genetic, geographical, and phenotypic data.

Regardless, coalescence of conspecific genetic samples often provides an unequivocal defence for distinguishing new species, cryptic or otherwise. However, we now set that issue aside and focus on the issue of what it potentially indicates, ecologically and evolutionarily, when a plant clade includes many species, each of which comprises coalescent conspecific genetic samples (i.e., monophyletic species).

8.1 Plant Studies Resolving Coalescent Conspecific Genetic Samples

Plant phylogenetic studies involving extensive conspecific genetic sampling are few, unfortunately. Jones et al. (2016) sampled 542 *Eucalyptus* trees representing 153 species, which represented many of the Australian subgroups of *Eucalyptus* that grow in the Currency Creek Arboretum. This area is located near the coast of South Australia. A majority of these species were resolved as not monophyletic. These trees come from throughout Australia, and Jones et al. do not report a relationship between habitat and degree of coalescence of conspecific samples. Studies reporting a relationship between habitat and degree of coalescence of conspecific samples come mostly from the legume family. One legume group particularly well sampled for conspecific genetic samples is the tribe Robinieae, where most of the 77 species in Robinieae are endemic to patches of seasonally dry neotropical dry forests and woodlands (Lavin 1988; Lavin 1993; Lavin and Sousa 1995; Särkinen et al. 2012; Lavin et al. 2018). Most of these species are represented by coalescing conspecific genetic samples, and the recent detection of additional monophyletic groups of genetic samples within a supposedly single species resulted in the recognition of newly described species (Duno de Stephano 2010; Pennington et al. 2011; Quieroz and Lavin 2011; Lavin et al. 2018; Figure 8.2). All these newly distinguished species had been relegated originally to environmental variants (Lavin 1988).

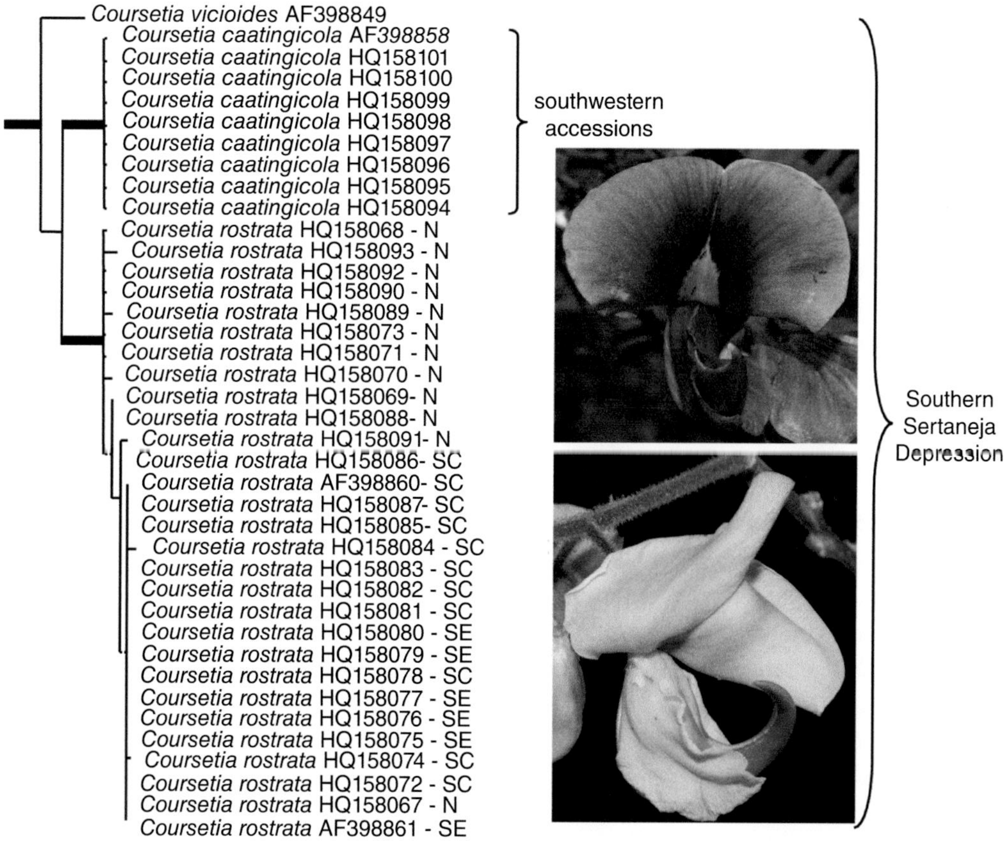

Figure 8.2 Phylogeny of the *Coursetia rostrata* group, which is endemic to the southern Sertaneja Depression of the caatinga (succulent biomes) in east central Brazil, where this lineage has been in residence for at least 17 Ma. *Coursetia caatingicola* specimens originally were named *C. rostrata*. The upper inset photo represents the flower of *Coursetia caatingicola*, whereas the lower photo represents the flower of *C. rostrata*. Both flowers are distinctively large and measure about 3 cm long and 3 cm wide but strongly contrast in colour. Sample labels comprise the Latin binomial followed by a GenBank accession number. N (northern), SC (southcentral), and SE (southeastern) indicate sampling region in the Southern Sertaneja Depression of the caatinga. Samples of *Coursetia caatingicola* occur to the southwest of the sample localities of *C. rostrata*. These phylogenetic results derive from the analysis of the nuclear ribosomal internal transcribed spacer region and the chloroplast *trnD-trnT* region. Thick branches indicate 100 per cent bootstrap values. Figure adapted from Queiroz and Lavin (2011).

Other examples of coalescing conspecific genetic samples taken from seasonally dry tropical forests and woodlands include *Wajira* (Thulin et al. 2004), *Mimosa* (Särkinen et al. 2011), diploid species of *Leucaena* (Govindarajulu et al. 2011), and *Arquita* (Gagnon et al. 2015). In the case of the legume tree species, *Cyathostegia mathewsii* (Benth.) Schery, Pennington et al. (2010) and Särkinen et al. (2012) found that in the inter-Andean Valleys in Ecuador and Peru with seasonally dry tropical forests and woodland, adjacent valleys harboured reciprocally monophyletic clades of conspecific genetic samples that range 2.5–5 Ma for the estimated ages of stem clades (Figure 8.3). Because these different ancient subclades of *Cyathostegia mathewsii* did not differ sufficiently ecologically and phenotypically, they were ranked as conspecific subclades (Figure 8.3).

Other plant families where coalescing conspecific genetic samples are commonly detected more in dry than wet environments include the genus *Manihot* (Euphorbiaceae; Duputie et al. 2011), where paraphyletic conspecific genetic samples are most common in a clade of cerrado (savanna) species than in clades of species from other dry areas, such as the caatinga (seasonally dry tropical forests and woodlands). A phylogenetic analysis of the trans-Atlantic Rutaceae genus *Thamnosma* (Thiv et al. 2011), which is confined to the succulent biome (i.e., seasonally dry tropical forests and woodlands), resolves monophyletic conspecific genetic samples, although conspecific sampling is limited. In the end, plant phylogenies representing dense conspecific genetic sampling are not common and should be a priority for future work (Pennington and Lavin 2016).

8.2 Explanations of Monophyly Versus Paraphyly of Conspecific Genetic Samples

In contrast to the general expectation in plants of paraphyly or non-coalescence of plant genetic samples, Pennington and Lavin (2016) suggested that the detection of coalescent conspecific plant genetic samples could be an indication of adaptation to dispersal limited ecological settings. They argued that seasonally dry tropical forests and woodlands (the succulent biome; Schrire et al. 2005) occur in a patchy manner throughout the Neotropics (Figure 8.4), and in the Paleotropics (Ringelberg et al. 2020). This biome has a harsh moisture regime with erratic intra- and inter-annual rainfall patterns (e.g., Martínez-Ramos et al. 2018), which commonly give rise to extended periods of drought. Resident plant lineages well adapted to these harsh moisture regimes are resistant to immigrants that are likely to be less well adapted to drought and therefore unlikely to colonise and establish except during infrequent wet intervals. The upshot is that well-adapted resident plant lineages can persist in small patches or nuclei of seasonally dry tropical forests and woodlands for potentially millions of years at low effective population sizes, a condition that increases the likelihood that conspecific genetic samples coalesce or are resolved as monophyletic (e.g., Särkinen et al. 2012; Pennington and Lavin 2016).

Adaptation of woody plant species to dispersal-limited seasonally dry tropical environments may only partly explain coalescence or monophyly of conspecific genetic samples. For example, geographical isolation of inter-Andean dry valleys can accentuate

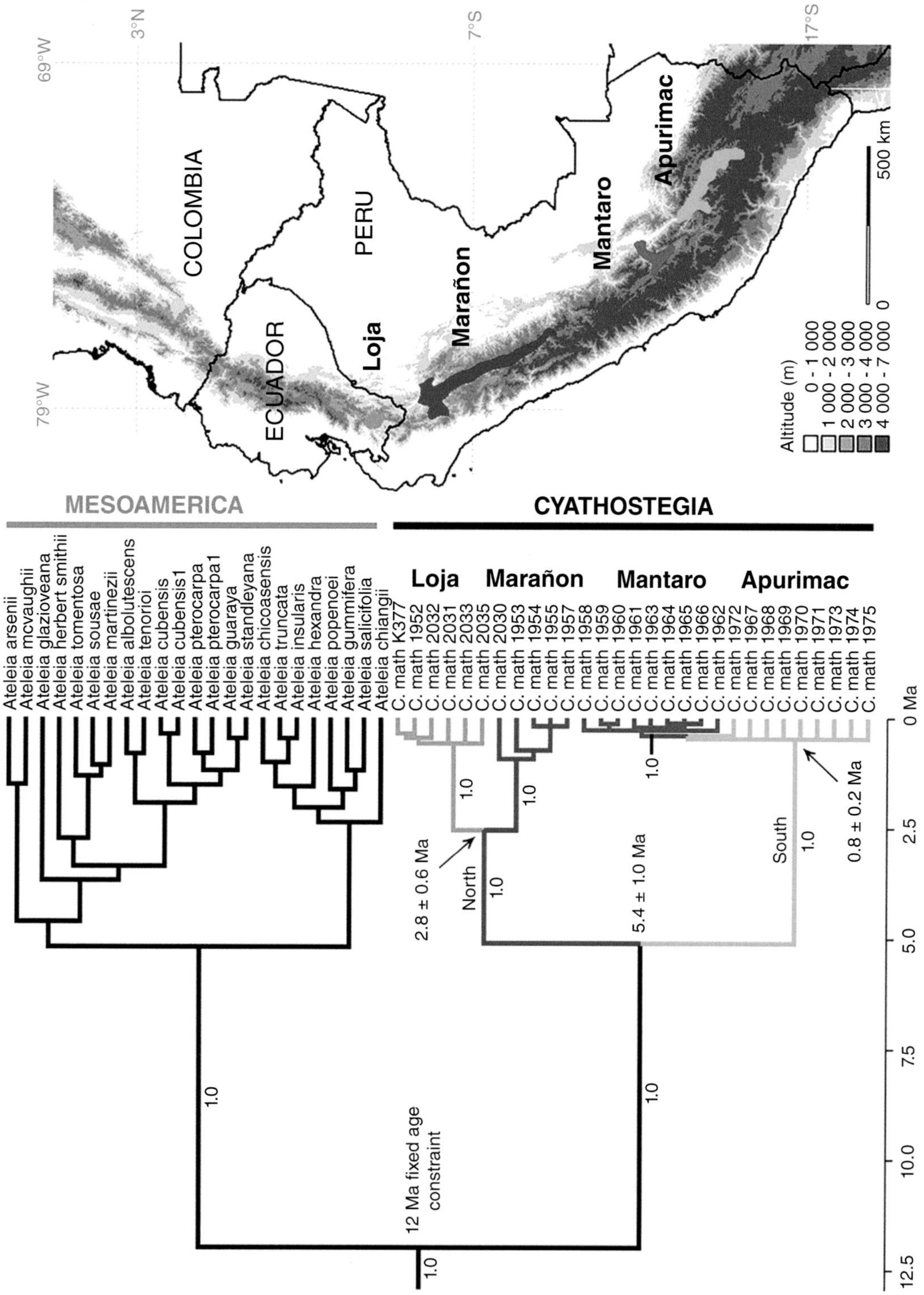

Figure 8.3 Chronogram for *Cyathostegia mathewsii* (C. math) derived from penalised likelihood rate smoothing of a 50 per cent majority rule Bayesian likelihood tree estimated from the ITS sequence data (a similar phylogeny was reproduced with an analysis of the chloroplast *matK* locus). Clade labels (Loja, Marañon, Mantaro, Apurimac) indicate the dry inter-Andean valleys where accessions were collected, the locations of which are shown on the map. Numbers (1.0) below branches indicate Bayesian posterior probabilities. Figure from Pennington et al. (2010).

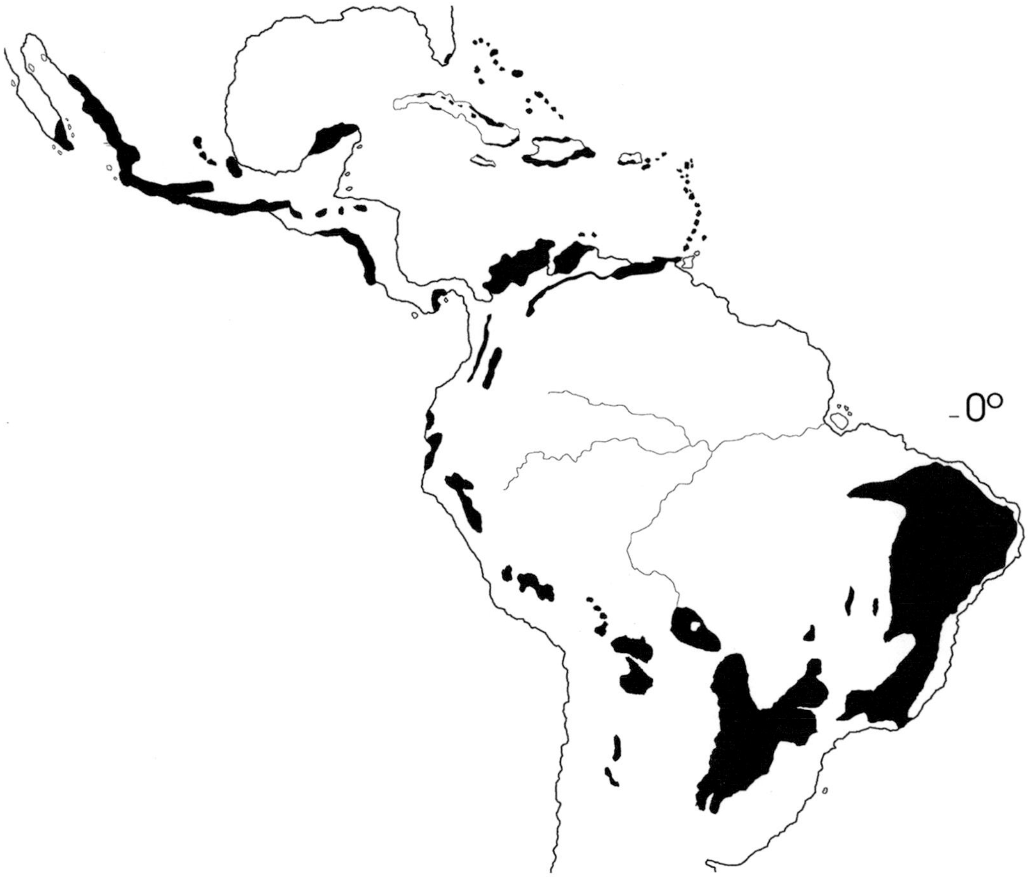

Figure 8.4 Schematic distribution of tropical dry forests in the Neotropics. Figure adapted from Pennington and Lavin (2017).

dispersal limitation already resulting from highly seasonal moisture regimes. This is the case of plant species endemic to nuclei or patches of seasonally dry tropical forests and woodlands from inter-Andean valleys (Figure 8.3; Pennington et al. 2011; Särkinen et al. 2012; Gagnon et al. 2015; Lavin et al. 2018). Plant populations isolated in such dry inter-Andean nuclei experience the ecological effect of lower effective population size, which results in the discovery of monophyletic conspecific genetic samples. This situation of plant species endemic to inter-Andean dry valleys greatly contrasts to the general finding for plants discussed by Fazekas et al. (2009) and Naciri and Linder (2015), which is the expectation that conspecific plant genetic samples are likely to be resolved as paraphyletic because they have young ages or large effective population sizes.

Naciri and Linder (2015) also point to polyploidy as increasing the effective population size of nuclear loci. Polyploid genomes therefore may explain plant clades endemic to seasonally dry tropical forests and woodlands that, in contrast to the previous examples, comprise species resolved as paraphyletic when represented by multiple conspecific genetic samples. This is likely the case for of the bombacoid tree genus *Ceiba* (Malvaceae), where recent evolution during the last few Ma within the same dry South American environments combined with palaeopolyploid genomes (2n = 86; Figueredo et al. 2016) and large effective population sizes may explain the paraphyly of conspecific genetic samples of about nine of the 16 *Ceiba* species (Pezzini 2019). In contrast to *Ceiba*, coalescent conspecific samples are ubiquitous in the legume clade Robinieae (e.g., Lavin et al. 2018), where reports of chromosome counts include only diploid numbers (Goldblatt 1981; Lavin and Sousa 1995). In addition, we infer low ploidy levels for species of Robinieae because direct sequencing of the nuclear ribosomal internal transcribed spacer region yields clean DNA sequence data (e.g., Lavin et al. 2003; Duno de Stephano 2010; Pennington et al. 2011; Quieroz and Lavin 2011; Lavin et al. 2018). According to Joly et al. (2004), a lack of nuclear ribosomal paralogs suggests low ploidy levels.

Recency of evolution by itself may explain plant clades endemic to seasonally dry tropical forests and woodlands that comprise species resolved as paraphyletic when represented by multiple conspecific genetic samples. This is likely the case for about seven of the 14 species of the legume genus *Luetzelburgia* (Leguminosae; Cardoso et al. 2014 and unpublished data), which shows evidence of having recently diversified mostly within the past one to two Ma within seasonally dry tropical forest and woodlands of South America (Cardoso et al. 2013; Trabuco da Cruz et al. 2018).

In sum, old clade ages, low ploidy levels, geographical isolation, and adaptation to dispersal-limited environments (e.g., Figures 8.2–8.4) could all work in concert to explain coalesced conspecific genetic samples of most Robinieae species. The same could be true for the *Caesalpinia* group, for which Souza et al. (2019) report polyploidy is rare. The *Caesalpinia* group includes such genera as *Arquita*, where conspecific plant genetic samples resolve as monophyletic (Gagnon et al. 2015). Here again, low ploidy levels, endemism to geographically localised inter-Andean dry valleys, and adaptation to dispersal-limited seasonally dry tropical forests and woodlands combine to explain why conspecific genetic samples of all *Arquita* species coalesce. This scenario of multiple ecological evolutionary processes working in concert to promote genetic coalescence is likely the same for species of other aforementioned legume genera. This includes *Wajira* (Thulin et al. 2004) with species endemic to small patches the succulent biome in Africa, *Mimosa* species endemic to inter-Andean dry valleys (Särkinen et al. 2011), the diploid species of *Leucaena* (Govindarajulu et al. 2011), and the ancient inter-Andean valley subclades of *Cyathostegia mathewsii* (Figure 3; Pennington et al. 2010).

8.3 Phylogenetic Niche Conservatism

Phylogenetic niche conservatism is the tendency of lineages to retain niche-related traits through speciation events (Crisp and Cook 2012). Adaptations to seasonally dry forests and woodlands appear to be phylogenetically niche conserved, as exemplified by the legume Caesalpinia group (Gagnon et al. 2019) and tribe Robinieae (Lavin 2006), where both of these studies formally tested phylogenetic niche conservatism or ecological phylogenetic structure with respect to seasonally dry forests and woodlands (the succulent biome).

Woody legumes, which dominate most types of lowland Neotropical forests and woodlands, as well as other arborescent plant groups, are phylogenetically structured along a moisture gradient. This gradient spans seasonally dry tropical forests and woodlands (i.e., the succulent biome) at the dry end, savanna woodlands at the middle of the gradient (e.g., cerrado and cerradão), and rain forests at the moist end of the gradient (Pennington et. al. 2009; Oliveira-Filho et al. 2013). Using a community phylogenetic structure analysis, Oliveira-Filho et al. (2013) found that this moisture gradient phylogenetically structures all arborescent taxa in southeastern Brazil in the caatinga region (seasonally dry tropical forests and woodlands) and the cerrado (savanna woodlands). This finding explains why the seasonally dry tropical forests and woodlands (e.g., caatinga) harbour an abundance and diversity of plant genera and families that are not abundant or diverse in the cerrado (e.g., Gagnon et al. 2019), and vice versa (e.g., Simon et al. 2009). The inference is that the ability of a species to inhabit just one end of the lowland tropical moisture gradient is a function of inheriting that ability. For reasons stated earlier, the dry end of the lowland Neotropical moisture gradient is highly dispersal limited, which favours the evolutionary persistence of localised drought-adapted residents and the concomitant higher likelihood of coalescence of conspecific genetic samples. Thus, coalescence of conspecific genetic samples can be a phylogenetically conserved trait because adaptation to seasonally dry tropical forests and woodlands is phylogenetically conserved.

Similarly, coalescence of conspecific genetic samples can be a phylogenetically conserved trait because the evolution of polyploidy often evolves in an evolutionarily conserved manner. For example, polyploidy is more common in plants than in animals (e.g., Bowers et al. 2003; Cui et al. 2006; Soltis et al. 2009; Van de Peer et al. 2009; Jiao et al. 2011; Kellogg 2016). Because of the effect of polyploidy on effective population size (e.g., Naciri & Linder 2015), this may explain, at least in part, why conspecific plant genetic samples coalesce less frequently than conspecific animal genetic samples. Within seed plants, certain plant families have higher incidences of polyploidy (e.g., the Brassicaceae and Rosaceae; Gaynor et al. 2018; reviewed in Persson et al. 2020). The legume family, Leguminosae is either paleo-polyploid or has multiple ancient polyploid events in its early lineages (Koenen et al. 2020). However, evolutionarily recent polyploidy is concentrated among certain genera (e.g., *Anarthrophyllum, Ateleia, Cologania, Dichrostachys, Dipteryx, Erythrina, Glycine*, and *Leucaena*), and especially among genera belonging to northern temperate Papilionoideae clades (e.g., *Astragalus, Lotus, Lupinus, Oxytropis*, and *Trifolium*; Goldblatt 1981). If recent polyploidy is more common among temperate legume clades, then coalescence of conspecific genetic samples may be less likely in temperate legume species and more common among species from tropical legume clades, especially those from seasonally dry tropical forests, as discussed earlier.

Conclusion

Conspecific plant genetic samples are expected to be often resolved as paraphyletic in phylogenies involving dense congeneric species-level sampling for reasons including recent evolution of species with large effective population sizes, introgressive hybridisation and polyploidy, and plastid loci not tracking species boundaries. We suggest that this will be the case even when large numbers of loci are available from next-generation sequencing studies, though we urge future workers to test this by dense species coverage, and sampling multiple individuals per species in their phylogenies, which is seldom accomplished. When genetic evidence reveals plant genera or other groups of plant species represented by coalescing conspecific genetic samples, which in some cases reveals cryptic and overlooked plant species, the ecological evolutionary implications explaining these patterns are as important as the issue of underestimated levels of plant species diversity.

Acknowledgements

We thank P. Muñoz-Rodríguez for constructive comments that improved the presentation of the ideas in this paper.

References

Bowers, J. E., Chapman, B. A., Rong, J., and Paterson, A. H. (2003) Unravelling angiosperm genome evolution by phylogenetic analysis of chromosomal duplication events. *Nature* 422: 433–438. doi:10.1038/nature01521

Cardoso, D., Queiroz, L. P., de Lima, H. C., Suganuma, E., van den Berg, C., and Lavin, M. (2013) A molecular phylogeny of the Vataireoid legumes underscores floral evolvability that is general to many early-branching papilionoid lineages. *American Journal of Botany* 100: 403–421. doi:10.3732/ajb.1200276.

Cardoso, D., Queiroz, L. P., and Lima, H. (2014) A taxonomic revision of the South American papilionoid genus *Luetzelburgia* (Fabaceae). *Botanical Journal of the Linnean Society* 175: 328–375. doi:10.1111/boj.12153.

Cornetti, L., Ficetola, G., Hoban, S., and Vernesi, C. (2015) Genetic and ecological data reveal species boundaries between viviparous and oviparous lizard lineages. *Heredity* 115: 517–526. doi:10.1038/hdy.2015.54.

Crisp, M. D. and Chandler, G. T. (1996) Paraphyletic species. *Telopea* 6(4): 813–844. doi:10.7751/telopea19963037

Crisp, M. D. and Cook, L. G. (2012) Phylogenetic niche conservatism: what are the underlying evolutionary and ecological causes? *New Phytologist* 2012(196): 681–694. doi:10.1111/j.1469-8137.2012.04298.x

Cui, L., Wall, P. K., Leebens-Mack, J., Lindsay, B. G., Soltis, D. E., Doyle, J. J. et al. (2006) Widespread genome duplications throughout the history of flowering plants. *Genome Research* 16: 738–749. doi:10.1101/gr.4825606

Delgado-Salinas, A., Thulin, M., Pasquet, R., Weeden, N., and Lavin, M. (2011) *Vigna* (Leguminosae) sensu lato: the names and identities of the American segregate genera. *American Journal of Botany* 98: 1694–1715. doi:10.3732/ajb.1100069.

de Queiroz, K. (2007) Species concepts and species delimitation. *Systematic Biology* 56 (6): 879–886. doi:10.1080/10635150701701083

Dexter, K. G., Lavin, M., Torke, B., Twyford, A., Kursar, T., Coley, P., Drake, C., Hollands, R., and Pennington, R.T. (2017) Dispersal assembly of rain forest tree communities across the Amazon basin. *Proceedings of the National Academy of Science* 114: 2645–2650. doi:10.1073/pnas.1613655114.

Duno de Stefano, R., Carnevali Fernández-Concha, G., Lorena Can-Itza, L., and Lavin, M. (2010) The morphological and phylogenetic distinctions of *Coursetia greenmanii* (Leguminosae): taxonomic and ecological implications. *Systematic Botany* 35: 289–295. doi:10.1600/036364410791638360

Duputie A., Salick, J., and McKey, D. (2011) Evolutionary biogeography of *Manihot* (Euphorbiaceae), a rapidly radiating Neotropical genus restricted to dry environments. *Journal of Biogeography* 38: 1033–1043. doi:10.1111/j.1365-2699.2011.02474.x.

Fazekas, A. J., Kesankurti, P. R., Burgess, K. S., Percy, D. M., Graham, S. W., Barrett, S. C., Newmaster, S. G., Hajibabei, M., and Husband, B. C. (2009) Are plant species inherently harder to discriminate than animal species using DNA barcoding markers? *Molecular Ecology Resources* 9: 130–139. doi:10.1111/j.1755-0998.2009.02652.x.

Figueredo, A., de Oliveira, A. W., Carvalho-Sobrinho, J. G., and Souza, G. (2016) Karyotypic stability in the paleopolyploid genus *Ceiba* Mill. (Bombacoideae, Malvaceae). *Brazilian Journal of Botany* 39: 1087–1093. doi:10.1007/s40415-016-0296-5.

Fišer, C., Robinson, C. T., and Malard, F. (2018) Cryptic species as a window into the paradigm shift of the species concept. *Molecular Ecology* 27: 613–635. doi:10.1111/mec.14486.

Freudenstein J. V., Broe, M. B., Folk, R. A., and Sinn, B. T. (2017) Biodiversity and the species concept: Lineages are not enough. *Systematic Biology* 66: 644–656. doi:10.1093/sysbio/syw098.

Gagnon, E., Hughes, C. E., Lewis, G. P., and Bruneau, A. (2015) A new cryptic species in a new cryptic genus in the *Caesalpinia* group (Leguminosae) from the seasonally dry inter-Andean valleys of South America. *Taxon* 64: 468–490. doi:10.12705/643.6.

Gagnon, E., Ringelberg, J. J., Bruneau, A., Lewis, G. P., and Hughes, C. E. (2019) Global Succulent Biome phylogenetic conservatism across the pantropical Caesalpinia Group (Leguminosae). *New Phytologist* 222: 1994–2008. doi:10.1111/nph.15633.

Gallardo, M. H. (2017) Phylogenetics, Reticulation and Evolution, chapter 3, In Phylogenetics, ed. Abdurakhmonov IY. IntechOpen doi:10.5772/intechopen.68564.

Gaynor, M. L., Ng, J., and Laport, R. G. (2018) Phylogenetic structure of plant communities: Are polyploids distantly related to co-occurring diploids? *Frontiers in Ecology and Evolution* 6: 52. doi:10.3389/fevo.2018.00052.

Gilbert, N. (2010) African elephants are two distinct species. *Nature* doi:10.1038/news.2010.691.

Goldblatt, P. (1981) Cytology and the phylogeny of Leguminosae, In *Advances in Legume Systematics*, Part 2. Ed. Polhill, R. M. and Raven, P. M. pp. 427–463. Royal Botanic Gardens, Kew.

Govindarajulu, R., Hughes, C. E., and Bailey, C. D. (2011) Phylogenetic and population genetic analyses of diploid *Leucaena* (Leguminosae–Mimosoideae) reveal cryptic species diversity and patterns of allopatric divergent speciation. *American Journal of Botany* 98: 2049–2063. doi:10.3732/ajb.1100259.

Groves, C. (2016) Two African elephant species, not just one. *Nature* 538: 317. doi:10.1038/538317a.

Heethoff, M. (2018) Cryptic species – conceptual or terminological chaos? A response to Struck et al. *Trends in Ecology & Evolution* 33: 310. doi:10.1016/j.tree.2018.02.006.

Hollingsworth P. M., Graham, S. W., and Little, D. P. (2011) Choosing and using a plant DNA barcode. *PLoS ONE* 6(5): e19254. doi:10.1371/journal.pone.0019254.

Hollingsworth, P. M., Li, D. Z., van der Bank, M., and Twyford, A. D. (2016) Telling plant species apart with DNA: from barcodes to genomes. *Philosophical Transactions of the Royal Society, London B* 371. doi:10.1098/rstb.2015.0338.

Jiao, Y., Wickett, N. J., Ayyampalayam, S., Chanderbali, A. S., Landherr, L., Ralph, P. E. et al. (2011) Ancestral polyploidy in seed plants and angiosperms. *Nature* 473: 97–100. doi:10.1038/nature09916

Joly, S., Rauscher, J. T., Sherman-Broyles, S. L., Brown, A. H. D., and Doyle, J. J. (2004) Evolutionary dynamics and preferential expression of homeologous 18S-5.8S-26S nuclear ribosomal genes in natural and artificial *Glycine* allopolyploids. *Molecular Biology and Evolution* 21: 1409–1421. doi:10.1093/molbev/msh140.

Jones, R. C., Nicolle, D., Steane, D. A., Vaillancourt, R. E., and Potts, B. M. (2016). High density, genome-wide markers and intra-specific replication yield an unprecedented phylogenetic reconstruction of a globally significant, speciose lineage of Eucalyptus. *Molecular Phylogenetics and Evolution* 105: 63–85. doi:10.1016/j.ympev.2016.08.009.

Kellogg, E. A. (2016) Has the connection between polyploidy and diversification actually been tested? *Current Opinion in Plant Biology* 30: 25–32. doi:10.1016/j.pbi.2016.01.002.

Koenen, E. J. M., Ojeda, D. I., Bakker, F. T., Wieringa, J. J., Kidner, C., Hardy, O. J., Pennington, R. T., Herendeen, P. S., Bruneau, A., and Hughes, C. E. (2020) The Origin of the Legumes is a Complex Paleopolyploid Phylogenomic Tangle Closely Associated with the Cretaceous–Paleogene (K–Pg) Mass Extinction Event, Systematic Biology syaa041. doi:10.1093/sysbio/syaa041.

Latta, R. G. and Mitton, J. B. (1999) Historical separation and present gene flow through a zone of secondary contact in Ponderosa pine. *Evolution*, 53: 769–776. doi:10.1111/j.1558-5646.1999.tb05371.x

Lavin, M. (1988) Systematics of *Coursetia* (Leguminosae-Papilionoideae). *Systematic Botany Monographs* 21: 1–167. doi:10.2307/25027701.

(1993) Systematics of the genus *Poitea* (Leguminosae): inferences from morphological and molecular data. *Systematic Botany Monographs* 37: 1–87. doi:10.2307/25027818.

(2006) Floristic and geographic stability of discontinuous seasonally dry tropical forests explains patterns of plant phylogeny and endemism, Chapter 19, In *Neotropical Savannas and Seasonally Dry Forests: Plant Biodiversity, Biogeographic Patterns and Conservation*. Ed. Pennington, R. T., Ratter, J. A., and Lewis, G. P. pp. 433–447. CRC Press, Boca Raton, FL. doi:10.1201/9781420004496.

Lavin, M. and Sousa-S., M. (1995) Phylogenetic systematics and biogeography of the tribe Robinieae. *Systematic Botany Monographs* 45: 1–165. doi:10.2307/25027850.

Lavin, M., Wojciechowski, M. F., Gasson, P., Hughes, C. H., and Wheeler, E. (2003) Phylogeny of robinioid legumes (Fabaceae) revisited: *Coursetia* and *Gliricidia* recircumscribed, and a biogeographical appraisal of the Caribbean endemics. *Systematic Botany* 28: 387–409. www.jstor.org/stable/3094008

Lavin, M., Pennington, R. T., Hughes, C. E., Lewis, G. P., Delgado Salinas, A., Duno de Stefano, R., Queiroz, L. P, Cardoso, D., and Wojciechowski, M. F. (2018) DNA sequence variation among conspecific accessions of the legume *Coursetia caribaea* reveal geographically localized clades here ranked as species. *Systematic Botany* 43: 664–675. doi:10.1600/036364418X697382.

Martínez-Ramos, M., Balvanera, P., Arreola Villa, F., Mora, F., Manuel Maass, J., and Maza-Villalobos Méndez, S. (2018) Effects of long-term inter-annual rainfall variation on the dynamics of regenerative communities during the old-field succession of a neotropical dry forest. *Forest Ecology and Management* 426: 91–100. doi:10.1016/j.foreco.2018.04.048.

Muñoz-Rodríguez, P., Carruthers, T., Wood, J. R. I., Williams, B. R. M., Weitemier, K., Kronmiller, B. Goodwin, Z., Sumadijaya, A., Anglin, N. L., Filer, D., Harris, D., Rausher, M. D., Kelly, S., Liston, A., and Scotland, R.W.

(2019) A taxonomic monograph of *Ipomoea* integrated across phylogenetic scales. *Nature Plants* 5: 1136–1144. doi:10.1038/s41477-019-0535-4.

Naciri Y. and Linder, H. P. (2015) Species delimitation and relationships: the dance of the seven veils. *Taxon* 64: 3–16. doi:10.12705/641.24.

Oliveira-Filho, A. T., Cardoso, D., Schrire, B. D., Lewis, G. P., Pennington, R. T., Brummer, T. J., Rotella, J., and Lavin, M. (2013) Stability structures tropical woody plant diversity more than seasonality: insights into the ecology of high legume-succulent-plant biodiversity. *South African Journal of Botany* 89: 42–57. doi:10.1016/j.sajb.2013.06.010.

Pennington, R. T. and Lavin, M. (2017) Dispersal, isolation and diversification with continued gene flow in an Andean tropical dry forest. *Molecular Ecology* 26: 3327–3329. doi:10.1111/mec.14182.

Pennington R. T., Lavin, M., and Oliveira-Filho, A. (2009) Woody plant diversity, evolution and ecology in the tropics: perspectives from seasonally dry tropical forests. *Annual Review of Ecology, Evolution, and Systematics* 40: 437–457. doi:10.1146/annurev.ecolsys.110308.120327.

Pennington, R. T., Lavin, M., Särkinen, T., Lewis, G. P., Klitgaard, B. B., and Hughes, C. E. (2010) Contrasting plant diversification histories within the Andean biodiversity hotspot. *Proceedings of the National Academy of Sciences, USA* 107 (31): 13783–13787. doi:10.1073/pnas.1001317107.

Pennington, R. T., Daza, A., Reynel, C., and Lavin, M. (2011) *Poissonia eriantha* (Leguminosae) from Cuzco, Peru: an overlooked species underscores a pattern of narrow endemism common to seasonally dry neotropical vegetation. *Systematic Botany* 36: 59–68. doi:10.1600/036364411X553135.

Pennington R. T. and Lavin, M. (2016) The contrasting nature of woody plant species in different neotropical forest biomes reflects differences in ecological stability. *The New Phytologist* 210: 25–37. doi:10.1111/nph.13724.

Persson, N. L., Eriksson, T., and Smedmark, J. E. E. (2020) Complex patterns of reticulate evolution in opportunistic weeds (*Potentilla* L., Rosaceae), as revealed by low-copy nuclear markers. *BMC Evol Biol* 20, 38. doi:10.1186/s12862-020-1597-7.

Pezzini, F. F. (2018) Phylogeny, taxonomy and biogeography of Ceiba Mill. (Malvaceae: Bombacoideae). PhD Thesis. The University of Edinburgh. 201 pp. https://hdl.handle.net/1842/36677.

Queiroz, L. P. and Lavin, M. (2011) *Coursetia* (Leguminosae) from eastern Brazil: nuclear ribosomal and chloroplast DNA sequence analysis reveal the monophyly of three caatinga-inhabiting species. *Systematic Botany* 36: 69–79. doi:10.1600/036364411X553144.

Rieseberg, L. H. and Brouillet, L., (1994) Are many plant species paraphyletic? *Taxon* 43: 21–32. doi:10.2307/1223457

Ringelberg, J. J., Zimmermann, N. E., Weeks, A., Lavin, M., and Hughes, C. E. (2020) Biomes as evolutionary arenas: convergence and conservatism in the trans-continental succulent biome. *Global Ecology and Biogeography* 29(7): 1100–1113. doi:10.1111/geb.13089.

Rohland N., Reich, D., Mallick, S., Meyer, M., Green, R. E., Georgiadis, N. J., Roca, A. L., and Hofreiter, M. (2010) Genomic DNA Sequences from Mastodon and Woolly Mammoth Reveal Deep Speciation of Forest and Savanna Elephants. *PLOS Biology* 8(12): e1000564. doi:10.1371/journal.pbio.1000564.

Särkinen, T. S., Marcelo Peña, J. L., Yomona, A. D., Simon, M. F., Pennington, R. T., and Hughes, C. E. (2011) Underestimated endemic species diversity in the Marañon seasonally dry tropical forests of Peru: An example from *Mimosa* (Leguminosae: Mimosoideae). *Taxon* 60: 139–150. doi:10.1002/tax.601012.

Särkinen, T., Pennington, R. T., Lavin, M., Simon, M. F., and Hughes, C. E. (2012) Evolutionary islands in the Andes:

persistence and isolation explains high endemism in Andean dry tropical forests. *Journal of Biogeography* 39: 884–900. doi:10.1111/j.1365-2699.2011.02644.x.

Schrire, B. D., Lavin, M. , and Lewis, G. P. (2005) Global distribution patterns of the Leguminosae: insights from recent phylogenies. In I. Friis & H. Balslev (eds.), Plant diversity and complexity patterns: local, regional and global dimensions. *Biologiske Skrifter* 55: 375–422.

Simon, M. F., Grether, R. Queiroz, L. P., Skema, C., Pennington, R. T., and Hughes, C. E. (2009) Recent assembly of the Cerrado, a neotropical plant diversity hotspot, by in situ evolution of adaptations to fire. *Proceedings of the National Academy of Sciences* 106: 20359–20364. doi:10.1073/pnas.0903410106.

Soltis, D. E., Albert, V. A., Leebens-Mack, J., Bell, C. D., Paterson, A. H., Zheng, C. et al. (2009) Polyploidy and angiosperm diversification. *Am. J. Bot.* 96: 336- 348. doi:/10.3732/ ajb.0800079

Souza, G., Costa, L., Guignard, M. S., Van-Lume, B., Pellicer, J., Gagnon, E., Leitch, I. J. , and Lewis, G. P. (2019) Do tropical plants have smaller genomes? Correlation between genome size and climatic variables in the Caesalpinia Group (Caesalpinioideae, Leguminosae). *Perspectives in Plant Ecology, Evolution and Systematics* 38: 13–23. doi:10.1016/j.ppees.2019.03.002.

Struck. T. H., Feder, J. L., Bendiksby, M., Birkeland, S., Cerca, J., Gusarov, V. I., Kistenich, S., Larsson, K. H., Liow, L. H., Nowak, M. D., Stedje, B., Bachmann, L., and Dimitrov. D. (2017) Finding evolutionary processes hidden in cryptic species. *Trends in Ecology & Evolution* 33: 153–163. doi:10.1016/j.tree.2017.11.007.

Struck. T. H., Feder, J. L., Bendiksby, M., Birkeland, S., Cerca, J., Gusarov, V. I., Kistenich, S., Larsson, K. H., Liow, L. H., Nowak, M. D., Stedje, B., Bachmann, (2018) Cryptic species – more than terminological chaos: A reply to Heethoff. *Trends in Ecology & Evolution* 33: 310–312. doi:10.1016/j. tree.2018.02.008.

Struck T. H. and Cerca, J. (2019) Cryptic species and their evolutionary significance. *eLS* (2019), pp. 1–9. doi:10.1002/9780470015902. a0028292.

ter Steege, H., Pitman, N. C. A., Sabatier, D., Baraloto, C., Salomão, R. P., Guevara, J. E., Phillips, O. L., Castilho, C. V., Magnusson, W. E., Molino, J. F. et al. (2013) Hyperdominance in the Amazonian Tree Flora. *Science* 342: 1243092. doi:10.1126/ science.1243092.

Thiv M., van der Niet, T., Rutschmann, F., Thulin, M., Brune, T., and Linder, H. P. (2011) Old—New World and trans-African disjunctions of *Thamnosma* (Rutaceae): intercontinental long-distance dispersal and local differentiation in the succulent biome. *American Journal of Botany* 98: 76–87. doi:10.3732/ajb.1000339.

Thulin, M., Lavin, M., Pasquet, R., and Delgado-Salinas, A. (2004) Phylogeny and biogeography of *Wajira* (Leguminosae): A monophyletic segregate of *Vigna* centered in the Horn of Africa region. *Systematic Botany* 29: 903–920. doi:10.1600/0363644042451035

Trabuco da Cruz, D., Idárraga, Á., Banda, K., van den Berg, C., Queiroz, L. P., Pennington, R. T., Lavin, M., and Cardoso, D. (2018) Ancient speciation of the papilionoid legume *Luetzelburgia jacana*, a newly discovered species in an inter-Andean seasonally dry valley of Colombia. *Taxon* 67: 931–943. doi:10.12705/675.6.

Van de Peer, Y., Fawcett, J. A., Proost, S., Sterck, L., and Vandepoele, K. (2009) The flowering world: a tale of duplications. *Trends Plant Sci.* 14: 680–688. doi:10.1016/j. tplants.2009.09.001.

Vargas O. M., Ortiz E. M., and Simpson B. B. (2017) Conflicting phylogenomic signals reveal a pattern of reticulate evolution in a recent high-Andean diversification (Asteraceae: Astereae: *Diplostephium*). *New Phytologist* 214: 1736–1750. doi:10.1111/ nph.14530.

Willyard, A., Gernandt, D. S., Potter, K., Hipkins, V., Marquardt, P., Mahalovich, M. F., Langer, S. K., Telewski, F. W., Cooper, B.,

Douglas, C., Finch, K., Karemera, H. H., Lefler, J., Lea, P., and Wofford, A. (2017) *Pinus ponderosa*: A checkered past obscured four species. *American Journal of Botany* 104: 161–181. doi:10.3732/ajb.1600336.

Wood, J. R. I., Muñoz-Rodríguez, P., Williams, B. R. M. , and Scotland, R. W. (2020) A foundation monograph of *Ipomoea* (Convolvulaceae) in the New World. *Phytokeys* 143: 1–843. doi:10.3897/phytokeys.143.32821.

9

Guerrilla Taxonomy and Discriminating Cryptic Species

Is Quick Also Dirty?

Paul H. Williams

Introduction

Can taxonomic revisions of even small groups of species be reliable from short projects? Increasingly, such projects are being proposed for MSc and PhD training for people previously unfamiliar with a group, presumably because supervisors reason that large amounts of data can now be obtained quickly from gene sequences. But while these data may represent a few individuals in some depth, is the representation of broad geographical variation from across all global populations now a greater challenge? Fundamental questions when designing revisionary studies are how much sampling effort is required to achieve reliable results and which factors affect the magnitude of this challenge? (Zhang et al. 2010; Luo et al. 2015; Phillips et al. 2018) Are the need for broad geographical coverage and the rarity (low encounter rates) of some taxa among the more serious constraints that work against achieving reliable results quickly, especially when dealing with poorly known cryptic species within short projects?

As far as possible, best practice in a taxonomic revision should include: (1) an explicit a priori statement of the theoretical species concept and appropriate practical criteria that need to be met for recognising species; (2) representation of all species worldwide from the entire monophyletic group; (3) representation of variation from across the entire geographical ranges of all of the included species, including potential cryptic species, covering all constituent taxa and representing all apparent gradients and clines; (4) iterative tree estimation (when trees are needed) as samples become available until convergence on robust trees and species assessments is attained; and (5) examination and comparison of type specimens for all of the included named taxa in order to apply names (Bolton 2007;

Table 9.1 GenBank COI sequences from Lecocq et al. (2015: their Supplementary Table S1) downloaded on 20 January 2018 and identified by searching the BOLD database (boldsystems.org/index.php/IDS_OpenIdEngine) and by including the sequences in a PTP analysis of the subgenus Melanobombus in Williams et al. 2020. Abbreviations for subgenera: Ml, *Melanobombus*; Pr, *Pyrobombus*; Bo, *Bombus* s. str.; Th, *Thoracobombus*

Given taxon name	GenBank numbers	BOLD identification[1]	PTP identification
Ml. alagesianus	KC915645	*Pr. brodmannicus*	
	KM458064	*Pr. brodmannicus*	
Ml. caucasicus[2]	KC915729	*Bo. lucorum*	
	KC915730	*Bo. patagiatus*	
	KC915731	*Bo. lucorum*	
	KC915732	*Bo. lucorum*	
Ml. erzurumensis	KC915646	*Ml. sichelii*	*Ml. incertus*[3]
	KC915647	*Ml. sichelii*	*Ml. incertus*[3]
	KM458065	*Ml. sichelii*	*Ml. incertus*[3]
	KM458066	*Ml. sichelii*	*Ml. incertus*[3]
Ml. incertus	KC915649	*Th. pascuorum*	

[1] BOLD identifications became complicated during 2017 because these GenBank sequences were uploaded into BOLD with names from Lecocq et al. 2013. Sequence identities have been checked using BOLD identification trees for all of the > 8,200 bumblebee sequences in BOLD in January 2018. The sequence names in BOLD are expected to be corrected soon.
[2] GenBank BLAST of a sequence emailed by T. Lecocq for the *caucasicus* specimen LaC001 also returned highest matches on 18 May 2018 for *B. lucorum* – the top match was for sequence JQ843494.
[3] Identification of reference sequence material in Williams et al. 2020 is based on original descriptions.

Williams et al. 2020). The process should be transparent and reproducible rather than based on unaccountable authority. Some recent high profile studies (of bumblebees) that seek to revise the status of species have not followed all of these principles (Koch et al. 2018), and in some cases this has resulted in taxon misidentification (Lecocq et al. 2015: see Table 9.1). At least for plants, a decline in the frequency with which taxonomic revisions cover the entire geographical distribution of taxa is suggested to be one major factor in current high rates of species misidentification (Goodwin et al. 2015).

DNA 'barcodes' have become popular as a data source for many more 'characters' to help in both species identification and (crucially here) in species discovery (Hebert et al. 2003). For insects, the standard DNA barcode is a sequence fragment (usually 657 base pairs) from the cytochrome c oxidase subunit 1 (COI or cox-1 gene, Hebert et al. 2004). This approach has inevitably raised the question of the extent of sampling required, particularly regarding the numbers of specimen sequences needed to recognise species (Matz and Nielsen 2005; Zhang et al. 2010; Bergsten et al. 2012; Lou and Golding 2012; Luo et al. 2015; Phillips et al. 2018). Simulation studies have considered sampling at random from individuals across the population, which should minimise the chances of bias. Unsurprisingly, inferences from these studies have highlighted the desirability of large samples per species and the need to represent the entire geographical range of each species (Zhang et al. 2010; Bergsten et al. 2012; Lou and Golding 2012; Luo et al. 2015; Phillips et al. 2018).

Coalescents in fast-evolving genes such as COI can provide one of the most direct sources of the evidence necessary for recognising the status of populations as evolutionarily independent lineages (EILs, De Queiroz 2007) and hence as separate species (Monaghan et al. 2005; Monaghan et al. 2009; Zhang et al. 2013). These coalescents represent the largest monophyletic groups, rather than merely an arbitrary threshold for degree of divergence or for degree of interbreeding (identifying lineages with shared genetic heritage, EILs), that crucially also correspond to a change in the pattern of relationships from rare speciation events to close intraspecific divergences. Coalescent species have also been found to show a 77 per cent correspondence to interbreeding species for *Drosophila* fruit flies, one of the groups studied most intensively in speciation studies (Campillo et al. 2020). COI barcodes match well with the assumptions required for identifying gene coalescents for species (Williams et al. 2020).

Bumblebees (genus *Bombus* Latreille) have long been claimed as a model group of organisms for the study of what species are in theory and how to recognise them in practice (Vogt 1909; Vogt 1911; Reinig 1939). There have been substantial changes in the preferred approach over the last two centuries (reviewed by Williams 1998; Williams et al. 2020). In 2008 the BEEBOL initiative began as a programme to barcode the bees of the world (Packer et al. 2009). COI barcoding has been successful in identifying bee species (Schmidt et al. 2015; Almeida et al. 2018). This rapidly evolving DNA fragment has also proved especially well-suited for use in assessing closely related and cryptic bumblebee species within revisionary taxonomy (Williams et al. 2016).

In 2011, two NHM projects began to revise worldwide the bumblebees of the contrasting subgenera *Alpinobombus* and *Melanobombus*, using integrative assessments that focus on evidence from morphology and from coalescents in the COI gene. These subgenera are each monophyletic based on evidence from five other genes (Cameron et al. 2007). The two groups are very different in that *Alpinobombus* species are associated primarily with arctic and subarctic environments, where many have large range sizes (Williams et al. 2019). In contrast, most *Melanobombus* species are associated primarily with high mountains in central and southern Eurasia, where many of these species have smaller range sizes (Williams et al. 2020). The difference in habitat is significant because, in consequence, the sampling programmes for the two studies are completely independent. Over the last century and including the last 20 years, much taxonomic uncertainty has remained in both groups, with *Alpinobombus* interpreted as consisting of 5–9 species and *Melanobombus* credited with 14–23 species (Williams 1998; Lecocq et al. 2013; Williams et al. 2015b; Martinet et al. 2018; Potapov et al. 2019), mostly by re-arrangements of previously known taxa. Both groups were known to include much variation of uncertain status, as indicated by the many published formal names (67 names in *Alpinobombus* and 186 names in *Melanobombus*). Nonetheless, both NHM projects have discovered new cryptic species that were unrecognised previously even as separate taxa at any rank, although all of these species were represented in collections for at least 50 years.

This chapter is a retrospective assessment that describes progress in attempts to revise the subgenera *Alpinobombus* and *Melanobombus* as re-sampling programmes grew over a period of approximately a decade (with analyses in the period 2011–2019). In particular, it examines the effects on the numbers of species recognised of: (1) under-sampling of some species; (2) ambiguous (missing) data; and (3) the over-sampling of some species (defined in Section 9.1). It considers the consequences of these challenges for recognising the most difficult species, the cryptic species.

9.1 Methods

9.1.1 Recognising Species

Species are considered here from recent revisions that cover all of the known species of the monophyletic subgenera *Alpinobombus* and *Melanobombus* (Williams et al. 2019, 2020). These revisions discuss the methods in more detail, but they both use integrative assessments to compare support from evidence of gene coalescents and from morphology. All samples and COI-barcode sequence data re-analysed here are taken from these earlier studies that compiled samples over the period 2008–2019. Species are recognised in an iterative three-step process.

9.1.2 Step 1: Representing Variation from Sampling

In the first step, the aim is to represent the principal variation across the entire global distributions to include all of the known taxa within the group and any obviously divergent individuals. Consequently, this is dependent on earlier published revisions (most based on morphology, but some based on genetic data) and on the museum collections of voucher specimens that were identified by their authors. There was not a substantial budget for a large systematic global re-sampling programme, so new samples had to be acquired as they became available through a global coalition of collaborators, to augment older collections within the constraints permitted by national policies. Nonetheless, sampling effort was deployed where possible in a directed and stratified way (in the sense of Neyman 1934; Guisan et al. 2006) to maximise the representation (1) of all species and then (2) of variation within each species across their entire distribution ranges. In practice this meant focussing increasingly on sampling geographical regions with concentrations of under-represented taxa based on the results of successive analyses.

The total number of *Alpinobombus* specimens examined for morphology was 4,622. It was not possible to publish a list of all the *Melanobombus* specimens because access to many of the data has become limited by national policies since the start of the project. Nonetheless, during the project substantially more than 14,000 *Melanobombus* preserved specimens (plus many thousands more field observations) were examined from the mountains of Central Asia, the Himalaya, and China alone (Williams 1991, 2011; Williams et al.

2009, 2010, 2015a; An et al. 2014), many through a ten-year project to enumerate the bumblebee species of China (Williams et al. 2017). Other large collections including from Mongolia, Iran, Russia, Europe, and North Africa were also examined.

9.1.3 Step 2: Identifying Candidate Taxa from Fast-Gene Coalescents

In the second step, 'candidate' species are identified (Vieites et al. 2009) from evidence of species coalescents in the fast-evolving mitochondrial COI gene from a representative subsample. The most important assumptions required by coalescent methods (Papadopoulou et al. 2008, 2009; Lohse 2009; Fujisawa and Barraclough 2013) are matched well by bumblebees, which should place bumblebees among the more suitable subjects for their application: (1) most bumblebees have a short (annual) generation time (Goulson 2010), which is short relative to the ages of most of their species (estimates in Hines 2008); (2) effective genetic population sizes may be low despite the abundance of foraging individuals that are sampled, because bumblebees are social and it is the colonies that are the genetic individuals (Zayed and Packer 2005), so these genetic individuals are rarer by 1–3 orders of magnitude than the foragers; and yet (3) bumblebee foragers tend to be relatively well sampled across their geographical ranges because bumblebees are large and brightly coloured, which attracts biologists (Williams 1998).

Barcode sequences are obtained using 'Lep' primers (Hebert et al. 2004) and standard protocols (Hebert et al. 2003) in the labs of: (1) the Canadian Centre for DNA Barcoding (CCDB) in Guelph (sequences now in the BOLD database, boldsystems.org); (2) the Natural History Museum in London (NHMUK); and (3) the Chinese Academy of Science's Institute of Botany in Kunming (CAS-IoB, IBK). To further reduce costs, for some European taxa of *Melanobombus* a subset of published sequences (Lecocq et al. 2015) is selected for sites dispersed widely across Europe and downloaded from GenBank for re-analysis (Williams et al. 2015a, 2020, excluding sequences in Table 9.1).

Evidence for species coalescents in the COI gene is obtained here using the Poisson-tree-process (PTP) technique as implemented in the online bPTP server (https://species.h-its.org/ accessed 2018; Zhang et al. 2013). The PTP method models branching events by relating them as rare events (hence Poisson-distributed) to the numbers of DNA-nucleotide substitutions, using the information on substitutions from the relative branch lengths in a metric gene tree. PTP models separately: (1) the infrequent among-species branching events; and (2) the more frequent within-species branching events. It then uses maximum likelihood to seek the best fit of these models and hence the transition points between them that are expected to indicate the speciation events. The online implementation of PTP includes, as an alternative, the results for solutions that use Bayesian support values to fit these models (which has the advantage of giving interval estimates of probability).

The newer multi-rate PTP (mPTP) method for species coalescents fits different models for branching rates within each species and is claimed to perform better at recognising candidate species that are more similar to those recognised from morphological taxonomy (Kapli et al. 2017). However, this version of PTP is not used here. Multi-rate PTP fails to discriminate between the *Alpinobombus* species *B. hyperboreus* Schönherr and *B. kluanensis* Williams & Cannings (Potapov et al. 2019; confirmed here using data from Williams et al.

2019), species that have been found to differ strongly in: (1) morphology, from the presence or absence of a male mandibular beard, and (2) behaviour, from the presence (or absence) of workers as females initiate their own colonies (rather than parasitising colonies initiated by other species). Crucially, these three species are all distinguished by the original PTP from their COI coalescents alone (Williams et al. 2015b, 2019). Similarly, for the subgenus *Melanobombus*, applying mPTP to the UAF data (a filter employed for reducing computation time by Kapli et al. 2017) also fails to distinguish *B. eximius* Smith from the morphologically distinct *B. rufipes* Lepeletier, or *B. pyrosoma* Morawitz from *B. formosellus* (Frison) (Williams et al. 2020). These *Melanobombus* species are also distinguished successfully from their COI coalescents by the original PTP. Consequently, when applied to bumblebee data, mPTP analysis appears to give inordinately conservative results, whereas the original PTP analysis yields coalescent candidate species that are consistent with the species also supported in integrative analyses by morphology and behaviour.

Metric gene trees are estimated here using the Bayesian procedure (Baum and Smith 2012) MRBAYES (version 3.1.2, Ronquist and Huelsenbeck 2003). The outgroup for rooting the *Alpinobombus* analyses is *B. ignitus* Smith and the outgroup for the *Melanobombus* analyses is *B. nobilis* Friese. The best available nucleotide-substitution model adopted for this gene fragment is selected using the Bayesian information criterion from MEGA (version 6.06, Tamura et al. 2013) as the general time-reversible model with a gamma-frequency distribution of changes among sites (GTR+Γ). Four Markov-chain Monte-Carlo chains with the temperature set to 0.2 are run for 10 million generations.

9.1.4 Step 3: Recognising Accepted Species from Integrative Assessment with Morphology

In the third step, the candidate species supported by gene coalescents are assessed within an integrative framework for the strength of corroboration by comparison with another source of evidence to identify 'accepted' species (Vieites et al. 2009; Padial et al. 2010; Schlick-Steiner et al. 2010). In the bumblebee analyses the second source of evidence is morphology: characters are identified either from earlier descriptions or from examining specimens to test for coincidence between different coalescent candidate species and different character states. Just as there are often more base-pair differences in fast-evolving genes than there are easily quantifiable morphological differences, so discovery of new easily quantifiable morphological differences at this stage is less common. Associations between genetic and morphological differences during integrative assessment are described in the bumblebee revisions (Williams et al. 2019, 2020) but a clearer way to document this corroboration would be desirable.

In the 'cumulative' approach to integrative assessment (Padial et al. 2010) any conflicting results with particular sources of evidence are not automatically taken as demonstrating that candidate species are conspecific but as an indication that an evolutionary explanation of the conflict is required (Schlick-Steiner et al. 2010). However, when candidate species are supported by only a single uncorroborated source of evidence with no clear explanation for the conflict, then the candidate species may be regarded provisionally as conspecific (Williams et al. 2019, 2020). Here, the numbers of these accepted species supported by both gene-coalescent and morphological evidence in the integrative assessments are the most likely numbers of species accepted at present as reliably corroborated by morphology in each subgenus (Williams et al. 2019, 2020).

9.1.5 Divergence and Geographical Distance

As a further assessment of the status of the candidate species, one of the simplest relationships expected of genetic variation can be examined: that within-species genetic divergence will increase with geographical distance (Dillon 1984; Jackson et al. 2018). There is no reason to expect this pattern to persist for long in fast-evolving genes among species after speciation, so finding the pattern may be evidence that populations are conspecific. A relationship between geographical and genetic distance is tested here with the Mantel test, which is used to assess the correlation between the pairwise distance matrices (Mantel 1967). This is implemented using the ADE4 package on the R software platform (Dray and Dufour 2007) and by drawing scatter plots using the GGPLOT2 package (Wickham 2011). Genetic divergence is measured using the maximum composite-likelihood model in MEGA, including all three codon positions of the sequence. Geographical (Great Circle) distances are approximated using an estimated mean radius of the Earth of 6,371 km for the WGS84 ellipsoid (program in Williams et al. 2020).

9.2 Sampling Bias

The effects of three sources of sampling bias on species recognition are considered in this section.

9.2.1 Under-Sampling

The most obvious source of bias for revisionary taxonomy arises from failing to include either all of the constituent species from within a monophyletic group or enough of the variation from across those species' populations. Although compiling data on bumblebee barcodes with BEEBOL began in 2008, NHM analyses that focussed on the subgenera *Alpinobombus* and *Melanobombus* did not begin until 2011. Since 2011, these projects have included a series of 30 separately analysed datasets of barcodes for the subgenus *Alpinobombus* and a series of 37 separately analysed datasets of barcodes for the subgenus *Melanobombus* as data were added. In this paper, five (nested) datasets are chosen as examples for each subgenus from the period 2011–2019 to represent the range of sample sizes as the barcode sequences were accumulated. These data sets are re-analysed using standardised settings and outgroups (see Section 9.1.3) so that the effects of sample size on the results for recognising species can be compared without being confounded by changes in analytical methods.

9.2.2 Ambiguous Data

The amount of ambiguous data arising from unsequenced nucleotide base pairs has been considered an important factor when identifying species. Nonetheless, sequences as short as 100 base pairs (bp) (15 per cent) have been found to be diagnostic for 90 per cent of species in other animal groups (Meusnier et al. 2008). When using barcodes to recognise known and new species, a threshold of 500 bp (76 per cent of the barcode fragment length used here) was recommended for use with the Barcode Index Number system (BINs: Ratnasingham and Hebert 2013). When using gene coalescents to recognise species within a revision for bumblebees of the subgenus *Mendacibombus*, a correlation was found between short barcode sequences and long terminal branches (which are then more likely to be interpreted as separate species) (Williams et al. 2016). That study preferred a threshold of 650 bp (99 per cent of barcode length) for accepting sequences when longer sequences were available for a taxon. Unfortunately, longer sequences were not always available, so the shortest sequence accepted was 603 bp (92 per cent of barcode length). Such a high threshold can be very restrictive when many of the available specimens are old, because older specimens tend to yield shorter sequences (Strange et al. 2009). Because of the difficulty of obtaining fresh material in some cases, thresholds may need to be relaxed, although the recent revision of the subgenus *Alpinobombus* was eventually able to use a threshold of 600 bp (91 per cent of barcode length) (Williams et al. 2019). For the present study, the simple PTP analyses are re-run with a fixed threshold, the 'long sequence filter' (LSF), which excludes sequences of less than 591 bp (90 per cent of barcode length). This allows comparison of the effects with and without short sequences on the results for recognising species.

9.2.3 Over-Sampling

Less intuitively obvious is the possible effect of including too many closely similar sequences from some of the more common species, a phenomenon described as 'over-sampling' (Zhang et al. 2013). The problem is that when sampling many closely related individuals within just some (not all) of the species, it can cause slightly less closely related individuals within other species to be interpreted falsely as several separate species. This is likely to be a particular problem when there are a few relatively invariant but particularly widespread and abundant species. For example, for the revision of *Melanobombus* there might be a risk of over-sampling *B. lapidarius* (Linnaeus), simply because it is by far the most accessible species to European students for sampling. The consequence could be to recognise false 'cryptic species' in other groups that are not true cryptic species but are artefacts of the method. To attempt to reduce this effect, all duplicate sequences can be removed from the analysis. Alleles that are unique for the sequenced fragment are identified here using COLLAPSE (version 1.2, darwin.uvigo.es/software/collapse.html), after ranking sequence fragments from longest to shortest. This ranking avoids matching longer to shorter sequences, which could lead to rejection of longer sequences that might otherwise obscure real differences (short sequences could omit un-sequenced differences). Applying this 'unique allele filter' (UAF) ensures a minimum (non-zero) divergence between all samples, which allows comparison of the effects with and without duplicate sequences on the results for recognising species.

9.3 Results

9.3.1 Stratified Sampling

For both of the subgenera *Alpinobombus* and *Melanobombus*, as the numbers of sample sequences increased in the two revisionary studies, so too did the numbers of longer sequences and the numbers of unique alleles (Tables 9.2 and 9.3).

As expected with the stratified sampling design, there is an increasing spatial bias in the pattern of barcode acquisition through time as described in the Methods, with sampling targeted increasingly towards Alaska and the Yukon for the subgenus *Alpinobombus* (red

Table 9.2 Numbers of candidate species from unfiltered Poisson-tree-process (PTP) analyses (with or without long sequence filtering, LSF, or unique allele filtering, UAF) of COI barcodes as sampling progressed for the subgenus *Alpinobombus*. All analyses were re-run retrospectively with the same outgroup and model settings. PTP analyses using maximum likelihood to fit the models.

Date of dataset	May 2011	Feb. 2012	Oct. 2014	Oct. 2016	Sep. 2018
Sequences	29	57	103	180	234
Final species represented[1]	7	9	9	9	9
Final coalescents represented[2]	7	9	9	10	10
Mean alleles/final species (9)	2.1	2.7	4.0	5.8	7.0
CV alleles/final species (9)	1.0	0.9	0.7	0.5	0.4
Unfiltered PTP candidate species	**8**	**10**	**12**	**10**	**10**
Error rate PTP/final (%)[3]	14	11	33	0	0
Long sequences ($\geq$ 591 bp)	23	50	95	164	204
LSF-PTP candidate species	**5**	**8**	**12**	**10**	**10**
Error rate LSF-PTP/final (%)[3]	29	11	33	0	0
Unique alleles	19	25	36	53	63
UAF-PTP candidate species	**7**	**9**	**9**	**10**	**10**
Error rate UAF-PTP/final (%)[3]	0	0	0	0	0

[1] The number of species included in the data that were recognised by Williams et al. 2019.

[2] The number of coalescent groups recognised by PTP analysis (to go forward to integrative assessment). In some cases, this includes not only final species from the integrative assessment but also a heteroplasmic paralogous copy of the COI gene.

[3] Error rates are calculated relative to the number of final coalescents (from the second row of results in the table) that are represented in the data.

Table 9.3 Numbers of candidate species from unfiltered Poisson-tree-process (PTP) analyses (with or without long sequence filtering, LSF, or unique allele filtering, UAF) of COI barcodes as sampling progressed for the subgenus *Melanobombus*. All analyses were re-run retrospectively with the same outgroup and model settings. PTP analyses using maximum likelihood to fit the models.

Date of dataset	Apr. 2011	Mar. 2014	Aug. 2018	Feb. 2019	Apr. 2019
Sequences	45	139	233	290	314
Final species represented[1]	14	19	24	24	25
Final coalescents represented[2]	14	19	25	25	26
Mean alleles/final species (25)	1.0	2.8	4.1	5.7	6.1
CV alleles/final species (25)	1.3	1.2	1.0	1.3	1.2
Unfiltered PTP candidate species	**17**	**22**	**27**	**30**	**31**
Error rate PTP/final (%)[3]	21	16	8	20	19
Long sequences ($\geq$ 591 bp)	21	126	157	215	218
LSF-PTP candidate species	**10**	**20**	**30**	**26**	**26**
Error rate LSF-PTP/final (%)[3]	29	5	20	4	0
Unique alleles	26	71	104	144	153
UAF-PTP candidate species	**16**	**21**	**26**	**25**	**26**
Error rate UAF-PTP/final (%)[3]	14	11	4	0	0

[1] The number of species included in the data that were recognised by Williams et al. 2020.
[2] The number of coalescent groups recognised by PTP analysis (to go forward to integrative assessment). In some cases, this includes not only final species from the integrative assessment but also a coalescent group within *B. lapidarius* that is uncorroborated by morphology.
[3] Error rates are calculated relative to the number of final coalescents (from the second row of results in the table) that are represented in the data.

spots in Figure 9.1), and targeted increasingly towards the east Qinghai-Tibetan Plateau for the subgenus *Melanobombus* (red spots in Figure 9.2). This resulted in the desired increasing mean numbers of alleles per accepted species (as interpreted in the integrated analyses: Tables 9.2 and 9.3). Initially, the coefficients of variation (CV) for the numbers of alleles per accepted species decreased for both subgenera, showing that the stratified sampling design had succeeded in making variation more evenly represented among species (Tables 9.2 and 9.3). Subsequently, as the numbers of sequences increased further, so the CV of the number of alleles per accepted species increased again for *Melanobombus*, as some species were found to be inherently more variable than others.

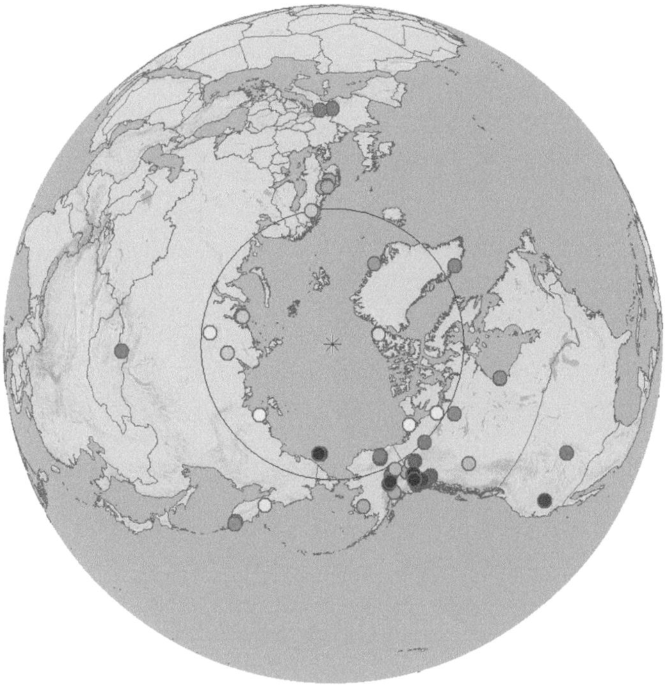

Figure 9.1 Sample sites for COI barcodes for the subgenus *Alpinobombus* from Williams et al. (2019), colour-coded for the order in which the samples were added to the analyses, from blue (2011) through green, yellow, and orange, to red (2019) (specimens may be much older). Spherical projection with the North Pole shown by a star, international boundaries and Arctic Circle shown as grey lines. Map projected in ArcGIS using the World_Shaded_Relief basemap © 2014 ESRI.

9.3.2 Under-Sampling

Under-sampling is most apparent in Table 9.2 in the initial samples for the subgenus *Alpinobombus*, because the number of accepted species represented initially in the barcode data (7) is less than the accepted total of nine species, as well as in the first two sets of samples in Table 9.3 for *Melanobombus*, for which the number of accepted species represented initially in the data (14 then 19) is again less than the accepted total of 25 species. As the numbers of sequences in the samples has increased in the two studies through time, so the numbers of candidate species obtained has tended to increase in the unfiltered PTP results (Figures 9.3 and 9.4). This shows some convergence in 2016 on ten candidate species for *Alpinobombus* but still no convergence on a consistent figure for *Melanobombus*. Initial under-sampling is biased against the rarer species with narrower range sizes (including some not recognised previously from morphology), as shown by the declining mean range size among included accepted species as sample sizes increase (Figure 9.5).

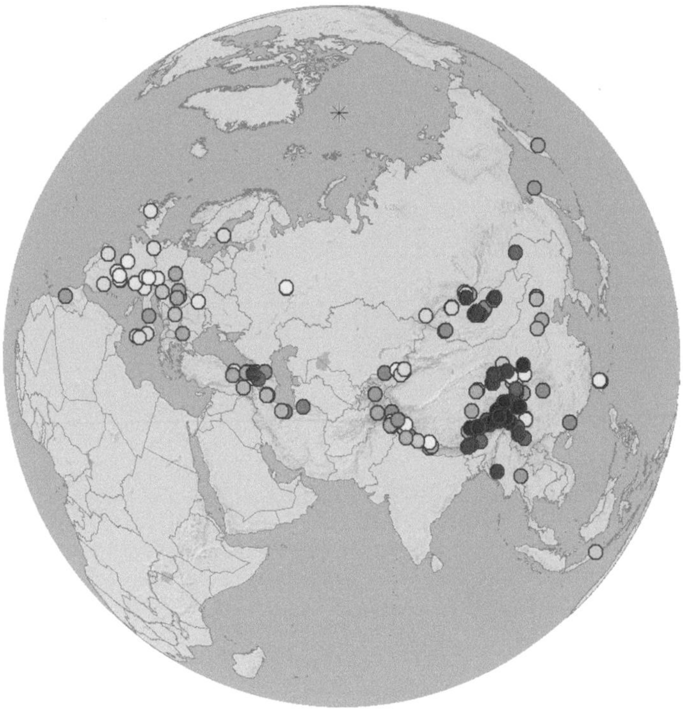

Figure 9.2 Sample sites for COI barcodes for (1) the subgenus *Melanobombus* from Williams et al. (submitted), colour-coded for the order in which the samples were added to the analyses, from blue (2011) through green, yellow, and orange, to red (2019) (specimens may be much older). Spherical projection with the North Pole shown by a star, international boundaries shown as grey lines. Map projected in ArcGIS using the World_Shaded_Relief basemap © 2014 ESRI.

9.3.3 Ambiguous Data

Filtering out short sequences (LSF) reduces the numbers of sequences admitted to the PTP analyses by a mean of 12 per cent for *Alpinobombus* and by 30 per cent for *Melanobombus*. Compared to the unfiltered PTP results (Tables 9.2 and 9.3), the LSF-PTP results show a reduction in the number of candidate species in seven out of ten cases (Figures 9.3 and 9.4). This is a difference (usually a reduction) of 12 per cent in candidate species for *Alpinobombus* and 18 per cent for *Melanobombus*. However, when viewed in terms of an 'error rate' relative to the number of accepted species represented in the data for each analysis, the LSF filter still results in a mean error rate of 15 per cent for *Alpinobombus* and 12 per cent for *Melanobombus*. This is interpreted as supporting the idea that the presence of short (ambiguous) sequences in unfiltered PTP analyses can increase the number of false candidate species obtained. With LSF, convergence in the numbers of candidate species is then achieved eventually for *Melanobombus* (Figure 9.4).

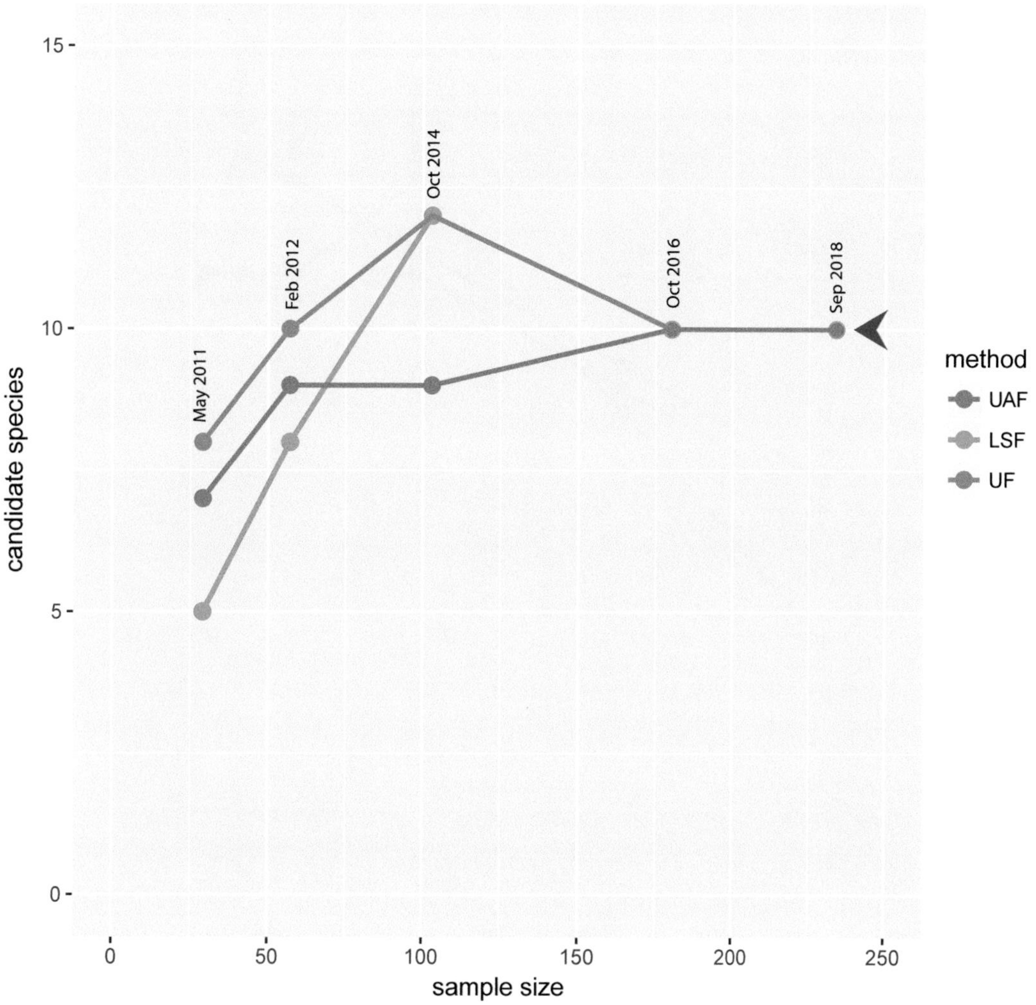

Figure 9.3 Numbers of candidate species for the subgenus *Alpinobombus* from Poisson-tree-process (PTP) analyses using either unfiltered (UF in blue), long sequence filtering (LSF in green), or unique allele filtering (UAF in red) from MrBayes trees from COI barcodes (y-axis) as sample sizes increased through time (x-axis). All analyses were re-run retrospectively with the same outgroup and model settings. PTP analyses using maximum likelihood to fit the models. Arrows indicate numbers of PTP candidate species accepted at the end of the revisionary studies.

9.3.4 Over-Sampling

Filtering out duplicate sequences (UAF) reduces the number of sequences admitted to the analyses by a mean of 60 per cent for *Alpinobombus* and 50 per cent for *Melanobombus* (Tables 9.2 and 9.3). In eight of the ten comparisons, the UAF-PTP analyses found fewer candidate species than the unfiltered PTP analyses (Figures 9.3 and 9.4). This is a smaller effect than for LSF-PTP as a mean reduction of 10 per cent for *Alpinobombus* and 9 per cent

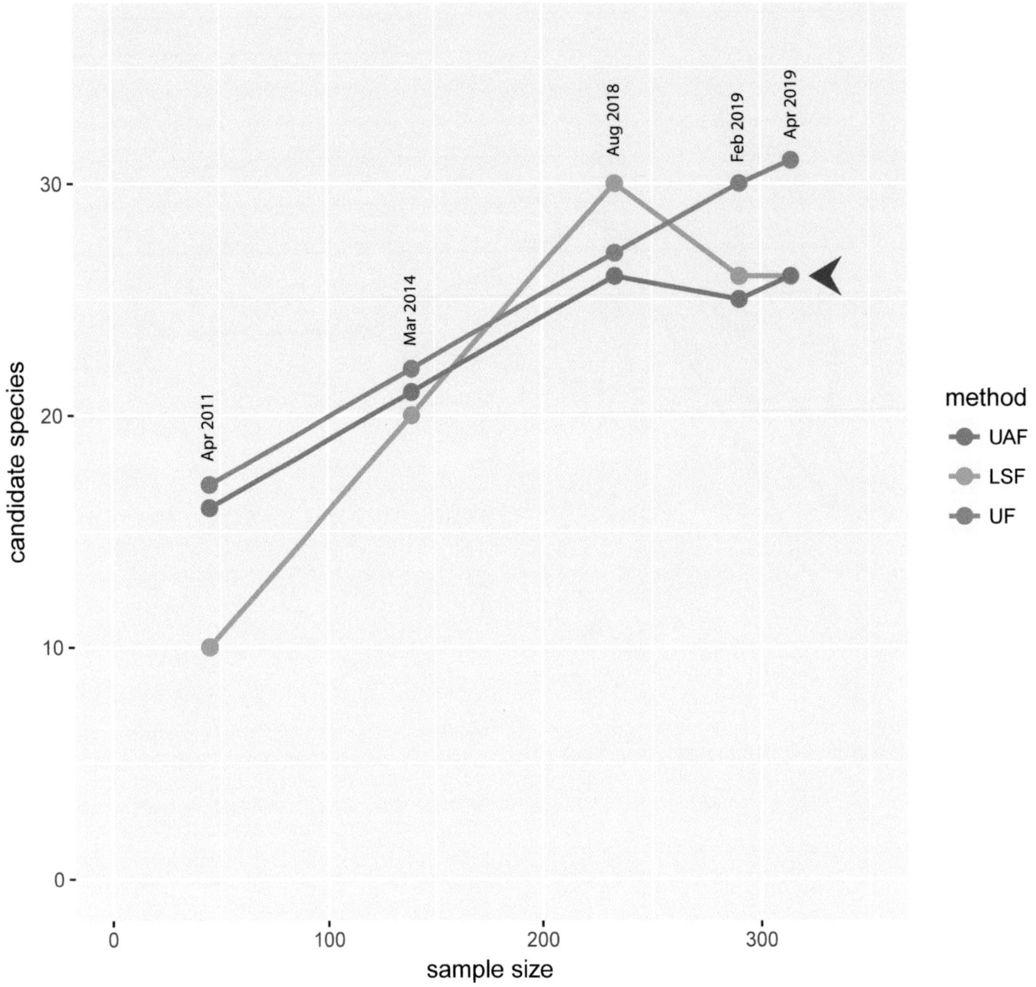

Figure 9.4 Numbers of candidate species for the subgenus *Melanobombus* from Poisson-tree-process (PTP) analyses using either unfiltered (UF in blue), long sequence filtering (LSF in green), or unique allele filtering (UAF in red) from MrBayes trees from COI barcodes (y-axis) as sample sizes increased through time (x-axis). All analyses were re-run retrospectively with the same outgroup and model settings. PTP analyses using maximum likelihood to fit the models. Arrows indicate numbers of PTP candidate species accepted at the end of the revisionary studies.

for *Melanobombus*. However, when viewed in terms of an error rate relative to the number of accepted species represented in the data for each analysis, the UAF results in a lower mean error rate than does the LSF filter of 0 per cent for *Alpinobombus* and 6 per cent for *Melanobombus*. This is interpreted as supporting the idea that the presence of duplicate sequences in the unfiltered PTP analyses can substantially increase the number of false candidate species obtained. The most promising aspect of these results is that as more samples are added, the UAF-PTP error rate is the most consistently low. The rate is zero for *Alpinobombus* and declines consistently with sample size for *Melanobombus*, so that

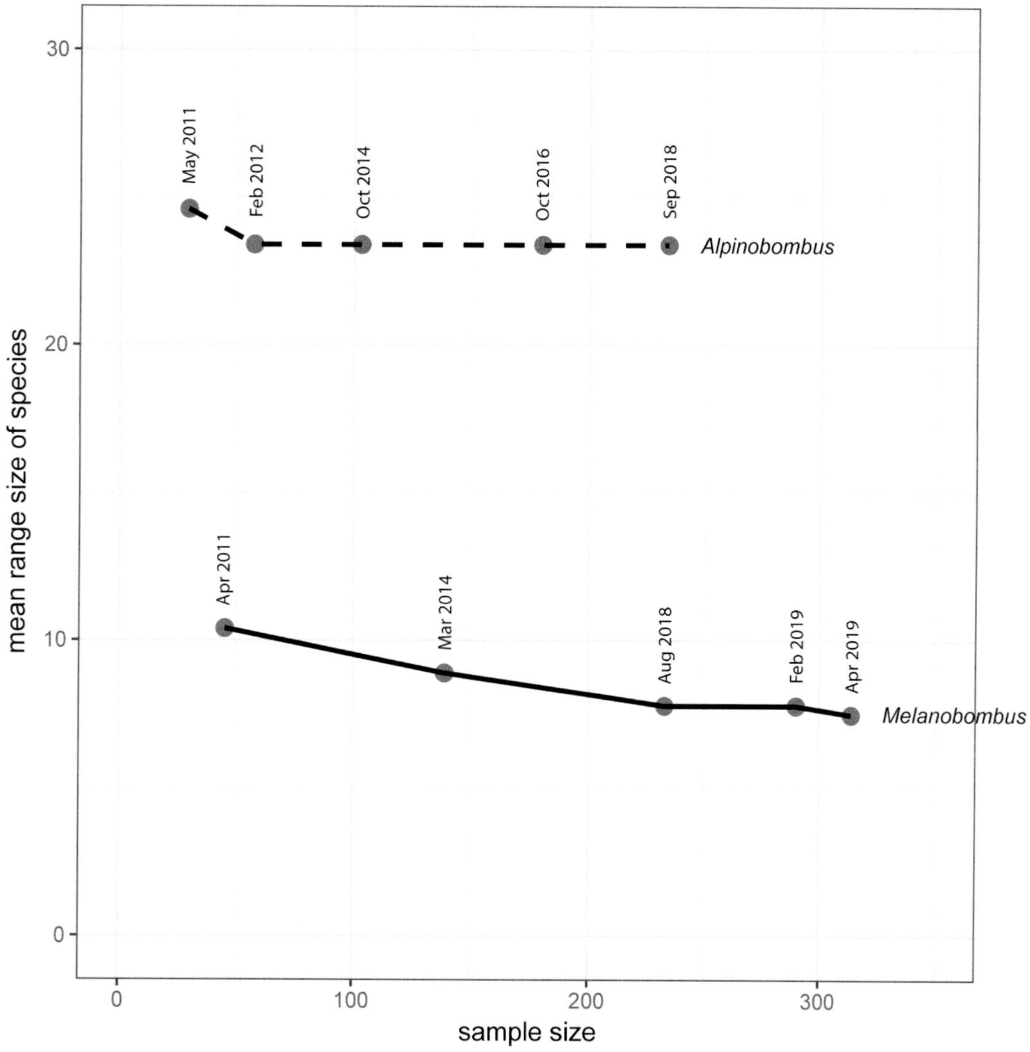

Figure 9.5 Plot of mean range sizes among the accepted species included (y axis) against the sample sizes (x axis) for the samples compiled for revisions of the subgenera *Alpinobombus* (dashed line) and *Melanobombus* (solid line) as in Tables 9.2 and 9.3. Range size is measured as the numbers of 611,000 km^2 (equal-area) grid cells with records (Williams 1998), and sample size is measured as the cumulative numbers of barcode sequences.

convergence is eventually achieved for number of candidate species, even if still only slowly for *Melanobombus* (Figures 9.3 and 9.4).

The online implementation of PTP also includes solutions with the highest Bayesian support values (rather than maximum likelihood) to fit the models (Tables 9.4 and 9.5). Unexpectedly, four results for *Alpinobombus* show a sharp jump in the numbers of candidate species from the unfiltered and from the LSF-PTP analyses in the last two time periods, a result confirmed by repeated analyses.

Table 9.4 Numbers of candidate species from unfiltered Poisson-tree-process (PTP) analyses (with or without long sequence filtering, LSF, or unique allele filtering, UAF) of COI barcodes as sampling progressed for the subgenus *Alpinobombus*. All analyses were re-run retrospectively with the same outgroup and model settings. PTP analyses using highest Bayesian support values to fit the models (95 per cent confidence intervals from PTP analyses are shown in parentheses).

Date of dataset	May 2011	Feb. 2012	Oct. 2014	Oct. 2016	Sep. 2018
Sequences	29	57	103	180	234
Final species represented[1]	7	9	9	9	9
Final coalescents represented[2]	7	9	9	10	10
Mean alleles/ final species (9)	2.1	2.7	4.0	5.8	7.0
CV alleles/final species (9)	1.0	0.9	0.7	0.5	0.4
Unfiltered PTP candidate species	**8** (6–16)	**11** (9–27)	**14** (10–62)	**120** (72–133)	**175** (104–175)
Error rate PTP/final (%)[3]	14	22	56	110	165
Long sequences (≥ 591 bp)	23	50	95	164	204
LSF-PTP candidate species	**5** (5–10)	**8** (7–17)	**12** (9–58)	**108** (44–119)	**142** (90–151)
Error rate LSF-PTP/ final (%)[3]	29	11	33	98	132
Unique alleles	19	25	36	53	63
UAF-PTP candidate species	**7** (4–14)	**10** (6–20)	**9** (7–21)	**10** (7–36)	**10** (7–44)
Error rate UAF-PTP/ final (%)[3]	0	11	0	0	0

[1] The number of species included in the data that were recognised by Williams et al. 2019.
[2] The number of coalescent groups recognised by PTP analysis (to go forward to integrative assessment). In some cases, this includes not only final species from the integrative assessment but also a heteroplasmic paralogous copy of the COI gene.
[3] Error rates are calculated relative to the number of final coalescents (from the second row of results in the table) that are represented in the data.

Table 9.5 Numbers of candidate species from unfiltered Poisson-tree-process (PTP) analyses (with or without long sequence filtering, LSF, or unique allele filtering, UAF) of COI barcodes as sampling progressed for the subgenus *Melanobombus*. All analyses were re-run retrospectively with the same outgroup and model settings. PTP analyses using highest Bayesian support values to fit the models (95 per cent confidence intervals from PTP analyses are shown in parentheses).

Date of dataset	Apr. 2011	Mar. 2014	Aug. 2018	Feb. 2019	Apr. 2019
Sequences	45	139	233	290	314
Final species represented[1]	14	19	24	24	25
Final coalescents represented[2]	14	19	25	25	26
Mean alleles/final species (25)	1.0	2.8	4.1	5.7	6.1
CV alleles/final species (25)	1.3	1.2	1.0	1.3	1.2
Unfiltered PTP candidate species	**17** (16–20)	**22** (20–30)	**27** (25–39)	**33** (27–44)	**33** (27–47)
Error rate PTP/final (%)[3]	21	16	8	32	27
Long sequences (≥ 591 bp)	21	126	157	215	218
LSF-PTP candidate species	**10** (9–12)	**20** (17–30)	**30** (25–39)	**28** (23–37)	**28** (24–37)
Error rate LSF-PTP/final (%)[3]	29	5	20	12	8
Unique alleles	26	71	104	144	153
UAF-PTP candidate species	**16** (13–19)	**21** (18–27)	**26** (22–41)	**25** (23–44)	**26** (24–44)
Error rate UAF-PTP/final (%)[3]	14	11	4	0	0

[1] The number of species included in the data that were recognised by Williams et al. 2020.
[2] The number of coalescent groups recognised by PTP analysis (to go forward to integrative assessment). In some cases, this includes not only final species from the integrative assessment but also a coalescent group within *B. lapidarius* that is uncorroborated by morphology.
[3] Error rates are calculated relative to the number of final coalescents (from the second row of results in the table) that are represented in the data.

9.3.5 Discerning Widespread Polytypic Species from Complexes of Species

The *sichelii*-complex of the subgenus *Melanobombus* shows weakly positive relationships in divergence with distance both within and among the seven candidate species from the Bayesian UF-PTP analysis (Figure 9.6: Mantel $r = 0.52$, $p < 0.001$; Figure 9.7: Mantel $r = 0.21$, $p < 0.001$), so that even among these candidate species the relationship is consistent with a

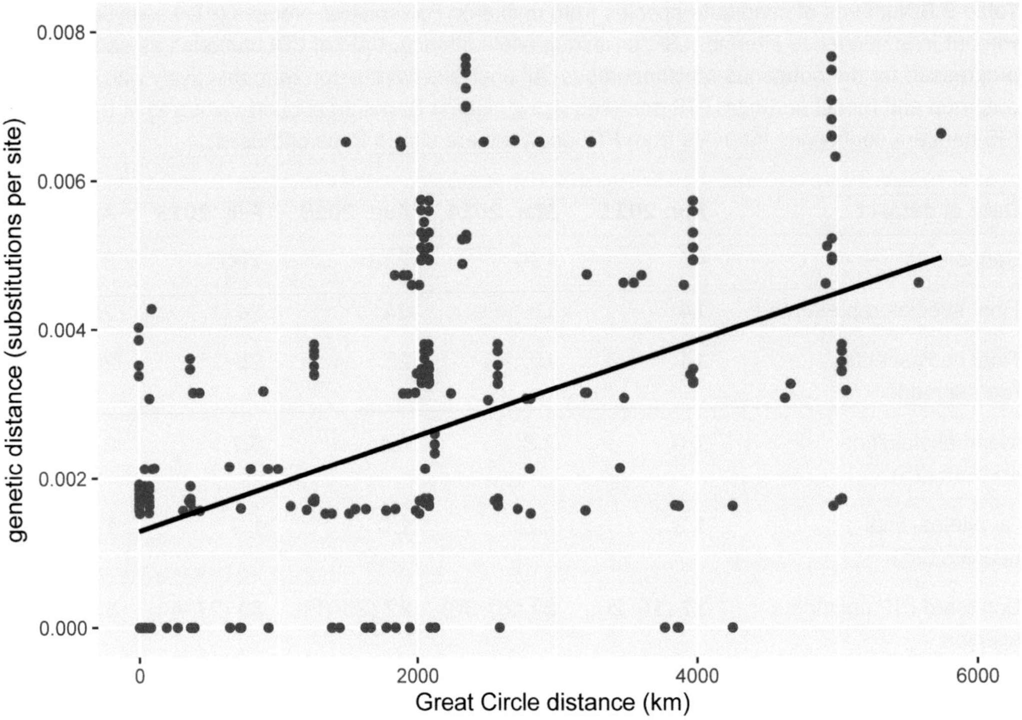

Figure 9.6 Plot of (y-axis) pairwise proportion of genetic divergence between COI barcode region sequences against (x-axis) pairwise Great Circle geographical distance in km between sample sites with linear trend lines (black) for genetic divergences within the candidate species for the *sichelii*-complex of the subgenus *Melanobombus* identified using Bayesian UF-PTP (Mantel r = 0.52).

single population model. Therefore, the *sichelii*-complex shows a pattern consistent with just one species being present. This result corroborates the UAF-PTP results that there is a single species in the *sichelii*-complex.

For the *keriensis*-complex of the subgenus *Melanobombus*, divergences within the six candidate species also appear to show a weakly positive relationship between divergence and distance from the Bayesian UF-PTP analysis (Figure 9.8: Mantel r = 0.16, $p < 0.001$), so these data are consistent with a single population model within each of the six candidate species. However, divergences among the six candidate species show no significant relationship between divergence and distance (Figure 9.9: Mantel r = −0.06, $p > 0.5$). Therefore, for the *keriensis*-complex, it is likely that there are multiple species present. This result corroborates the UAF-PTP results that there are multiple species in the *keriensis*-complex.

Comparing the results for the two species complexes supports a better performance for the UAF-PTP procedure in discerning between widespread polytypic species (*B. sichelii*) and complexes of species (the *keriensis*-complex), for which the species are near-cryptic.

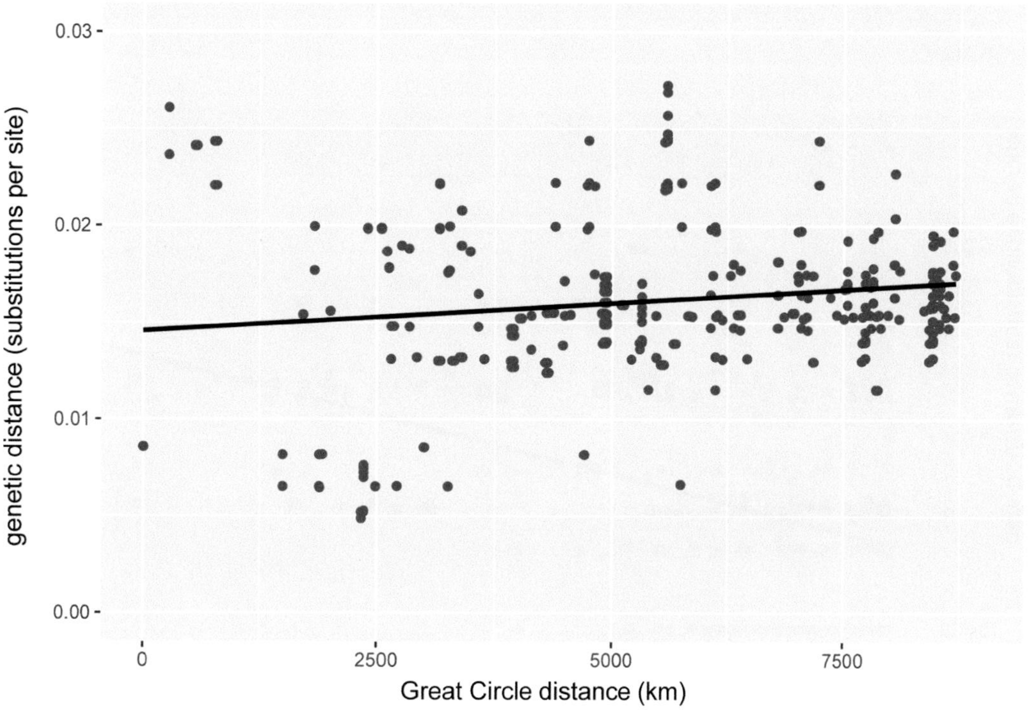

Figure 9.7 Plot of (y-axis) pairwise proportion of genetic divergence between COI barcode region sequences against (x-axis) pairwise Great Circle geographical distance in km between sample sites with linear trend lines (black) for genetic divergences among the candidate species for the *sichelii*-complex identified using Bayesian UF-PTP (Mantel r = 0.21).

9.4 Discussion

9.4.1 Stratified Sampling

This empirical study attempted to improve sampling by incorporating knowledge of the existing taxonomic hierarchy (both from species and from lower-rank taxa representing putative population structure) in order to provide a more even representation of the known species and of their intraspecific diversity. Coefficients of variation (Tables 9.2 and 9.3) show some progress towards achieving this, although inherent differences in intra-specific diversity among species within the subgenus *Melanobombus* became apparent with larger sample sizes. The best chance for discovery of any completely unknown species (from genes or morphology) would still require random sampling in order to reduce bias, because the prior taxonomic hierarchy cannot account for any unknown species, except through misclassified lower-rank taxa.

9.4.2 Reducing Bias

Compared to the results for accepted species from the integrative analyses, the closest and most consistent results for PTP candidate species follow from excluding duplicate

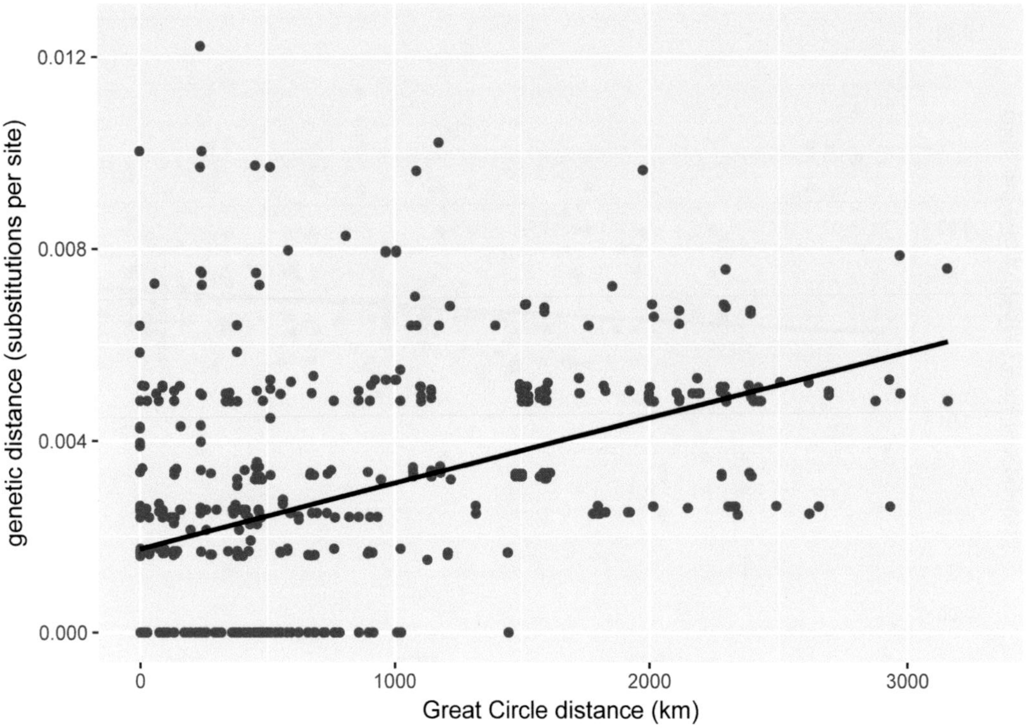

Figure 9.8 Plot of (y-axis) pairwise proportion of genetic divergence between COI barcode region sequences against (x-axis) pairwise Great Circle geographical distance in km between sample sites with linear trend lines (black) for genetic divergences within the candidate species for the *keriensis*-complex of the subgenus *Melanobombus* identified using either Bayesian UF-PTP or UAF-PTP (Mantel r = 0.16).

sequences (the UAF-PTP analyses: Tables 9.2 and 9.3; Figures 9.3 and 9.4). The most consistently declining error rate was found with increasing sample sizes when using the UAF-PTP analysis for *Melanobombus*. In contrast, the continuing monotonic increase in numbers of candidate species for *Melanobombus* from the unfiltered PTP analysis as the numbers of sequences increased (Table 9.3; Figure 9.4) was unexpected. Using the UAF procedure to exclude duplicate sequences could be combined with excluding short sequences as in some previous studies, although UAF will tend to do this anyway (Williams et al. 2016, 2019). However, care has to be taken not to exclude short sequences when these are all that are available for some genuinely separate species (sometimes discernible as strongly divergent sequences).

False species in the results of some unfiltered analyses, perhaps arising from short sequences or from over-sampling other species, might be mistaken for fully cryptic species because they would not be supported by morphological characters, either known or discoverable. This is not to deny that genuine cryptic species exist, which they undoubtedly do even among these bumblebees (Williams et al. 2015b, 2020). Near-cryptic species are especially well studied among bumblebees in the subgenus *Bombus* s. str., but subtle

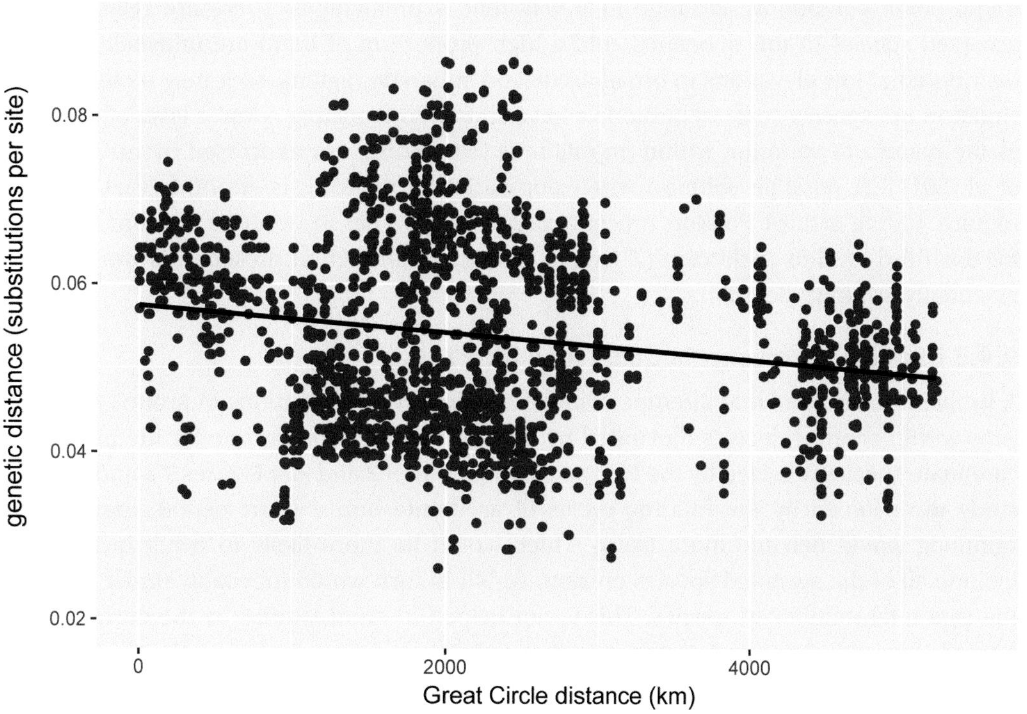

Figure 9.9 Plot of (y-axis) pairwise proportion of genetic divergence between COI barcode region sequences against (x-axis) pairwise Great Circle geographical distance in km between sample sites with linear trend lines (black) for genetic divergences among the candidate species for the *keriensis*-complex identified using either Bayesian UF-PTP or UAF-PTP (positive relationship rejected). For details of the measurements see the text, data from Williams et al. 2020.

coincident morphological differences could be found (Rasmont 1984; Bertsch et al. 2005; Williams et al. 2012). Discerning false from true cryptic species might depend not only on applying filters to exclude some sequences as suggested previously but also on formulating appropriate stopping rules for studies, which here depended on checking for repeatability (stability) in the results as more sequences are added.

The online implementation of PTP includes solutions with the highest Bayesian support values (rather than maximum likelihood) to fit the models (Tables 9.4 and 9.5). With this Bayesian method, four results for the subgenus *Alpinobombus* show sharp jumps in the numbers of candidate species by more than an order of magnitude (Table 9.4). This results from the peaks in probability for 'species' coalescents shifting much closer to the terminals of the tree. Previously, there had been little agreement on the number of species recognised in *Alpinobombus* (Williams et al. 2015b: their Table 1), and although the nine species accepted here come closest to being supported by morphology (Williams et al. 2019), even this figure is considered too high by some (Martinet et al. 2018; Potapov et al. 2019). The dramatic jumps in Table 9.4 raise the possibility that PTP may fail to give reliable results under some conditions, although in a way that is predictable. There are relatively few accepted species in this subgenus, and a high proportion of them are unusually broadly distributed at low elevations in broad arctic and subarctic regions, so it may be that in this situation the species coalescent models have shifted to detecting a lower level of structure in the pattern of variation within populations (something not addressed by mPTP: Kapli et al. 2017). A possible solution when population structuring is strong is that it might require adding a third Poisson process class to the model in order to account explicitly for this third level of coalescent (Zhang et al. 2013), although at present this would be an essentially arbitrary decision.

9.4.3 Guerrilla Taxonomy: Is Quick Also Dirty?

A problem for studies that attempt to make revisionary investigations of groups of organisms within short projects is illustrated by the trajectories through time for the number of candidate species detected by the PTP analyses (Tables 9.2 and 9.3; Figures 9.3 and 9.4). If a study were forced by the funding cycles of academia into a short period, then under-sampling would become more likely, which would be more likely to result in failing to include all of the accepted species present, which in turn would inevitably under-estimate the true total number of species. This is not helped in short studies by the current trend towards increasing national restrictions that reduce access to specimens (Prathapan et al. 2018) and even data from some key countries. Attempts to overcome the time problem by sampling more intensively but constrained to just some larger geographical regions might then run the risk of the opposite problem: that of over-sampling a few widespread, abundant, and in some cases less variable species. This context-sensitivity could bias analyses in favour of over-estimating the number of species present by promoting false 'cryptic species'. The solution has to be to explicitly address two realities: (1) that sampling has to represent the entire geographical distributions of all of the species involved; and (2) that some species are simply fundamentally genuinely rare (Gaston 1994), so that their low encounter rates when sampling means that studies cannot be rushed if these species are to be included with adequate representation of their variation. Because this is a fundamental problem in gathering information about individuals around the world, it is likely to persist independently of which approach to genes or any other characters is adopted for recognising species.

Some new morphologically cryptic or near-cryptic candidate species were discovered in the course of revising the two bumblebee subgenera *Alpinobombus* and *Melanobombus*. In the subgenus *Alpinobombus*, one candidate was recognised initially from its COI coalescent and was then found to have support from subtle morphological characters in females (Williams et al. 2016). More obvious characters were later found in the rarer males (Williams et al. 2019), so that it became possible to describe the new species with confidence, as *B. kluanensis* Williams & Cannings. Subsequently, a second new closely related coalescent group was discovered within the same species complex. This candidate was subsequently shown to be likely to be based on a paralogous copy of the COI gene, as an instance of mitochondrial heteroplasmy (Williams et al. 2019). Heteroplasmy is a phenomenon that has been reported for other bees (Magnacca and Brown 2010), including some bumblebees (Francoso et al. 2016), but so far it appears to be uncommon among bumblebees. In the subgenus *Melanobombus*, two candidates were recognised from COI coalescents within each of the former 'species' *B. miniatus* Bingham and *B. rufofasciatus* Smith, although diagnostic characters in morphology to support their status as separate species are subtle (Williams et al. 2020). In contrast, so far there is no consistent diagnostic morphological or pheromonal evidence to support two candidate coalescent groups detected within the European *B. lapidarius* (Lecocq et al. 2015, 2019; Williams et al. 2020). Cryptic species were also recognised within the *keriensis*-complex of *Melanobombus* (see Section 9.3).

In the past, revisionary taxonomy has been largely dependent on museum specialists, who have often maintained an interest in studying particular groups over long careers, accumulating detailed worldwide experience of species' characters over many years. More recently, especially with the escalating biodiversity crisis and the decreasing number of taxonomists, there has been pressure to find short cuts, in order to broaden and fast-track the process of discovery (Packer et al. 2009). DNA analysis has promised a 'Brave New World' with more data and quicker answers, tempting people to attempt revisionary taxonomy within the scope of short projects. But even with DNA, global sampling is still necessary for revisionary studies if the groups have a global distribution. Consequently, as shown here, even DNA studies may still require many years to assemble data, particularly in order to encompass larger geographic ranges and specially to improve the representation of the rarer taxa. The concern about short studies will be all the greater if revisionary taxonomy were judged by some to be more successful when it makes changes to the established taxonomy. Change for its own sake is detrimental as instability, and 'guerrilla taxonomy', in which a short study of an unfamiliar group makes substantial but in some cases poorly supported changes before the authors move on to work in other subject areas (and before any errors become apparent), should be discouraged. This accelerated approach may have the potential to contribute easy citations for the authors, but it risks chaos for the users of taxonomy, which is everyone. It could constitute a perverse incentive if authors were rewarded simply for changing the taxonomy of a group, especially if this were more likely to result from short projects. We would hope that reviewers would catch these problems, although this might be reliable only if reviewers were always familiar with the group in question.

If good revisionary taxonomy depends on experience, or at least on accumulating large samples, which often requires a long period, how can people be expected to make progress when long-term funding is unrealistic? One possibility for small projects might be to begin with shorter local field-survey projects, for example, of the kind that are needed for pollinator conservation (Cameron et al. 2011; Holland et al. 2015). For those who need to find short and publishable projects in a related subject area, one of many better uses for student-project resources currently might be to develop new genomic methods for recognising species. Science urgently needs more taxonomists (Drew 2011; Paknia et al. 2015), but until there is more funding to support people in this role for the long term, it may be a double disservice to seek to train more taxonomists for whom there are no jobs when also asking them to work on short revisionary projects that might add to taxonomic confusion. The solution more broadly for revisionary studies has to be for people to collaborate more in order to combine existing specialist experience with new capacity for sampling and analysis.

9.4.4 Scaling Up for the Species-Discovery Problem

With the growing concern over the extent and severity of threats to biodiversity, and with much of that biodiversity as yet undescribed, let alone critically revised, there is an urgent need to develop techniques for recognising species quickly and reliably on a large scale (Packer et al. 2009). One example where this challenge has been taken up is within the International Barcode of Life's (iBOL) online barcode of life data (BOLD) platform (Ratnasingham and Hebert 2007). An automated COI barcode index number (BIN) tool uses an empirically-justified approximation to discovering species (Ratnasingham and Hebert 2013; Meierotto et al. 2019), rather than a theoretically based approach such as gene coalescents. When applied to bumblebees across North America, the BIN system seems to work reasonably well, although this may be in large part because the North American bumblebee fauna is already relatively thoroughly sampled across all of the species' ranges (as species are reasonably well understood from morphology and barcodes: Williams et al. 2014) in the BOLD database.

Could automated procedures such as the BIN tool provide a good approximation when applied to all bumblebees? In principle they might, once sufficient samples from across the ranges of all species were in place – but the results of the present analysis show that achieving adequate sampling is the larger part of the problem. Most of the world's bumblebee species are either restricted to, or occur within, Asia (Williams 1998). Representation of most groups of bumblebees from the many species-rich mountain ranges across Asia in BOLD is currently especially thin and patchy (among the 8,200 BOLD barcodes for bumblebees accessed by the author in 2019). When low sampling is combined with the current national restrictions on access for further sampling in much of Asia (Prathapan et al. 2018), as well as with the complexity of bumblebee population structures within Asian mountain systems (Williams et al. 2020), this makes reliable automated discovery of Asian bumblebee species impractical at present. In consequence, most of the Asian species do not yet have BIN assignments and it will require broader permission and more funding for a very substantial survey effort in order to sample and represent them all adequately in BOLD. If this were achieved, then re-assessment with coalescent methods for more reliable interpretations would also become quick and easy. In summary, if we are going to reap the full benefits of quantitative methods for recognising species with minimum bias, then we will need to assess data globally, so we will have to find ways to collaborate across international boundaries in order to examine as much of the global variety as we can, a broad message that is not new for taxonomy (de Candolle 1880).

Acknowledgements

Thanks for arranging collecting trips to J. An, S. Cameron, J. Huang, and Y. Tang; for providing additional specimens to M. Berezin, A. Byvaltsev, S. Cannings, B. Cederberg, R. De Jonghe, A. Dorjsuren, S. Jaffar, G. Japoshvili, S. Kahono, H. Liang, M. Mei, A. Monfared, F. Ødegaard, M. Orr, R. Raina, C. Rasmussen, Z. Ren, L. Richardson,

J. Rykken, C. Sheffield, C. Starr, C. Thanoosing, and Y. Zhao; for access to specimens in collections to J. An, Y. Astarova, M. Aytekin, F. Bakker, S. Belokobylskij, L. Best, W. van Bohemen, A. Byvaltsev, S. Cameron, S. Cardinale, S. Colla, S. Droege, B. Harris, J. Hogan, M. Kalnins, F. Koch, T. Levchenko, D. Notton, M. Orr, L. Packer, M. Proshchalykin, P. Rasmont, V. Richter, D. Sheppard, D. Sikes, J. Strange, J. Thomas, A. Tripodi, E. Wyman, J. Yao, and C. Zhu; for obtaining DNA sequences to the Canadian Centre for DNA Barcoding, R. Canty, T. Lecocq, L. Packer, S. Russell, and C. Thanoosing; for funding to SAFEA High-End Foreign Experts Programme of China, Canadian Centre for DNA Barcoding; and for discussion to L. Bailey and S. Mayo.

References

Almeida, E. A. B., Packer, L., Melo, G. A. R. et al. (2018) The diversification of neopasiphaeine bees during the Cenozoic (Hymenoptera: Colletidae). *Zoologica Scripta* 12333: 1–17. https://doi.org/10.1111/zsc.12333

An, J.-D., Huang, J.-X., Shao, Y.-Q. et al. (2014) The bumblebees of North China (Apidae, *Bombus* Latreille). *Zootaxa* 3830: 1–89. https://doi.org/10.11646/zootaxa.3830.1.1

Baum, D. and Smith, S. (2012) *Tree Thinking: An Introduction to Phylogenetic Biology*. Roberts and Company, Greenwood Village.

Bergsten, J., Bilton, D. T., Fujisawa, T. et al. (2012) The effect of geographical scale of sampling on DNA barcoding. *Systematic Biology* 61: 851–869. https://doi.org/10.1093/sysbio/sys037

Bertsch, A., Schweer, H., Titze, A., and Tanaka, H. (2005) Male labial gland secretions and mitochondrial DNA markers support species status of *Bombus cryptarum* and *B. magnus* (Hymenoptera, Apidae). *Insectes Sociaux* 52: 45–54.

Bolton, B. (2007) How to conduct large-scale taxonomic revisions in Formicidae. *Memoirs of the American Entomological Institute* 80: 52–71.

Cameron, S. A., Hines, H. M., and Williams, P. H. (2007) A comprehensive phylogeny of the bumble bees (*Bombus*). *Biological Journal of the Linnean Society* 91: 161–188.

Cameron, S. A., Lozier, J. D., Strange, J. P. et al. (2011) Patterns of widespread decline in North American bumble bees. *Proceedings of the National Academy of Sciences of the United States of America* 108: 662–667.

Campillo, L. C., Barley, A. J., and Thomson, R. C. (2020) Model-based species delimitation: Are coalescent species reproductively isolated? *Systematic Biology* 69: 708–721. https://doi.org/10.1093/sysbio/syz072

de Candolle, A. (1880) *La Phytographie ou l'art de décrire les végétaux considérés sous différents points de vue*. G. Masson, Paris.

De Queiroz, K. (2007) Species concepts and species delimitation. *Systematic Biology* 56: 879–886. https://doi.org/10.1080/10635150701701083

Dillon, R. T. (1984) Geographic distance, environmental difference, and divergence between isolated populations. *Systematic Zoology* 33: 69–82.

Dray, S. and Dufour, A.-B. (2007) The ade4 package: Implementing the duality diagram for ecologists. *Journal of Statistical Software* 22: 1–20.

Drew, L. (2011) Are we losing the science of taxonomy? *BioScience* 61: 942–946. https://doi.org/10.1525/bio.2011.61.12.4

Francoso, E., Zuntini, A. R., Carnaval, A. C., and Arias, M. C. (2016) Comparative phylogeography in the Atlantic forest and Brazilian savannas: Pleistocene fluctuations and dispersal shape spatial patterns in two bumblebees. *BMC Evolutionary Biology*: 16. https://doi.org/10.1186/s12862-016-0803-0

Fujisawa, T. and Barraclough, T. G. (2013) Delimiting species using single-locus data and the Generalized Mixed Yule Coalescent approach: A revised method and evaluation on simulated data sets. *Systematic Biology* 62: 707–724.

Gaston, K. J. (1994) *Rarity*. Springer, London.

Goodwin, Z. A., Harris, D. J., Filer, D., Wood, J. R. I., and Scotland, R. W. (2015) Widespread mistaken identity in tropical plant collections. *Current Biology* 25: R1066–1067. https://doi.org/10.1016/j.cub.2015.10.002

Goulson, D. (2010) *Bumblebees, Behaviour, Ecology, and Conservation* (2nd ed.). Oxford University Press, Oxford.

Guisan, A., Broennimann, O., Engler, R. et al. (2006) Using niche-based models to improve the sampling of rare species. *Conservation Biology* 20: 501–511. https://doi.org/10.1111/j.1523-1739.2006.00354.x

Hebert, P. D. N., Ratnasingham, S., and deWaard, J. R. (2003) Barcoding animal life: cytochrome *c* oxidase subunit 1 divergences among closely related species. *Proceedings of the Royal Society of London (B)* 270: S96–S99. https://doi.org/10.1098/rsbl.2003.0025

Hebert, P. D. N., Cywinska, A., Ball, S. L. et al. (2003) Biological identifications through DNA barcodes. *Proceedings of the Royal Society of London B, Biological Sciences* 270: 313–321.

Hebert, P. D. N., Penton, E. H., Burns, J. M., Janzen, D. H., and Hallwachs, W. (2004) Ten species in one: DNA barcoding reveals cryptic species in the neotropical skipper butterfly *Astraptes fulgerator*. *Proceedings of the National Academy of Sciences* 101: 14812–14817. https://doi.org/10.1073/pnas.0406166101

Hines, H. M. (2008) Historical biogeography, divergence times, and diversification patterns of bumble bees (Hymenoptera: Apidae: *Bombus*). *Systematic Biology* 57: 58-75.

Holland, J. M., Smith, B. M., Storkey, J., Lutman, P. J. W., and Aebischer, N. J. (2015) Managing habitats on English farmland for insect pollinator conservation. *Biological Conservation* 182: 21–222. https://goi.org/10.1016/j.biocon.2014.12.009

Jackson, J. M., Pimsler, M. L., Oyen, K. J. et al. (2018) Distance, elevation and environment as drivers of diversity and divergence in bumble bees across latitude and altitude. *Molecular Ecology* 27: 2926–2942.

Kapli, P., Lutteropp, S., Zhang, J. et al. (2017) Multi-rate Poisson tree processes for single-locus species delimitation under maximum likelihood and Markov chain Monte Carlo. *Bioinformatics* 33: 1630–1638. https://doi.org/10.1093/bioinformatics/btx025

Koch, J. B., Rodriguez, J., Pitts, J. P., and Strange, J. (2018) Phylogeny and population genetic analyses reveals cryptic speciation in the *Bombus fervidus* species complex (Hymenoptera: Apidae). *PLoS ONE* 13. https://doi.org/10.1371/journal.pone.0207080

Lecocq, T., Dellicour, S., Michez, D. et al. (2013) Scent of a break-up: Phylogeography and reproductive trait divergences in the red-tailed bumblebee (*Bombus lapidarius*). *BMC Evolutionary Biology* 13: 263. https://doi.org/10.1186/1471-2148-13-263

(2015) Methods for species delimitation in bumblebees (Hymenoptera, Apidae, Bombus): Towards an integrative approach. *Zoologica Scripta* 44: 281–297. https://doi.org/10.1111/zsc.12107

Lecocq, T., Biella, P., Martinet, B., and Rasmont, P. (2019) Too strict or too loose? Integrative taxonomic assessment of Bombus lapidarius complex (Hymenoptera: Apidae). *Zoologica Scripta*. https://doi.org/10.1111/zsc.12402

Lohse, K. (2009) Can mtDNA barcodes be used to delimit species? A response to Pons et al. (2006). *Systematic Biology* 58: 439–442.

Lou, M. and Golding, G. B. (2012) The effect of sampling from subdivided populations on species identification with DNA barcodes using a Bayesian statistical approach. *Molecular Phylogenetics and Evolution* 65: 765–773. https://doi.org/10.1016/j.ympev.2012.07.033

Luo, A., Lan, H.-Q., Ling, C. et al. (2015) A simulation study of sample size for DNA barcoding. *Ecology and Evolution* 5:

5869–5879. https://doi.org/10.1002/ece3.1846

Magnacca, K. N. and Brown, M. J. F. (2010) Mitochondrial heteroplasmy and DNA barcoding in Hawaiian *Hylaeus (Nesoprosopis)* bees (Hymenoptera: Colletidae). *BMC Evolutionary Biology* 10: 1–16. https://doi.org/10.1186/1471-2148-10-174

Mantel, N. (1967) The detection of disease clustering and a generalized regression approach. *Cancer Research* 27: 209–220.

Martinet, B., Brasero, N., Lecocq, T. et al. (2018) Adding attractive semio-chemical trait refines the taxonomy of *Alpinobombus* (Hymenoptera: Apidae). *Apidologie*. https://goi.org/10.1007/s13592-018-0611-1

Matz, M. V. and Nielsen, R. (2005) A likelihood ratio test for species membership based on DNA sequence data. *Transactions of the Royal Society (Biological Sciences)* 360: 1969–1974.

Meierotto, S., Sharkey, M. J., Janzen, D. H. et al. (2019) A revolutionary protocol to describe understudied hyperdiverse taxa and overcome the taxonomic impediment. *Deutsche Entomologische Zeitschrift* 66: 119–145. https://doi.org/10.3897/dez.66.34683

Meusnier, I., Singer, G. A. C., Landry, J.-F. et al. (2008) A universal DNA mini-barcode for biodiversity analysis. BioMed Central *Genomics* 9: 214[4 pp.]. https://doi.org/10.1186/1471-2164-9-214

Monaghan, M. T., Balke, M., Gregory, T. R., and Vogler, A. P. (2005) DNA-based species delineation in tropical beetles using mitochondrial and nuclear markers. *Philosophical Transactions of the Royal Society (B)* 360: 1925–1933.

Monaghan, M. T., Wild, R., Elliot, M. et al. (2009) Accelerated species inventory on Madagascar using coalescent-based models of species delineation. *Systematic Biology* 58: 298–311.

Neyman, J. (1934) On the two different aspects of the representative method: The method of stratified sampling and the method of purposive sampling. *Journal of the Royal Statistical Society* 97: 558–625.

Packer, L., Sheffield, C. S., Gibbs, J. et al. (2009) The campaign to barcode the bees of the world: Progress, problems, prognosis. In: C. L. Yurrita (ed.) *Memorias: VI Congresso Mesoamericano Sobre Abejas Nativas*. Universidad de San Carlos de Guatemala, Centro de Estudios Conservacionistas, Guatemala, pp. 178–180.

Padial, J. M., Miralles, A., De La Riva, I., and Vences, M. (2010) The integrative future of taxonomy. *Frontiers in Zoology* 7: 16. https://doi.org/10.1186/1742-9994-7-16

Paknia, O., Rajaei, H., and Koch, A. (2015) Lack of well-maintained natural history collections and taxonomists in megadiverse developing countries hampers global biodiversity exploration. *Organisms Diversity and Evolution* 15: 619–629. https://doi.org/10.1007/s13127-015-0202-1

Papadopoulou, A., Bergsten, J., Fujisawa, T. et al. (2008) Speciation and DNA barcodes: Testing the effects of dispersal on the formation of discrete sequence clusters. *Philosophical Transactions of the Royal Society (B)* 363: 2987–2996.

Papadopoulou, A., Monaghan, M. T., Barraclough, T. G., and Vogler, A. P. (2009) Sampling error does not invalidate the Yule-coalescent model for species delimitation: A response to Lohse (2009). *Systematic Biology* 58: 442–444.

Phillips, J. D., Gillis, D. J., and Hanner, R. H. (2018) Incomplete estimates of genetic diversity within species: Implications for DNA barcoding. *Ecology and Evolution*: 1–15. https://doi.org/10.1002/ece3.4757

Potapov, G. S., Kondakov, A. V., Filippov, B .Y. et al. (2019) Pollinators on the polar edge of the Ecumene: Taxonomy, phylogeography, and ecology of bumble bees from Novaya Zemlya. *Zookeys* 866: 85–115. https://doi.org/10.3897/zookeys.866.355084

Prathapan, K. D., Pethiyagoda, R., Bawa, K. S. et al. (2018) When the cure kills: CBD limits biodiversity research. *Science* 360(6396): 1405–1406. https://10.1126/science.aat9844

Rasmont, P. (1984) Les bourdons du genre *Bombus* Latreille sensu stricto en Europe occidentale et centrale (Hymenoptera, Apidae). *Spixiana* 7: 135–160.

Ratnasingham, S. and Hebert, P. D. N. (2007) BOLD: The Barcode Of Life Data system (www.barcodinglife.org). *Molecular Ecology Notes* 2007: 1–10. 10.1111/j.1471-8286.2006.01678.x

(2013) A DNA-based registry for all animal species: The barcode index number (BIN) system. *PLoS ONE* 8: 1–16. https://doi.org/10.1371/journal.pone.0066213

Reinig, W. F. (1939) Die Evolutionsmechanismen, erläutert an den Hummeln. *Verhandlungen der Deutschen zoologischen Gesellschaft (supplement)* 12: 170–206.

Ronquist, F. and Huelsenbeck, J. P. (2003) MRBAYES 3: Bayesian phylogenetic inference under mixed models. *Bioinformatics* 19: 1572–1574.

Schlick-Steiner, B. C., Steiner, F. M., Seifert, B. et al. (2010) Integrative taxonomy: A multisource approach to exploring biodiversity. *Annual Review of Entomology* 55: 421–438. https://doi.org/10.116/annurev-ento-112408-085432

Schmidt, S., Schmid-Egger, C., Morniere, J., Haszprunar, G., and Hebert, P. D. N. (2015) DNA barcoding largely supports 250 years of classical taxonomy: Identifications for Central European bees (Hymenoptera, Apoidea partim). *Molecular Ecology Resources* 15: 985–1000. https://doi.org/10.1111/1755-0998.12363

Strange, J. P., Knoblett, J., and Griswold, T. (2009) DNA amplification from pin-mounted bumble bees (*Bombus*) in a museum collection: Effects of fragment size and specimen age on successful PCR. *Apidologie* 40: 134–139. https://doi.org/10.1051/apido/2008070

Tamura, K., Stecher, G., Peterson, D., Filipski, A., and Kumar, S. (2013) MEGA6: Molecular evolutionary genetics analysis version 6.0. *Molecular Biology and Evolution* 30: 2725–2729.

Vieites, D. R., Wollenberg, K. C., Andreone, F. et al. (2009) Vast underestimation of Madagascar's biodiversity evidenced by an integrative amphibian inventory. *PNAS* 106: 8267–8272. https://doi.org/10.1073/pnas.0810821106

Vogt, O. (1909) Studien über das Artproblem. 1. Mitteilung. Über das Variieren der Hummeln. 1. Teil. *Sitzungsberichte der Gesellschaft naturforschender Freunde zu Berlin* 1909: 28–84.

(1911) Studien über das Artproblem. 2. Mitteilung. Über das Variieren der Hummeln. 2. Teil. (Schluss). *Sitzungsberichte der Gesellschaft naturforschender Freunde zu Berlin* 1911: 31–74.

Wickham, H. (2011) GGPLOT2. *WIREs Computational Statistics* 3: 180–185. https://doi.org/10.1002/wics.147

Williams, P. H. (1991) The bumble bees of the Kashmir Himalaya (Hymenoptera: Apidae, Bombini). *Bulletin of the British Museum (Natural History) (Entomology)* 60: 1–204.

(1998) An annotated checklist of bumble bees with an analysis of patterns of description (Hymenoptera: Apidae, Bombini). *Bulletin of The Natural History Museum (Entomology)* 67: 79–152 [updated at www.nhm.ac.uk/bombus/ accessed 2015].

(2011) Bumblebees collected by the Kyushu University Expeditions to Central Asia (Hymenoptera, Apidae, genus *Bombus*). *Esakia* 50: 27–36.

Williams, P. H., Altanchimeg, D., Byvaltsev, A. et al. (2020) Widespread polytypic species or complexes of local species? Revising bumblebees of the subgenus *Melanobombus* world-wide (Hymenoptera, Apidae, Bombus). *European Journal of Taxonomy* 719: 1–120. https://doi.org/10.5852/ejt.2020.719.1107

Williams, P. H., Berezin, M. V., Cannings, S. G. et al. (2019) The arctic and alpine bumblebees of the subgenus *Alpinobombus* revised from integrative assessment of species' gene coalescents and morphology (Hymenoptera, Apidae, *Bombus*). *Zootaxa*

4625: 1–68. https://doi.org/10.11646/zootaxa.4625.1.1

Williams, P. H., Brown, M. J. F., Carolan, J. C. et al. (2012) Unveiling cryptic species of the bumblebee subgenus *Bombus s. str.* worldwide with COI barcodes (Hymenoptera: Apidae). *Systematics and Biodiversity* 10: 21–56. https://doi.org/10.1080/14772000.2012.664574

Williams, P. H., Bystriakova, N., Huang, J.-X. et al. (2015a) Bumblebees, climate and glaciers across the Tibetan plateau (Apidae: *Bombus* Latreille). *Systematics and Biodiversity* 13: 164–181. https://doi.org/10.1080/14772000.2014.982228

Williams, P. H., Byvaltsev, A. M., Cederberg, B. et al. (2015b) Genes suggest ancestral colour polymorphisms are shared across morphologically cryptic species in arctic bumblebees. *PLoS ONE* 10: 1–26. https://doi.org/10.1371/journal.pone.0144544

Williams, P. H., Cannings, S. G., and Sheffield, C. S. (2016) Cryptic subarctic diversity: A new bumblebee species from the Yukon and Alaska (Hymenoptera, Apidae). *Journal of Natural History* 50: 1–13. https://doi.org/10.1080/00222933.2016.1214294

Williams, P. H., Huang, J.-X., and An, J.-D. (2017) Bear wasps of the Middle Kingdom: A decade of discovering China's bumblebees. *Antenna* 41: 21–24.

Williams, P. H., Huang, J.-X., Rasmont, P. et al. (2016) Early-diverging bumblebees from across the roof of the world: The high-mountain subgenus *Mendacibombus* revised from species' gene coalescents and morphology (Hymenoptera, Apidae). *Zootaxa* 4204: 1–72. https://doi.org/10.11646/zootaxa.4204.1.1

Williams, P. H., Ito, M., Matsumura, T. et al. (2010) The bumblebees of the Nepal Himalaya (Hymenoptera: Apidae). *Insecta Matsumurana* 66: 115–151.

Williams, P. H., Tang, Y., Yao, J. et al. (2009) The bumblebees of Sichuan (Hymenoptera: Apidae, Bombini). *Systematics and Biodiversity* 7: 101–190. https://doi.org/10.1017/S1477200008002843

Williams, P. H., Thorp, R. W., Richardson, L. L., and Colla, S. R. (2014) *Bumble Bees of North America: An Identification Guide.* Princeton University Press, Princeton, NJ.

Zayed, A. and Packer, L. (2005) Complementary sex determination substantially increases extinction proneness of haplodiploid populations. *Proceedings of the National Academy of Sciences* 102: 10742–10746.

Zhang, A.-B., He, L. J., Crozier, R. H., Muster, C., and Zhu, C.-D. (2010) Estimating sample sizes for DNA barcoding. *Molecular Phylogenetics and Evolution* 54: 1035–1039. https://doi.org/10.1016/.ympev.2009.09.014

Zhang, J.-J., Kapli, P., Pavlidis, P., and Stamatakis, A. (2013) A general species delimitation method with applications to phylogenetic placements. *Bioinformatics* 29: 2869–2876. https://doi.org/10.1093/bioinformatics/btt499

10

Cryptic Lineages among Seychelles Herpetofauna

JIM LABISKO, SIMON T. MADDOCK, SARA ROCHA, AND DAVID J. GOWER

Introduction

Amphibian and reptile systematics have been in a state of great flux in the twenty-first century. This flux is not only because of the expansion of the use of molecular genetic data to underpin classifications but also due to substantial increases in the number of described species. For example, the number of recognised extant squamate reptile (lizard and snake) species increased by 27.4 per cent (from 8,396 to 10,954) between February 2008 and August 2020 (Uetz et al. 2020), and 2,646 new species of amphibians have been described in the 16 years 2004–2020; close to one-third (32 per cent) of the 8,276 extant species currently recognised (AmphibiaWeb 2021). The proposed main causes of this increase in recognised species-level diversity are numerous and include: more collaborations; increased in-country expertise and capacity in regions with naturally high diversity (especially Asia and South America); extended spatio-temporal sampling of individuals and populations; new fieldwork; increased molecular data and new analytical tools; new acoustic and behavioural data; development of species distribution modelling techniques; increased application of phenotypic and genetic data; and increased willingness to consider relatively small phenotypic differences (especially where accompanied by molecular genetic disparity) a sufficient basis for the description or elevation of species (e.g. Padial et al. 2010; Hillis 2019; Streicher et al. 2020). Further, molecular genetic data alone have been used (not without debate) in recent years for diagnosis of some new species (e.g. Leache and Fujita 2010; Bauer et al. 2011; Wagner et al. 2014).

The great increase in the number of recognised species of amphibians and reptiles (herpetofauna) has been accompanied by widespread statements about some of this newly discovered diversity being 'cryptic' (e.g. Vieites et al. 2009; Makhubo et al. 2015). Proposed causes of instances of 'cryptic species' (resulting in phenotypically similar but genetically distinct lineages) in herpetofauna include both intrinsic and extrinsic factors such as low dispersal abilities, morphological conservatism, high degrees of homoplasy, small body size, the influence of climate, and physical separation of populations via barriers or vicariance (Stuart et al. 2006; Florio et al. 2012; Fusinatto et al. 2013; Arteaga et al. 2016; Priti et al. 2016; Liu et al. 2018). In this chapter, we consider the 'cryptic species issue' for terrestrial herpetofauna of the Seychelles archipelago (Figure 10.1), a region where divergent genetic lineages (including potentially cryptic taxa) have also been discovered in other faunal groups, including freshwater decapods (Daniels 2011; Cumberlidge and Daniels 2014) and intertidal isopods (Santamaria et al. 2017).

There are several reasons why islands (and island-like biogeographic systems) might be considered worthy of closer examination with respect to cryptic species. Islands often harbour a disproportionate number of endemic taxa, so we might also expect them to frequently harbour cryptic species. Studies of island biodiversity have had a disproportionate impact on the understanding of organismal evolution, including knowledge of phenomena such as speciation, random genetic drift, founder effects, and local adaptation (e.g. Losos 1992; Hartl and Clark 1997; Kolbe et al. 2012; Malenda et al. 2012; O'Neill et al. 2012; Spurgin et al. 2014). In addition, island systems are often held as exemplars of allopatry, and cryptic species are often allopatrically distributed (e.g. Stuart et al. 2006; Vieites et al. 2009; Forlani et al. 2017; Simo-Riudalbas et al. 2017). Allopatry represents a pervasive challenge in systematics, raising questions about how much geno- and phenotypic divergence is enough to warrant specific status, and how much time is required for incipient allopatric species to become full species that would be maintained even on secondary contact. On the other hand, populations on small and/or environmentally heterogenous islands might be especially prone to local adaptation, perhaps making morphologically cryptic species less likely.

Most of the studies of evolution on islands have been on emergent islands and/or of rapidly evolving, often relatively recent radiations (Wallace 1880; Grant 1999; Malhotra and Thorpe 1991; Gillespie 2004; Thorpe et al. 2005; Grant and Grant 2014; Lamichhaney et al. 2015; Thorpe et al. 2015). In contrast, the Seychelles archipelago includes ancient landmasses of continental origin that have been repeatedly (dis)connected. The Seychelles harbour diverse and endemic reptile and amphibian fauna, with differing historical origins and ecologies, including dispersal abilities. In this chapter, we examine intraspecific diversity in native, terrestrial amphibians and reptiles of the Inner Islands of the Seychelles Archipelago (Figure 10.1). We review their taxonomic history and knowledge of their diversity and distribution from phenotypic and molecular genetic perspectives. Based on this review, we ask such questions as: How has species discovery changed through time and with the onset of molecular data? Does the Seychelles harbour cryptic reptile and amphibian lineages? Can patterns of the degree of crypsis be explained by ecology and/or longevity of lineages on the Seychelles?

We use the term 'cryptic species' generally to refer to species-level taxa that were discovered through analyses of molecular rather than phenotypic data. Under this usage, 'cryptic species' mostly lack notable phenotypic divergence from their closest relatives, but there is a continuum, and some might display low to moderate morphological

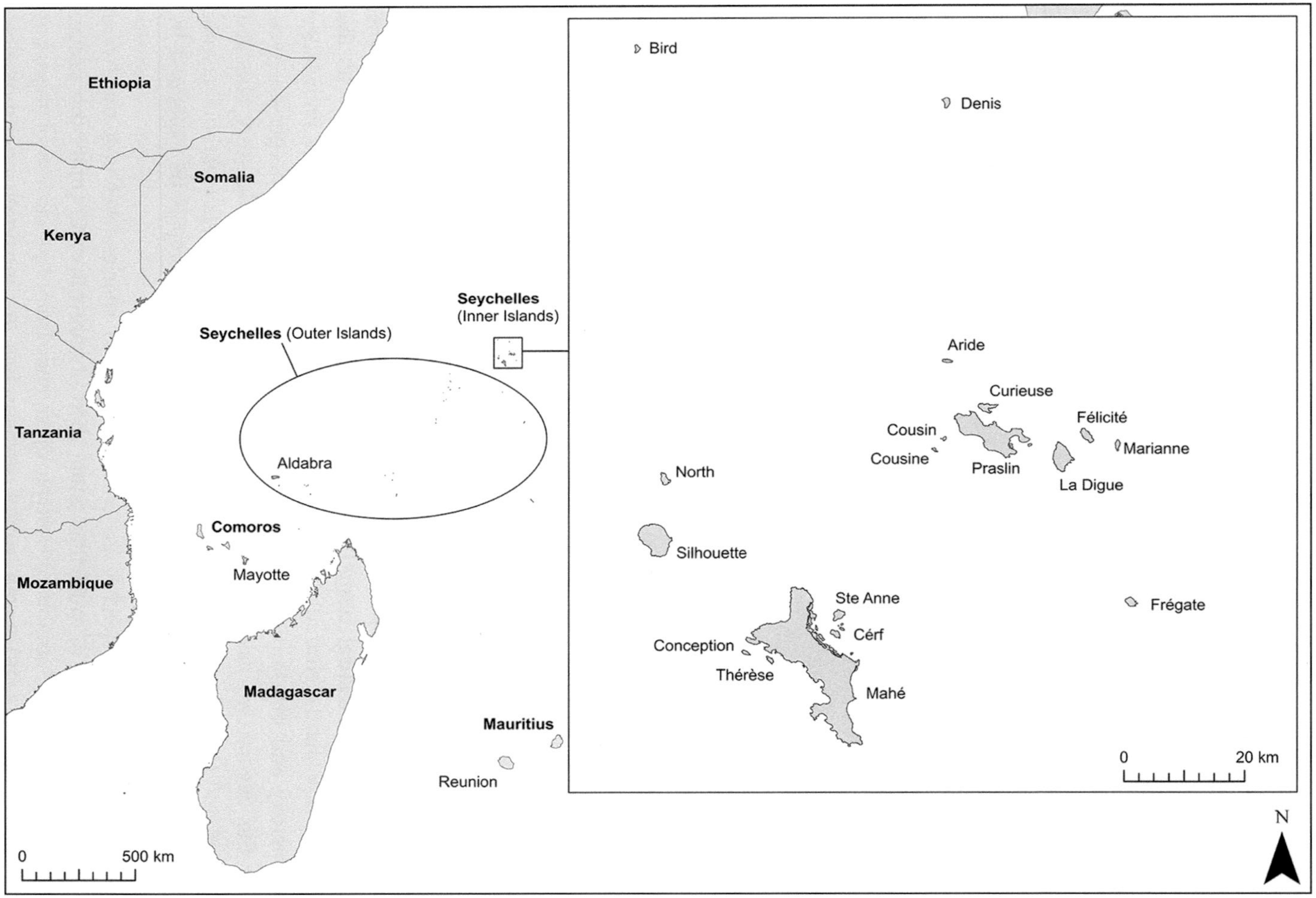

Figure 10.1 Western Indian Ocean region and Seychelles Inner and Outer Islands. The Inner Islands (inset), also include Bird and Denis from the Northern Coral Group.

distinctiveness that was overlooked due to lack of study. We leave further debate about definitions to other contributions to this volume.

10.1 Geophysical and Environmental Setting

The granitic (inner) islands of the Seychelles are of continental origin, once part of the supercontinent Gondwana. Gondwana began to fragment during the Middle Jurassic (~170 million years ago: Ma), splitting first into West and East Gondwana (Coffin and Rabinowitz 1987; Chatterjee et al. 2013). In a second splitting phase during the Early Cretaceous (~118 Ma), a rift formed in East Gondwana, separating what would become Antarctica and Australia, and releasing a smaller fragment composed of Sri Lanka, India, the Laxmi Ridge (a now submarine feature underlain possibly by a sliver of continental crust that once lay between Seychelles and India: see e.g. Yatheesh 2020), Seychelles, and Madagascar (Chatterjee et al. 2013). A third phase of fragmentation (Late Cretaceous, ~88 Ma) saw the separation of Africa from Madagascar, and of the Indian subcontinent from what became the Seychelles microcontinent, the latter, along with the Laxmi Ridge, having been originally situated between Madagascar and India prior to the onset of Gondwanan breakup (Chatterjee et al. 2013). During this period (~65 Ma), eruption of the Deccan Traps plumed basalt across large portions of India (Courtillot et al. 1986; Courtillot et al. 1988; Duncan and Pyle 1988), and the K-Pg boundary mass-extinction event occurred. The separation of Seychelles from the Indian plate possibly started with separation of India from Seychelles and the Laxmi Ridge ca. 71–56 Ma (see summary of estimated timing of opening of the Gop Rift presented by Gower et al. 2016), with the split between Seychelles and the Laxmi Ridge more precisely dated as beginning by 63.4 Ma (Armitage et al. 2011), with unambiguous separation by 62–61.7 Ma (Collier et al. 2008). India continued its journey northward, colliding with Asia during the Early Eocene (~50 Ma), by which time Seychelles had already been isolated in the Western Indian Ocean for ~13 million years (Chatterjee et al. 2013; Collier et al. 2008).

Consideration of the origin of Seychelles aided Alfred Wegener (1924) in developing his landmark theory of continental drift. It is now recognised that isolated blocks of islands such as Seychelles are rare, with only two (the other being Madagascar) formally diagnosed in the strictest sense (Ali 2018). During the region's separation from India, volcanic activity from the Deccan Traps might have made large parts of the Seychelles microcontinent uninhabitable, evidenced by basalt plumes as well as extensive felsic and mafic aspects in Seychelles geology (Ganerød et al. 2011; Owen-Smith et al. 2013; Shellnutt et al. 2017). While rifting from India, Seychelles is likely to have transported numerous Gondwanan lineages with it but volcanism may have had a heavy toll on biota in affected regions (Thewissen and McKenna 1992; Bossuyt and Milinkovitch 2001).

Today, Seychelles is tectonically stable (Kopp et al. 2009; Ali 2018). The main, Inner Islands of the archipelago—a group of 45, also known as the Granitic Seychelles—are situated 4–5°S to 55–56°E in the western Indian Ocean, and form a terrestrial area <250 km^2, emerging from the ocean atop a now submerged plateau of some 44,000 km^2. This submerged plateau would reconnect all of the inner islands with a sea-level drop of only 60 m (Ryan et al. 2009; Rocha et al. 2013; Ali 2018). The archipelago further includes ca. 70 low-lying coralline islands (the Outer Islands) that lie in a southerly distribution from the Inner Islands, stretching to within ca. 260 km of the northernmost tip of Madagascar. This low-lying territory includes the Aldabra Atoll UNESCO World Heritage Site, positioned ca. 625 km from the East African coast of Mozambique.

Notwithstanding the arrival on Seychelles of various flora and fauna via transoceanic dispersal, which notably includes strong Asian as well as African affinities (Cogan 1984; Procter 1984a; Warren et al. 2010), since the end of the Eocene (~34 Ma), a primary influence on the region's biota is likely to have been eustatic fluctuations in sea level driven by variations in the volume of polar ice-sheets (Miller et al. 2005). Combined with regional effects of hydro-isostatic uplift and volcanic subsidence (Montaggioni & Hoang 1988; Colonna et al. 1996; Camoin et al. 2004), these factors will have driven irregular cycles of isolation and reconnection between the Inner Islands. At the commencement of the last deglaciation (18–17 thousand years ago), the western Indian Ocean experienced a gradual rise in sea level from ~110 m below present (Camoin et al. 2004) to today's levels, with little fluctuation beyond 2 m during the last two thousand years (Woodroffe et al. 2015). Therefore, over both recent and deep-time, geologic and environmental conditions have likely influenced a variety of phylogeographic patterns in Seychelles biota, which will have also been subject to immigration, extinction, and speciation as expected to some extent by island biogeography theory (MacArthur and Wilson 1967).

Seychelles today has a tropical climate, with a mean annual temperature of ~26.5°C. Two distinct wind-dominated seasonal patterns occur: the southeast monsoon (May to September) and the wetter northwest monsoon (November to March). The first human settlers arrived on the islands in the mid-eighteenth Century, and within 100 years had cleared much of the forests and mangroves, and replaced native and endemic vegetation with spice, cotton, and coconut plantations (Baker 1877; Vesey-Fitzgerald 1940; Kueffer et al. 2013). By the late 1870s the only large trees remaining were found in inaccessible areas (Baker 1877), and many animal species, including crocodiles and giant tortoises, had been exterminated (Stoddart 1984b), although a relict population of tortoises survives on Aldabra (Stoddart 1984a).

Despite the devastating impacts of anthropogenic activity on the islands, losses may have been limited due to the dispersed nature of the Inner Islands, protecting some from introduced predators (Simberloff and Cox 1987; Rocamora and Henriette 2015). Similarly, habitat destruction left refugia (Baker 1877; Wallace 1892; Simberloff and Cox 1987; Kueffer et al. 2013), and some native flora and fauna have since recolonised the largely mountainous terrain and long-abandoned plantations. Other organisms seemingly adapted to new environmental conditions – tree species formerly present in lowland forest survive as miniaturised forms in marginal habitats (Kueffer et al. 2010; Kueffer and Kaiser-Bunbury 2014).

Alfred Russel Wallace (1880) identified Seychelles as an ancient continental island group, characterised by the degree of separation from the nearest neighbouring landmass, the surrounding deep ocean, and by possessing a biota with biological links to both the closest (i.e. Africa, Madagascar) and more distant (i.e. India) landmasses. This characterisation was recently reappraised, and physical island types redefined, with Seychelles classed as *isolated block islands*, with a biota falling into two defined categories: (1) *deep time vicariant* and (2) *overwater dispersed* forms (Ali 2017, 2018). Although the origins of most of Seychelles' endemic vertebrate fauna can be ascribed to transoceanic dispersal, at least two amphibian clades with Gondwanan origins are truly deep-time vicariants (Ali 2018).

10.2. Amphibian and Reptile Taxon Accounts

In this section we summarise knowledge on the diversity of Seychelles herpetofauna, paying special attention to studies relevant to cryptic species: taxonomy, variation within and among islands, and phylogeography. We here focus on extant, native, terrestrial species. We exclude marine species (some turtles and snakes), probable aliens, species that do/did not naturally occur on the Inner (granitic) Islands, and taxa that are widespread beyond Seychelles. Data are from the literature, including online resources (especially Amphibian Species of the World, the IUCN Red List, and the Reptile Database). We summarise information on distribution across Seychelles granitic islands in Table 10.1, and taxonomic history of species and subspecies in Online Appendix 1 available via the Data Portal of the Natural History Museum, London (data.nhm.ac.uk: https://doi.org/10.5519/007423wv). Taxa representing examples of the main taxonomic groups and/or those most frequently encountered species are shown in Figure 10.2.

10.2.1 Amphibians

Anura, Sooglossidae

Sechellophryne gardineri (Boulenger, 1911)
Sechellophryne pipilodryas (Gerlach and Willi, 2002)
Sooglossus sechellensis (Boettger, 1896)
Sooglossus thomasseti (Boulenger, 1909)

Seychellois: Pti grenwir

Two genera and four species comprise the Seychelles endemic family Sooglossidae, the taxonomy of which has varied greatly in both familial and generic assignment, although no subspecies have ever been described, and no species names are currently in synonymy. Until recently, evolutionary relationships of Sooglossidae remained unclear but multi-gene and phylogenomic studies have since reached solid agreement that they are sister to the peninsular Indian Nasikabatrachidae Biju & Bossuyt, 2003 together comprising a sister group to Ranoidea Rafinesque, 1814 (Pyron 2014; Frazão et al. 2015; Feng et al. 2017; Hime et al. 2021). The divergence between sooglossids and their Indian sister taxon is ancient, which is consistent with the hypothesis that the presence of sooglossids on Seychelles is best explained by Gondwanan vicariance.

Sooglossids—one of only two amphibian families endemic to an archipelago—are restricted to the three largest inner islands; Mahé, Praslin, and Silhouette, but are not evenly distributed, with only *Sooglossus sechellensis* occurring on all three (Taylor et al. 2012; Labisko et al. 2015; Labisko et al. 2019) (Figure 10.3). Both *Sooglossus thomasseti*

Table 10.1 Distribution of endemic Seychelles herpetofauna across the Inner Islands. Islands and islets are presented in approximate size order by group (see also Figures 10.1 and 10.3). The giant tortoise *Aldabrachelys gigantea* is not included because the only wild population is found on the Outer Island of Aldabra (see main text for details). * includes subspecies. Key: Y = present; Y? = genus present (species undetermined: see main text); ? = unconfirmed. NB. There are two Round Islands: one northeast of Mahé, one southeast of Praslin.

	Northern group																Southern group									
	Praslin	**La Digue**	**Curieuse**	**Felicité**	**Frégate**	**Marianne**	**Grande Soeur**	**Aride**	**Petite Soeur**	**Cousin**	**Cousine**	**Recifs**	**Round**	**Booby**	**St. Pierre**	**Cocos**	**Mahé**	**Silhouette**	**St. Anne**	**North**	**Cerf**	**Thérèse**	**Conception**	**Anonyme**	**Mammelles**	**Round**
Amphibians																										
Sechellophryne gardineri																	Y	Y								
Sechellophryne pipilodryas																		Y								
Sooglossus sechellensis	Y																Y	Y								
Sooglossus thomasseti																	Y	Y								
Tachycnemis seychellensis	Y	Y															Y	Y								
Grandisonia alternans	Y	Y		Y	Y												Y	Y								
Grandisonia larvata	Y	Y		Y													Y	Y	Y							
Grandisonia seychellensis	Y																Y	Y								
Hypogeophis brevis																	Y									
Hypogeophis montanus																	Y									
Hypogeophis pti	Y																									
Hypogeophis rostratus	Y	Y	Y	Y	Y		Y										Y	Y	Y		Y					
Praslinia cooperi																	Y	Y								

	Northern group																Southern group									
	Praslin	**La Digue**	**Curieuse**	**Felicité**	**Frégate**	**Marianne**	**Grande Soeur**	**Aride**	**Petite Soeur**	**Cousin**	**Cousine**	**Recifs**	**Round**	**Booby**	**St. Pierre**	**Cocos**	**Mahé**	**Silhouette**	**St. Anne**	**North**	**Cerf**	**Thérèse**	**Conception**	**Anonyme**	**Mammelles**	**Round**
Reptiles																										
Ailuronyx seychellensis	Y	Y?	Y?	Y?	Y	Y?	Y	Y		Y	Y		Y				Y	Y	Y			Y		Y		Y?
Ailuronyx tachyscopaeus	Y	Y	Y	Y			Y										Y	Y			Y		Y			
Ailuronyx trachygaster	Y																									
*Phelsuma astriata**	Y	Y	Y		Y		Y	Y		Y	Y						Y	Y	Y		Y					
*Phelsuma sundbergi**	Y	Y	Y	Y	Y	Y	Y		Y				Y			Y	Y	Y	Y	Y	Y	Y	Y			Y
Urocotyledon inexpectata	Y	Y	Y	Y	Y		Y	Y	Y	Y	Y			Y		Y	Y	Y		Y						
Archaius tigris	Y																Y	Y								
Janetaescincus braueri																	Y?	Y?								
Janetaescincus veseyfitzgeraldi	Y?	Y?	Y?	Y?	Y?												Y?	Y?								
Pamelaescincus gardineri	Y	Y	Y		Y		Y	Y		Y	Y		Y				Y	Y			Y					
Trachylepis sechellensis	Y	Y	Y	Y	Y	Y	Y	Y	Y	Y	Y		Y			Y	Y	Y	Y	Y	Y	Y	Y	Y		Y
Trachylepis wrightii					Y			Y		Y	Y	Y		?	Y										Y	
Lycognathophis seychellensis	Y	Y			Y			Y		Y	Y						Y	Y								
Lamprophis geometricus	Y	Y			Y												Y	Y								
Pelusios castanoides intergularis	Y	Y			Y					Y							Y	Y			Y					
Pelusios subniger parietalis	Y	Y			Y					Y							Y	Y			Y					

Figure 10.2 Representative examples of Seychelles' amphibian and reptile taxa. A: Sooglossidae – *Sooglossus thomasseti* (J. Labisko); B: Hyperoliidae – *Tachycnemis seychellensis* (S. T. Maddock); C: Indotyphlidae – *Praslinia cooperi* (D. J. Gower & S. T. Maddock); D: Gekkonidae – *Phelsuma astriata* (J. Labisko); E: Gekkonidae – *Urocotyledon inexpectata* (J. Labisko); F: Chamaeleonidae – *Archaius tigris* (J. Labisko); G: Scincidae – *Trachylepis sechellensis* (J. Labisko); H: Colubridae – *Lycognathophis seychellensis* (S. T. Maddock).

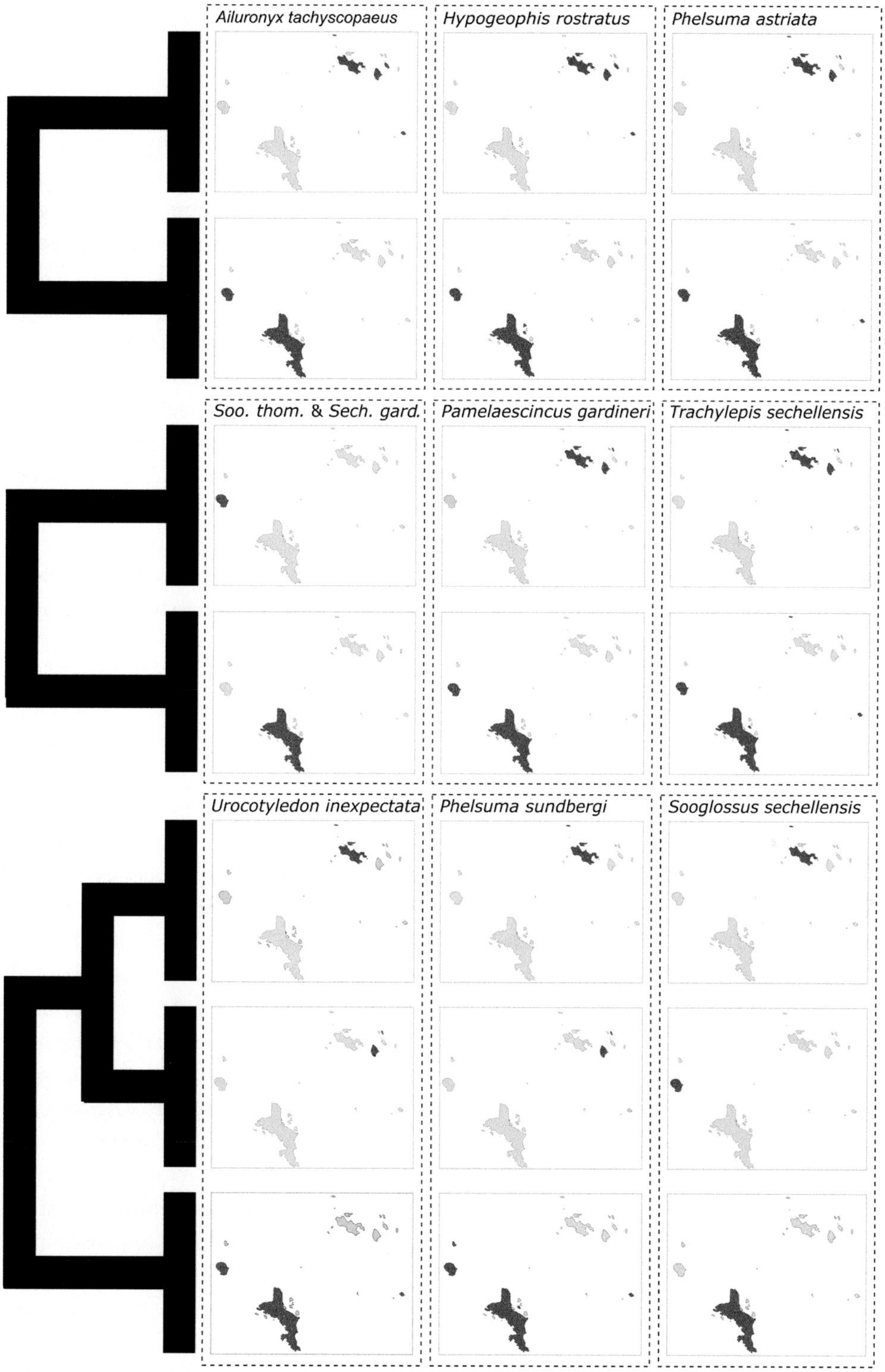

Figure 10.3 A schematic illustrating the north-south pattern of evolutionary relationships among some Seychelles reptile and amphibian species. Dashed boxes contain multiple maps indicating genetic variation for the taxa shown; each map depicts distinct clades with dark grey islands referring to populations belonging to that clade. Pale grey islands indicate either unsampled localities (or that the species is not present) or that the taxa present are a different clade shown on the corresponding map(s) for that group. See main text and Table 10.1 for further details relating to the phylogenetic patterns depicted. *Soo. thom.* and *Sech. gard.* are *Sooglossus thomasseti* and *Sechellophryne gardineri*, respectively.

(Figure 10.2A) and *Sechellophryne gardineri* are found on Mahé and Silhouette (Nussbaum 1984a; Labisko et al. 2015; Labisko et al. 2019), while *Sechellophryne pipilodryas* occurs only on Silhouette (Gerlach and Willi 2002). The presence of sooglossid frogs on Praslin is a relatively recent discovery (Taylor et al. 2012) and subsequent assessment confirmed that *Sooglossus sechellensis* on Praslin are an evolutionarily distinct lineage and not a recent introduction, with 2.0–4.5 per cent sequence divergence in mitochondrial *16s rRNA* between the three island populations (Labisko et al. 2019). The same study also revealed that Mahé and Silhouette populations of *Sooglossus thomasseti* and *Sechellophryne gardineri* are reciprocally monophyletic and notably divergent (2.0 per cent and 3.6 per cent sequence divergence in *16s*, respectively). The validity of each island-specific lineage was further supported by numerous geographically explicit haplotypes across four nuDNA markers, in addition to independent analyses using multispecies coalescent models. Pending assessment of other commonly used means for population- and species-level differentiation in anurans (e.g. morphology, bioacoustics), each island lineage of *Sooglossus sechellensis*, *Sooglossus thomasseti*, and *Sechellophryne gardineri* has been ascribed as candidate species, further assessment of which may yet necessitate taxonomic review (Labisko et al. 2019).

Anura, Hyperoliidae

Tachycnemis seychellensis (Duméril and Bibron, 1841)

Seychellois: Krapo

The Seychelles treefrog, *Tachycnemis seychellensis* (Figure 10.2B), is the only treefrog species found within Seychelles and belongs to the African and Malagasy family Hyperoliidae. *Tachycnemis* Fitzinger, 1843 is a monotypic genus and the single recognised species is distributed across the four largest islands of the granitic Seychelles: Mahé, Praslin, Silhouette, and La Digue. On Mahé and Praslin the species is common and widespread, whereas on Silhouette and La Digue very few populations exist (pers. obs.). The taxonomic history of the species is complicated, with the original description of the species by Tschudi (1838) considered unavailable, and consequently the species name is attributed to Duméril and Bibron (1841). Only in 1981 (Dubois 1981) was *T. seychellensis* moved into *Tachycnemis* from *Eucnemis* to distinguish this distinct hyperoliid. Much earlier, *Megalixalus infrarufurus* Günther, 1869 was synonymised with *T. seychellensis* by Peters (1881).

Tachycnemis is thought to have colonised the Seychelles via transoceanic dispersal, splitting from its sister genus (*Heterixalus* Laurent, 1944) in Madagascar approximately 35–10 Ma (Crottini et al. 2012; Maddock et al. 2014). Nussbaum and Wu (1995) identified morphological differences between distinct populations of *T. seychellensis*, both between and within islands (Mahé). They refrained from making any taxonomic changes despite these morphological differences because variation between populations within Mahé were almost as high as between island populations. Maddock et al. (2014) applied a molecular genetic approach to four (all islands) of the five populations investigated by Nussbaum and Wu (1995) and identified a lack of major mtDNA and nuDNA geographic structuring in 52 samples, supporting the morphological results that a single species should be recognised.

As some substructure was observed, despite multiple clades for each island population (except La Digue), the results were interpreted as indicating past rapid range expansion or multiple admixture events (Maddock et al. 2014).

Gymnophiona, Indotyphlidae

Grandisonia alternans (Stejneger, 1893)
Grandisonia larvata (Ahl, 1934)
Grandisonia seychellensis (Boulenger, 1911)
Hypogeophis brevis Boulenger, 1911
Hypogeophis montanus Maddock, Wilkinson and Gower, 2018
Hypogeophis pti Maddock, Wilkinson, Nussbaum and Gower, 2017
Hypogeophis rostratus (Cuvier, 1829)
Praslinia cooperi Boulenger, 1909

Seychellois: Leverter nwar

The Seychelles caecilians are monophyletic, comprising eight currently recognised species in three genera endemic to Seychelles: the monotypic *Praslinia* Boulenger, 1909, three species of *Grandisonia* Taylor, 1968, and four species of *Hypogeophis* Peters, 1880 (Nussbaum 1984a; Hass et al. 1993; Hedges et al. 1993; Loader et al. 2007; Roelants et al. 2007; Gower et al. 2011; Kamei et al. 2012; San Mauro et al. 2014; Maddock et al. 2017; Maddock et al. 2018). The Seychelles caecilians are thought to have remained isolated from confamilials since Seychelles broke away from India ca. 63 Ma (Collier et al. 2008; Armitage et al. 2011; Chatterjee et al. 2013). As with other amphibians within the Seychelles, they are restricted to the ancient granitic islands, with any single species occurring on a maximum of ten islands (*H. rostratus*; Nussbaum 1984a). All but one species (*Hypogeophis pti*, which is restricted to Praslin: Maddock et al. 2017) can be found on the largest island, Mahé.

Although few in number, the small radiation of Seychelles caecilians constitutes 4 per cent of extant global caecilian diversity (214 species: Frost 2020; AmphibiaWeb 2021). The Seychelles taxa include the smallest known extant caecilians (*H. brevis*, *H. pti* and *H. montanus*) (Maddock et al. 2017, 2018) and a taxon with one of the smallest known geographic ranges of any caecilian (*H. montanus*, $<2\,km^2$: Maddock et al. 2018). In addition, Seychelles caecilians have diverse life-history strategies, including possibly the most extended and aquatic larval stage of any teresomatan caecilian (*Praslinia cooperi*; Figure 10.2C), species with highly abbreviated larval stages (*Grandisonia larvata* and *G. sechellensis*), and species with no larval stage (*H. rostratus*) (San Mauro et al. 2014). Most species are seemingly obligatorily fossorial as adults, but at least *H. rostratus* is somewhat semiaquatic (Nussbaum 1984a).

In addition to the eight species of Seychelles caecilian currently recognised, four have been synonymised: *Hypogeophis guentheri* was described by Boulenger (1882) as being from 'Zanzibar' (probably in error: Parker 1958) and relegated to a subspecies of *H. rostratus* by Taylor (1968); *Dermophis flaviventer* was described as being from Seychelles (no island reported) by Ahl (1926) and synonymised with *G. sechellensis* by Parker (1941); *Hypogeophis angusticeps* was described as being from Mahé by Parker (1941) and synonymised with *G. larvata* by Taylor (1968); and *Grandisonia diminutiva* was described as being from Praslin by Taylor (1968) and synonymised with *G. sechellensis* by Wilkinson and Nussbaum (2006).

The most common and widespread Seychelles caecilian, *H. rostratus*, has been subject to the greatest level of taxonomic analysis for the clade, with up to four subspecies recognised. Using genetic (mtDNA and nuDNA sequences, and AFLPs) and morphological data, Maddock et al. (2020) found notable geographic structuring that is largely congruent with the subspecies *H. r. praslini* Parker 1958 for a northern island lineage and *H. r. rostratus* Parker 1958 for a southern island lineage (with mean between-group p-distances of 3.7 per cent) (Figure 10.3). The lack of precise locality data for the types of the remaining two subspecies (*H. r. lionneti* Taylor 1969 and *H. r. guentheri* Taylor 1968) (Online Appendix 1) further complicates attempts to establish correspondence between named subspecies and the genetic lineages and morphological clusters identified by Maddock et al. (2020). For this reason, and because individuals from the eastern island of Frégate are most similar morphologically to the southern island group, but genetically to the northern island group, Maddock et al. (2020) recommended that no subspecies should currently be recognised.

Most molecular genetic work on Seychelles caecilians to date has focussed on inferring interspecific relationships, which remain incompletely resolved (Hedges et al. 1993; Loader et al. 2007; Roelants et al. 2007; Gower et al. 2011; Kamei et al. 2012; Pyron 2014; San Mauro et al. 2014; Maddock et al. 2016; Maddock et al. 2018). Few studies have quantitatively assessed intraspecific morphological or genetic variation in caecilians with intense sampling. The tools are now available for detailed studies (e.g. Adamson et al. 2016; Maddock et al. 2016; Marshall et al. 2019; Maddock et al. 2020), and we anticipate that thorough analyses of variation will see renewed interest in taxonomy, possibly leading to the identification of subspecies or previously unrecognised species within currently recognised taxa.

10.2.2 Reptiles

Squamata, Gekkonidae, *Ailuronyx*

Ailuronyx seychellensis (Duméril & Bibron, 1834)
Ailuronyx tachyscopaeus Gerlach & Canning, 1996
Ailuronyx trachygaster (Duméril & Bibron, 1851)

Seychellois: Lezar bronz

The bronze geckos, *Ailuronyx* Fitzinger, 1843 comprise three species, all endemic to the Seychelles Inner Islands. No subspecies have been described and there are no junior synonyms. Dating analyses have yet to be carried out, but closest living relatives of *Ailuronyx* are within sub-Saharan Africa (Gamble et al. 2012). The most widespread species, *A. seychellensis*, is reported from at least 12 islands, *A. tachyscopeus* is reported from nine islands (sympatric with *A. seychellensis* on at least four), while *A. trachygaster* is endemic to Praslin, and restricted to a single area of less than 10 km^2 (reports from Silhouette have never been verified and are currently best considered as questionable). Rocha et al. (2017) used mtDNA and nuDNA sequence data to assess molecular genetic variation in 47 *A. seychellensis* (from six islands), 47 *A. tachyscopaeus* (eight islands), and 11 *A. trachygaster* (one island). MtDNA variation within *A. trachygaster* is very low but is considerable and strongly geographically structured in *A. seychellensis* and *A. tachyscopaeus*. Within *A. seychellensis*, two deeply divergent clades were discovered; from northeastern islands (Praslin, Aride, Cousine, Frégate) and southwestern islands (Mahé, Silhouette). Within *A. tachyscopaeus*, three major haplogroups were identified; from southern Mahé, from other southwestern island localities (Mahé, Silhouette, Cerf), and from northern islands (Praslin, Curieuse, La Digue, Grande Soeur, Felicité) (Figure 10.3). NuDNA is less variable intraspecifically, and incompletely sorted with respect to mtDNA groups. Species delimitation analyses found strong support for *A. seychellensis* containing at least one currently unrecognised species; support for the same within *A. tachyscopaeus* was also found, but less strong.

Rocha et al. (2017) also found some indications of morphological intraspecific variation being congruent with their molecular data, but firm conclusions were constrained by limited sampling for morphological characters: six variables in adults of 14 *A. seychellensis* (from two islands), 15 *A. tachyscopaeus* (three islands), and three *A. trachygaster* (one island). Rocha et al. (2017) discussed limitations in their sampling of some populations and of nuDNA, and also stressed that additional data generation and analyses are required to determine whether the lineages they discovered are morphologically distinct.

Squamata, Gekkonidae, *Phelsuma*

Phelsuma astriata Tornier, 1901
Phelsuma sundbergi Rendahl, 1939

Seychellois: Lezar ver

The day geckos *Phelsuma* Gray, 1825 comprise a widespread and common genus restricted to the western Indian Ocean, with the exception of one species that occurs on the Andaman Islands. Of the ca. 52 currently recognised species—most of which occur on Madagascar and Mauritius—Seychelles is home to four, distributed across the granitic and coralline islands. As the common name suggests, *Phelsuma* are diurnal, unlike many other geckos. Two of the Seychelles species, *P. abbotti* Stejneger, 1893 and *P. laticauda* (Boettger, 1880), are distributed more broadly across the western Indian Ocean (Raxworthy et al. 2007). In the Seychelles, *P. abbotti* is found on the coralline islands of Aldabra and Assumption. The native status of *P. laticauda* within the Seychelles is unclear (Gerlach 2008). Seychelles *Phelsuma* are more closely related to western Indian Ocean than to Andaman island congeners, though dating analyses have yet to be carried out (Rocha et al. 2010).

The two endemic *Phelsuma* in the Seychelles, *P. astriata* (Figure 10.2D) and *P. sundbergi*, are widespread, and like *Trachylepis* (see next subsection) are the most frequently encountered native lizards, occurring extensively around human habitation on granitic as well as coralline islands. These two endemics are sister species (Rocha et al. 2010) and have been the subject of extensive morphological (Cheke 1982; Gardner 1984, 1986b, 1987; Radtkey 1996; Rocha et al. 2010) and genetic studies (Rocha et al. 2010, 2013). Within each of the nominal species, numerous subspecies have been described: three for *P. astriata* and six for *P. sundbergi*. By analysing mtDNA and nuDNA for 73 *P. astriata* and 126 *P. sundbergi* from across their known ranges, including all known morphological variation (subspecies) in each species, Rocha et al. (2013) identified two distinct geographically structured clades in each species: a northern and a southern island lineage (Figure 10.3). The genetic data support the recognition of two subspecies for *P. astriata* (*P. a. astriata* Tornier 1901 from the southern islands and *P. a. semicarinata* Cheke 1982 from the northern islands) and three for *P. sundbergi* (*P. s. sundbergi* Rendahl 1939, *P. s. longinsulae* Rendahl 1939 and *P. s. ladiguensis* Böhm & Meier 1981). *Phelsuma s. longinsulae* refers to lizards within the southern granitic islands. Within the northern group Rocha et al. (2013) identified low levels of substructuring between a clade consisting of lizards from Praslin + Curieuse (*P. s. sundbergi*) and from La Digue + Grande Soeur (*P. s. ladiguensis*), which is broadly concordant with morphological analyses (Gardner 1986a, 1987). Both *P. astriata* and *P. sundbergi* share haplotypes between Inner and Outer Island populations, suggesting recent introduction or dispersal to the latter (Rocha et al. 2013).

Squamata, Gekkonidae, *Urocotyledon*

Urocotyledon inexpectata (Stejneger, 1893)

Seychellois: Lezar disik

The Seychelles prehensile- or sucker-tailed gecko (Figure 10.2E) is endemic to the Seychelles; its four currently recognised extant congeners are restricted to tropical mainland Africa (e.g. Uetz et al. 2020), though comprehensive, dated analyses of phylogeny have yet to be carried out. It appears to be largely a specialist of rock crevices, and thus has been proposed to have a relatively low dispersal capacity (Rocha et al. 2011). It has been recorded from 15 Seychelles islands (Gerlach and Ineich 2006e). No subspecies have been described, and there are no junior synonyms. Rocha et al. (2011) generated and analysed mtDNA and nuDNA sequence data from 51 individuals of *U. inexpectata* sampled from eight islands. They found high levels of genetic diversity partitioned into two deeply divergent (95 per cent posterior density interval ca. 13.9–3.4 My) lineages; from northeastern islands (Praslin, Curieuse, Aride, Grande Soeur, La Digue, Frégate) and from southwestern islands (Mahé, Silhouette) (Figure 10.3). Within each of these two main lineages, there was additional genetic structure, again substantially geographically partitioned by island; these splits were estimated to be much younger (posterior density intervals to as low as 0.18 Ma). Rocha et al. (2011) considered the two main genetic lineages within *U. inexpectata* to be cryptic and 'good candidates to be considered as full species'. A lack of detailed morphological data prevented further conclusions or formal taxonomic action (DNA data were generated from non-lethal tail tip samples).

Squamata, Iguania, Chamaeleonidae

Archaius tigris (Kuhl, 1820)

Seychellois: Kameleon

Two chameleon taxa—*Chamaeleo scychellensis* (Kuhl, 1820) and *C. tigris* (Kuhl, 1820) (Figure 10.2F)—were described from Seychelles, but Duméril and Bibron (1836) proposed their synonymy. Placed in the genus *Calumma* (from Madagascar) by Klaver and Böhme (1986), subsequent molecular phylogenetic work found *C. tigris* to be sister taxon to *Rieppeleon* (from mainland Africa), resulting in resurrection of the monotypic genus

Archaius Gray, 1865 (Townsend et al. 2011), and establishing that the ancestral *Archaius* likely arrived in Seychelles via transoceanic dispersal from Africa during the Eocene-Oligocene (~34 Ma) (Raxworthy et al. 2002; Townsend et al. 2011; Tolley et al. 2013).

Seychelles chameleons are known to occur only on the three largest islands, Mahé, Silhouette, and Praslin (Rendahl 1939; Gerlach 2008; Townsend et al. 2011) despite recent, extensive surveys across the archipelago (C. Raxworthy pers. comm). An introduced population was thought to occur on Zanzibar (Rendahl 1939), although this is more likely the result of incorrect labelling during specimen transportation (Gerlach 1999). Variation in DNA, morphology, and population ecology may exist between islands (C. Raxworthy pers. comm.) but it is currently unclear whether this will translate to recognition of more (sub)specific taxa.

Squamata, Scincidae, *Janetaescincus*

Janetaescincus braueri (Boettger 1896)
Janetaescincus veseyfitzgeraldi (Parker 1947)

Seychellois: Lezar fen

Janetaescincus Greer, 1970 is a genus of (semi)fossorial skink endemic to the Seychelles comprising two species (*J. braueri* and *J. veseyfitzgeraldi*), neither of which contain junior synonyms or have had subspecies described. Although these two species are currently recognised as valid (e.g. Uetz et al. 2020), the status of *J. veseyfitzgeraldi* as a distinct species has been questioned on the basis of morphology (e.g. Cheke 1984; Bowler 2006). Given the taxonomic uncertainty, it is difficult to verify distributional records but *J. braueri* has most recently been considered to occur on as few as two or three islands and *J. veseyfitzgeraldi* on as many as six or seven (Gerlach and Ineich 2006a, 2006b; Rocha et al. 2009; Harris et al. 2015) (Figure 10.3). *Janetaescincus* is sister to the Seychelles endemic *Pamalaescincus* (see next section), with their closest living relatives thought to be among African and/or Malagasy scincines, but available phylogenies (e.g. Brandley et al. 2005; Pyron et al. 2013) do not include some Asian lineages, and dating analyses have yet to be carried out.

Harris et al. (2015) generated and analysed mtDNA and nuDNA from 75 *Janetaescincus* individuals from six of the seven islands from which the genus has been recorded. For mtDNA, four distinct, deeply divergent (uncorrected p-distances for *cytb* ca. 4.0–16 per cent) clades were found. There is some geographic structure, with all 13 samples from La Digue and all but one of 12 from Frégate constituting one of the four major lineages. Silhouette samples comprised a further two (not sister) clades, and the fourth lineage was a mixture of Mahé, Silhouette, Praslin, and Curieuse samples, plus one from Frégate. There was less variation in the two nuDNA markers, but the mtDNA lineages could partly be detected among haplotype clusters. Harris et al.'s (2015) DNA data found partial support (on Silhouette) for at least some of the major clades being elevationally restricted, which is relevant taxonomically because Gerlach and Ineich (2006a, 2006b) reported that *J. veseyfitzgeraldi* and *J. braueri* were low- and high-elevation species, respectively. Harris et al. (2015) inferred that the divergences among the four major clades they identified occurred between ca. 19–1 Ma, most much older than Pleistocene sea-level fluctuations.

Harris et al. (2015) concluded that the genetic variation they discovered within *Janetaescincus* is 'considerably more than expected for one or two species', thus proposing the existence of at least one unrecognised species within the genus. As well as sampling more specimens from the largest and highest island of Mahé (only three were included in their study), Harris et al. (2015) recognised the need for morphological data (their DNA data were generated from non-lethal tail tip samples), and proposed that *Janetaescincus* spp. is currently best referred to as a species complex.

Squamata, Scincidae, *Pamelaescincus*

Pamelaescincus gardineri (Boulenger, 1909)

Seychellois: Lezar gra

The monotypic Seychelles endemic (semi)fossorial skink *Pamelaescincus* Greer, 1970 is known from at least 12 granitic islands (Gerlach and Ineich 2006d; Valente et al. 2014) and the taxon contains no synonyms or subspecies. Valente et al. (2014) generated and analysed mtDNA and nuDNA from 89 individuals from nine islands. They found strong evidence for a deep mtDNA divergence (ca. 8.0 per cent for *cytb*) between two main lineages, from southwestern (Mahé, Silhouette) and northeastern (Frégate, Praslin, La Digue, Curieuse, Grande Soeur, Aride, Cousine) islands, and with moderate additional geographic substructure for the mtDNA data within these two clades (Figure 10.3). This pattern was not entirely consistent, with five Mahé samples grouping with the northeastern island lineage, which was interpreted as possibly indicative of recent introductions and/or gene flow during low sea-level stands. Variation in nuDNA was low and not fully concordant with that for mtDNA. Valente et al. (2014) considered the molecular genetic evidence for *P. gardineri* less indicative of the presence of a cryptic species than for the gecko *Urocotyledon inexpectata*, which they attributed to lower philopatry and likely greater dispersal ability in the former. A lack of detailed morphological data prevented further conclusions or formal taxonomic action (DNA data were generated from non-lethal tail tip samples). *Pamelaescincus* is most closely related to the Seychelles endemic *Janetaescincus* (see previous section).

Squamata, Scincidae, *Trachylepis*

Trachylepis sechellensis (Duméril & Bibron, 1839)
Trachylepis wrightii (Boulenger, 1887)

Seychellois: Lezar mangouya (*T. sechellensis*), lezar tengteng (*T. wrightii*)

The Afro-Malagasy mabuyas, *Trachylepis* Fitzinger, 1843, occur across the western Indian Ocean and Africa, with at least one species (*T. atlantica* (Schimidt 1945)) occurring in South America. The Seychelles *Trachylepis* are monophyletic and estimated to have arrived from Africa by overseas dispersal ca. 25 Ma (Weinell et al. 2019). There are currently two *Trachylepis* species recognised within Seychelles: *T. sechellensis* (Figure 10.2G) and *T. wrightii*, which can be found in sympatry on four islands. *Trachylepis sechellensis* is the most widely distributed endemic reptile species across the granitic Seychelles, occurring on 22 islands and islets (Figure 10.3). The much larger (often considered 'gigantic' form) *T. wrightii* is restricted to six small islands and islets (Frégate, Cousine, Aride, St. Pierre, Mammelles, and Recifs), but there are other islands on which the populations are presumed extinct or for which only isolated records exist (Rendahl 1939; Gerlach 2007). One subspecies of *T. wrightii* was described as *Mabouia wrightii ilotensis* Rendahl, 1939 from 'L'ilot' (which could be one of two named sites: a small islet off the northwest tip of Mahé, or more likely a larger islet ~3 km southwest of Frégate) based on midbody scale row counts that differ from those in *T. wrightii* and overlap with those in *T. sechellensis*. To our knowledge *T. w. ilotensis* has not been documented since its description and its validity remains unclear. Molecular analyses have found that the two Seychelles *Trachylepis* are sister species (Lima et al. 2013; Rocha et al. 2016).

Investigations into mtDNA variation and phylogenetic relationships within Seychelles *Trachylepis* found *T. sechellensis* to be non-monophyletic with respect to *T. wrightii* (Rocha, Harris, and Carretero 2010, 2016). However, nuDNA clearly differentiates the two species (Rocha et al. 2016). Rocha et al. (2016) analysed genetic data for 117 *T. sechellensis* and 27 *T. wrightii*, finding low levels of mtDNA and nuDNA variation within *T. wrightii* but substantial, geographically structured (northern island vs southern island populations) variation within *T. sechellensis* (Figure 10.3). All sampled nuclear genes supported *T. wrightii* as a distinct species from a seemingly monophyletic *T. sechellensis*, leading the authors to conclude that the mito-nuclear discordance is likely due to historic introgressive hybridisation and alien mtDNA fixation. Morphological analyses also distinguish the two species, although the colour patterns of *T. sechellensis* from Cousine are clearly distinct from other *T. sechellensis* populations (an observation not reflected in the genetic data). The only southern island population of *T. sechellensis* that was examined (Mahé) was also distinct (Rocha et al. 2016).

Squamata, Serpentes, Natricidae

Lycognathophis seychellensis (Schlegel, 1837)

Seychellois: Koulev zonn

The Seychelles tree-snake, wolf snake, or cliff snake (Nussbaum 1984b; Dowling 1990) (Figure 10.2H) is the only member of its genus, and the only Seychelles natricid (or natricine, depending on classification), a group of snakes with ca. 250 extant species globally (Uetz et al. 2020), only 14 of which occur in sub-Saharan Africa with no taxa found elsewhere in the Indian Ocean region). *Lycognathophis seychellensis* has been reported from eight of the granitic Seychelles islands (Nussbaum 1984b; Gerlach and Ineich 2006c; Rocha et al. 2009). Detailed studies of morphological intraspecific variation have not been published, though note has been made of distinct colour morphs (e.g. Nussbaum 1984b) albeit without evidence that this variation is geographically structured or representative of unrecognised (sub)specific diversity. No subspecies have been described. Recent examination of molecular genetic variation (mtDNA) in *L. seychellensis* individuals from five islands found very little variation, consistent with *L. seychellensis* being a single species (Deepak et al. 2021). The same molecular study found strong support for *L. seychellensis* being a member of a sub-Saharan African radiation and inferred that the Seychelles lineage probably arrived via overseas dispersal ca. 43–25 Ma.

Squamata, Serpentes, Lamprophiidae

Lamprophis geometricus (Schlegel, 1837)

Seychellois: Koulev gri

The Seychelles house snake is the only Seychelles lamprophiid, a group of snakes with ca. 82 extant species globally (Uetz et al. 2020). Beyond the Seychelles, *Lamprophis* comprises six species restricted to sub-Saharan Africa. *Lamprophis geometricus* is known from five of the granitic Seychelles islands (Nussbaum 1984b; Rocha et al. 2009; Kopeczky and Gemel 2016). There is disagreement about the generic classification of the Seychelles house snake. For example, Kelly et al. (2011) suggested possible transferral to the then newly resurrected *Boaedon*, pending generation of DNA sequence data, while both Wallach et al. (2014) and Trape and Mediannikov (2016) preferred the combination *B. geometricus* even without such data. *Boaedon* is a lamprophiid closely related to *Lamprophis*, comprising ca. 18 species restricted to sub-Saharan Africa and the southernmost Arabian Peninsula (Uetz et al. 2020). No molecular genetic data have been published for *L. geometricus*. Rendahl (1939) presented information on morphological variation within *L. seychellensis* but detailed some of these data under entries for *Lycognathophis seychellensis*, thereby negating their usefulness. Nussbaum (1984b) considered *L. geometricus* to be very similar to African congeners and thus likely a relatively recent overseas arrival in the Seychelles; Dowling (1990) even suggested it might be introduced, an idea that has apparently not gained traction.

Testudines, Cryptodira, Testudinoidea, Testudinidae

Aldabrachelys gigantea (Schweigger, 1812)

Seychellois: Torti-d-ter

One extant species of Seychelles giant tortoise (*Aldabrachelys gigantea*) is currently recognised, though this has a contentious and confusing taxonomic history (see Braithwaite (2016) for a summary). Until recently, four species were suggested based on morphology, two of which were considered extant (Gerlach and Canning 1998). Subsequent DNA and microsatellite analysis found no notable genetic variation or geographic structuring in the sampled material, indicating a single extant lineage (Palkovacs et al. 2002, 2003; Le et al. 2006). Giant tortoises were present on at least 23 of the granitic islands between 1609 and 1850 (Gerlach et al. 2013) although it remains unclear whether multiple taxa existed across the archipelago. Initial eighteenth-century reports suggest there were size differences between islands (Bour 1984) but contemporary understanding is clouded by confounding factors. First, giant tortoises were a prized resource of transportable, readily available food for whalers, fishermen, and naval and merchant vessels visiting Indian Ocean Islands (Stoddart and Peake 1997). Animals were frequently moved between islands, including imports from Aldabra when no more tortoises could be found on the Inner Islands, and also shipped to the Mascarenes, when endemic giant tortoises had become extinct there (Stoddart and Peake 1997). Second, available (sub)fossils are fragmentary and scarce

(Arnold 1979; Gerlach et al. 2013), especially from the Inner Islands, and analyses of their degraded DNA are challenging (Caccone et al. 2005) despite advancing techniques (e.g. Shapiro et al. 2019) that offer hope for further assessments of Indian Ocean giant tortoise diversity and phylogeography. However, carbon-dating verifies the natural occurrence of giant tortoises on the Inner Islands prior to the arrival of humans, placing them on North Island between 764 and 395 BCE (Caccone et al. 2005) and Denis in CE 680 (Stoddart and Peake 1997), and combined with molecular dating provides strong evidence that *A. gigantea* on Aldabra resulted from repeated colonisations since the Early Oligocene until at least the Late Pleistocene (Braithwaite et al. 1973; Taylor et al. 1997; Braithwaite 2016; Le and Raxworthy 2017).

Testudines, Pleurodira, Pelomedusidae

Pelusios castanoides intergularis Bour, 1983
Pelusios subniger parietalis Bour, 1983

Seychellois: Torti soupap

Freshwater turtles of the genus *Pelusios* Wagler 1830, often referred to as hinged terrapins (Bramble and Hutchison 1981; Bour 1984; Stuckas et al. 2013; Kindler et al. 2016), are found across sub-Saharan Africa, Madagascar, and Seychelles. Bour (1984) reports a lack of records for *Pelusios* in Seychelles until the late nineteenth century but three endemic taxa were considered to have been present on the inner islands; *P. seychellensis* (Siebenrock, 1906) (of unknown distribution), *P. castanoides intergularis* Bour, 1983, and *P. subniger parietalis* Bour, 1983, of which the latter two are extant and recorded from Cerf, Cousin, Frégate, La Digue, Mahé, Praslin, and Silhouette (Buitelaar and de Pous 2011; Gerlach et al. 2008a, 2008b). *Pelusios seychellensis* was known only from three specimens but the provenance of this type series lacks decisive collection data (Kindler et al. 2016), and recent molecular work found this taxon embedded within the West African *P. castaneus* (Schweigger 1812) (Stuckas et al. 2013; Kindler et al. 2016) making *P. seychellensis* a junior synonym. Human-mediated introduction of *P. castaneus* to the Seychelles may also explain previous recognition of this taxon as a Seychelles endemic (if the specimens were indeed collected there), a dispersal mechanism that has not been discounted for *P. c. intergularis* and *P. s. parietalis*, questioning the validity of their subspecific status (Bour 1984; Silva et al. 2010; Fritz et al. 2012; Rhodin et al. 2017). Morphological differentiation between *P. subniger parietalis* and the East African *P. subniger* appears to be insufficient, and their synonymy is further supported by molecular analyses (Fritz et al. 2012). Yet despite weak molecular differentiation between Malagasy *P. c. castanoides* and Seychellois *P. c. intergularis* (Fritz et al. 2010, 2012; Silva et al. 2010) suggesting these taxa are also synonymous, both are differentiated from mainland *P. castanoides* Hewitt 1931. Resolution of the affinity of the Seychelles population has been hampered by limited sampling of *P. castanoides* from both Madagascar and mainland (west) Africa (Fritz et al. 2010, 2012; Silva et al. 2010; Stuckas et al. 2013), indicating the need for further work to resolve the affinity of the Seychelles population.

10.3 Discussion

Over the past 50 years, the ways in which many species have been discovered and described have changed. The advent of molecular data and development of new analytical tools modified some species concepts and prompted empirical studies of the extent to which populations might represent distinct species. Systematic studies that incorporate molecular data continue to reveal cases of morphologically similar, supposedly conspecific populations that are discovered to comprise highly divergent, evolutionarily distinct but phenotypically more or less cryptic lineages. Such work has resulted in amphibians being highlighted as reservoirs of hitherto unrecognised diversity (Bickford et al. 2007; Vieites et al. 2009; Funk et al. 2012) even for large, charismatic, seemingly well-known taxa (Yan et al. 2018). Similar observations are also widely reported for reptiles (Oliver et al. 2009; Domingos et al. 2014; Makhubo et al. 2015; Perez-Ponce de Leon and Poulin 2016). Yet, a pervasive bias exists in studies of herpetological diversity, with taxa and (notably) those studying them more likely to originate from regions such as Europe and North America (Gardner et al. 2007; Brito 2008; Winter et al. 2016; Titley et al. 2017), with global centres of herpetological diversity—the Neotropical, Afrotropical, Indomalayan, and Australasian regions—remaining understudied (Gardner et al. 2007; Brito 2008; Böhm et al. 2013; Winter et al. 2016; Roll et al. 2017; Titley et al. 2017).

As demonstrated by several recent studies, multiple, more or less cryptic lineages have been discovered within the Seychelles herpetofauna (e.g. Rocha et al. 2017; Labisko et al. 2019; Maddock et al. 2020) (Figure 10.3). Indeed, all but the house snake have received at least some attention resulting in published molecular data, with cryptic lineages being revealed in the majority of recent studies. This prompts several questions, including some that we address here: (1) Why have high levels of previously undiscovered diversity remained in what is, in relative terms, a small number of taxa spread across a few small islands in a relatively well-studied region? (2) Are these lineages truly phenotypically cryptic and do they represent distinct taxa? (3) Can differences in the extent of lineage crypsis among different amphibian and reptile groups be explained by their ecology and/or duration of residence in the Seychelles?

10.3.1 Why, until Recently, Have So Many Seychelles Amphibian and Reptile Lineages Remained Cryptic?

Being approximately 1,000 km distant from Madagascar, 1,300 km from Africa, 1,700 km from the Mascarenes, and 2,800 km from India, the Granitic Seychelles are more distant from their nearest neighbour than Madagascar is from Africa (ca. 420 km) or from the Mascarenes (ca. 680 km). During the eighteenth century, the trade route from East Africa to Seychelles would take four to five months to complete (Benedict 1984), so despite the arrival of the first permanent settlers, and subsequent replacement of natural vegetation with subsistence farming and plantations in the 1700s, formal exploration of the islands yielding biological data or specimens may have been more punctuated than to other, more accessible Indian Ocean ports. Even so, given that the terrestrial area of the Inner Islands is

< 250 km^2, it is remarkable that the rate of deforestation was not as extensive as in (for example) the Mascarenes. Approximately 60 per cent of native forest cover on Mauritius was lost between 1773 and 1872 (Norder et al. 2017), an estimated area of more than 1,200 km^2, nearly five times the combined area of Seychelles' Inner Islands. Following this deforestation much of the Mascarene biota was rendered extinct before they could be subjected to scientific study (Cheke and Hume 2010).

The most historically important contributions to Seychelles' natural history did not take place until the late 1890s and early 1900s (Stoddart 1984b). Yet despite the loss and replacement of forests with plantations and the undoubted declines (and extinctions) of many taxa, remnant native vegetation patches and faunal populations survived (Baker 1877; Wallace 1892; Stoddart 1984a; Simberloff and Cox 1987; Kueffer et al. 2013). The dispersed nature of the Inner Islands, their geographic position (relative to nearest landfall), later settlement, and availability of all-important refugia may therefore have provided enough opportunity for today's extant taxa to persist, often maintaining seemingly natural, notably geographically structured populations, all the while remaining relatively unstudied.

Visitors to the Seychelles' have of course increased since the advent of accessible air-travel but the archipelago has maintained a less anthropogenically impacted state than many other island destinations, no doubt aided by its position in the western Indian Ocean. Although some taxa may be more suited to persist in anthropogenically impacted habitat (e.g. *Tachycnemis seychellensis* and *Phelsuma* spp.) this has yet to be formally tested. Nevertheless, such qualities, as well as recolonisation from refugia, and the remoteness of the region are likely important factors that have both ensured the survival of Seychelles herpetofauna, and also until recently limited its exposure to modern research techniques.

10.3.2 Are Recently Discovered Genetic Lineages of Seychelles Herpetofauna Truly Phenotypically Cryptic, and Do They Represent Distinct Taxa?

In most instances, the requisite studies have not been carried out to answer this question, except to the extent that such Seychelles herpetofauna lineages are routinely so phenotypically similar as to be indistinguishable to the naked eye. In several cases, especially for lizards (see taxon accounts in Section 10.2), collection of tissues for genetic analysis was non-lethal, thereby precluding subsequent morphological comparisons of genetically divergent lineages beyond gross appearance from photographs and/or field notes. In some cases where newly discovered genetic lineages are allopatrically distributed, historical morphological vouchers are accompanied by at least island-level collection data but work to generate and analyse detailed morphological data has not yet been undertaken. In this sense, a thorough examination of the 'cryptic species issue' for Seychelles herpetofauna is not yet possible.

Despite multiple discoveries of distinct lineages during the past decade, very few taxonomic changes have taken place. Only a single recent study has assigned populations to previously reported subspecies (day geckos in the genus *Phelsuma*: Rocha et al. 2013), and two recently described amphibian species could be considered somewhat cryptic (*Hypogeophis montanus* and *Sechellophryne pipilodryas*) due to their populations being sympatric or parapatric with generally similar and most-closely related congeners. The general lack of taxonomic (re)assessment thus far is likely due to one or more of: (i) morphological investigation of phenotypically similar lineages often taking place some time after discovery of substantial molecular variation; (ii) lack of sufficient geno- and/or phenotypic differentiation to present a compelling case for recognition of discrete taxa; (iii) historic descriptions of (sub)species containing erroneous and/or inaccurate (e.g. 'Seychelles') locality data making clear assignation of provenance difficult; and (iv) lack of available names from historical studies for some lineages. The need for further study has been often highlighted, for example the proposal to consider *Janetaescincus* spp. a species complex pending further investigation (Harris et al. 2015).

There are several interconnected issues in determining whether divergent genetic lineages warrant taxonomic recognition. Choice of operational species concept is not addressed here but it impacts how this question is answered in individual cases. In those where (sub)species names already exist in the literature that are currently synonyms, but which might apply to divergent genetic lineages, allopatry of those genetic lineages allows greater confidence in matching morphological vouchers (including name-bearing types) that lack genetic data with genetic lineages (based on samples that might lack morphological vouchers). However, even deeply divergent allopatric lineages lacking evidence of inter-lineage gene flow might simply be (temporarily) physically separated incipient species (e.g. Rocha et al. 2011; Valente et al. 2013) rather than being unable to successfully interbreed, and in this regard allopatry is a confounding factor. We know of no experimental data for captive breeding of allopatric, genetically divergent, Seychelles reptile or amphibian species.

In some cases, potential assignment of names currently in synonymy to divergent genetic lineages is impeded by mismatches between phenotypic and genotypic patterns of variation and/or lack of even island-level precision for collection locality of name-bearing types (e.g. see Maddock et al.'s 2020 discussion of variation and taxonomy of *Hypogeophis rostratus*). Even where phenotypic and genetic divergence patterns are broadly congruent, using only these data to hypothesise species boundaries among allopatric lineages can be challenging, especially given that populations on small islands are generally small and might be especially prone to rapid local adaptation (e.g. Baxter-Gilbert et al. 2020; Donihue et al. 2020a, 2020b; Velo-Antón et al. 2012), perhaps as has occurred within what appears otherwise to be a single species of Seychelles treefrog, *Tachycnemis seychellensis* (Maddock et al. 2014). Nonetheless, well-sampled instances of deep genetic divergence without evidence of introgression or hybridisation, such as within *Urocotyledon inexpectata* (candidate species as proposed by Rocha et al. 2011), are leading candidates for cryptic taxa if congruent phenotypic data are discovered. Molecular genetic species delimitation methods such as the generalised mixed Yule-coalescent (GMYC) (Fujisawa and Barraclough 2013) and Poisson tree processes (PTP) (Zhang et al. 2013) have not been extensively applied across the Seychelles herpetofauna, but we question their ability to solve decisively the question of potential boundaries among allopatric populations that we already know are substantially divergent.

Whether notably divergent genetic Seychelles amphibian and reptile lineages represent distinct taxa is likely a somewhat arbitrary decision in at least some cases. Even so, divergent genetic lineages rightly have implications for conservation biology. Island populations are generally small and especially vulnerable to threats such as climatic fluctuations, disease, and invasive alien species (e.g. Bullock 2009; Bickford et al. 2010; Smith et al. 2012;

Hudson et al. 2016). Endemic Seychelles vertebrates include several threatened taxa that have been subject to management programmes including inter-island translocations (e.g. Hambler 1994; López-Sepulcre et al. 2008; Wright et al. 2014). Such management interventions benefit from understanding spatial patterns of phenotypic and genetic variation, and the ecological and evolutionary processes that generated them (Sodhi and Ehrlich 2010; IUCN 2013).

10.4.3 Does Ecology and/or Duration of Residence Explain Differences in Degree of Lineage Crypsis among Seychelles Amphibians and Reptiles?

Thorough studies have not yet been conducted, but preliminary qualitative assessment indicates plausible links in some cases. One ecological trait worth consideration is vagility, itself influenced by multiple factors including reproductive mode, fecundity, environmental tolerance, and habitat preference. Vagility in the case of Seychelles gene flow relates not only to transmarine dispersal but also to dispersal across ephemeral land bridges that emerged between islands at low sea-level stands. Among non-aquatic reptiles, giant tortoises are perhaps best-suited to transmarine dispersal (e.g. Cheke et al. 2017; Hansen et al. 2017; Hawlitschek et al. 2017), having large lungs that keep them buoyant, and the ability to survive up to six months without sustenance (Caccone et al. 2002). This might be expected to maintain gene flow among populations on different Seychelles islands – though verification of that causal link in the present and recent past is complicated by human-mediated translocation of these reptiles (e.g. Stoddart and Peake 1997) and the absence of remnant, wild populations on any of the Inner Islands. At the other end of the spectrum, very small sooglossid frogs and mostly soil-dwelling caecilians might be expected to have generally low dispersal abilities, with the osmotic properties of amphibian skin (e.g. Balinsky 1981) making it intolerant of salt water (e.g. Myers 1953). This expectation is matched by molecular genetic data showing little to no evidence of recent gene flow among intraspecific populations occurring on separate islands or proximate island groups (Labisko et al. 2019; Maddock et al. 2020).

Duration of residence on the Seychelles is partly linked with vagility. All but one of the native major amphibian lineages (*Tachycnemis seychellensis*) are likely to be Gondwanan relics, whereas most, or perhaps all, of the major native reptile lineages are the product of transmarine colonisation. For example, the Seychelles natricine snake lineage (solely represented by *Lycognathophis seychellensis*) likely reached Seychelles by transmarine dispersal from Africa sometime in the last ca. 43 Ma, and this ability to disperse is likely to have contributed (possibly with a founder effect) to the low genetic diversity and lack of geographical structure within this species across multiple islands (Deepak et al. 2021). The presence of squamates but not amphibians on non-granitic Seychelles islands and the greater likelihood of major Seychelles squamate lineages being transmarine arrivals instead of Gondwanan relics is also consistent with their expected greater vagility. However, lizard vagility on Seychelles is not so great in most cases to have prevented the origin and maintenance of allopatric, divergent, seemingly phenotypically cryptic genetic lineages on the major islands or proximate island groups. The pervasiveness of the pattern of multiple, (currently intraspecific) allopatric genetically divergent lineages across a wide range of taxa, demonstrate that geographic distance and/or marine barriers and/or ephemerally emergent land habitat barriers have had a major influence on Seychelles biodiversity patterns, despite cyclical reconnection of the granitic islands during sea-level changes (e.g. Maddock et al. 2020). Among Seychelles herpetofauna, the tree frog *Tachycnemis seychellensis* is an outlier, in that geographically isolated populations within and among islands are morphologically distinct yet genetically variable without obvious geographic structure (Maddock et al. 2014). More detailed population genetic and ecological data might shed light on this anomaly.

Conclusion*

If the recently discovered allopatric (currently intraspecific) genetic lineages within many taxa prove to be largely phenotypically cryptic, that might be explained by environmental homogeneity across the archipelago and/or large populations now and/or in the recent past. Regarding the former, currently even the granitic islands are physically disparate in size and topography, though detailed long-term climate data are not available and human activities since the mid-eighteenth century have substantially eroded many natural vegetation patterns (Procter 1984b; Kueffer et al. 2013; Rocamora and Henriette 2015). We are uncertain of the potential for palynological studies to provide island-scale paleoenvironmental data.

The Seychelles herpetofauna remains little explored concerning cryptic species questions, though the system offers much potential in this field. Most studies investigating intraspecific variation in Seychelles amphibians and reptiles have identified (mostly geographically structured) genetically divergent populations, some of which are potentially cryptic (sub)species. More research is required to test or verify hypotheses of phenotypic crypsis, and to test hypotheses of the causes of (non-) crypsis by generating data on the ecology of these taxa (including habitat preferences and tolerances, plasticity, fecundity, behaviour, population genetics, and demography), ideally to inform an ecological niche-modelling approach, as has proved informative in delimiting closely related, locally endemic Malagasy *Phelsuma* spp. (Raxworthy et al. 2007). Precise and detailed environmental data are also needed to understand abiotic variation across the archipelago. Comparative studies would also benefit from additional work on phylogeography, molecular dating of divergences, and historical biogeography, as well as integration of phenotypic trait data (e.g. see Papadopoulou and Knowles 2016; Zamudio et al. 2016). The Seychelles herpetofauna is ripe for exploration of cryptic lineage questions. There are significant opportunities for continued study, not only to more fully understand patterns and processes but also to inform and ensure appropriate measures of protection in the face of ongoing and increasing anthropogenic threats.

* Lobón-Rovira et al. (in press) report small differences in morphology between *Uroctyledon* from the northern and southern Seychelles islands, and describe the northern lineage as a new species.

Lobón-Rovira, J., Rocha, S., Gower, D. J., Perera, A. and Harris, D. J. (in press) The unexpected gecko: A new cryptic species within *Urocotyledon inexpectata* (Stejneger, 1893) from the northern granitic Seychelles. *Zootaxa*.

References

Adamson, E. A. S., Saha, A., Maddock, S. T. et al. (2016) Microsatellite discovery in an insular amphibian (*Grandisonia alternans*) with comments on cross-species utility and the accuracy of locus identification from unassembled Illumina data. *Conservation Genetics Resources* 8(4): 541–551. https://doi.org/10.1007/s12686-016-0580-5

Ahl, E. (1926) Neue Eidechsen und Amphibien. *Zoologischer Anzeiger* 67: 186–192.

Ali, J. R. (2017) Islands as biological substrates: Classification of the biological assemblage components and the physical island types. *Journal of Biogeography* 44(5): 984–994. http://dx.doi.org/10.1111/jbi.12872

(2018) Islands as biological substrates: Continental. *Journal of Biogeography* 45(5): 1003–1018. https://doi.org/10.1111/jbi.13186

AmphibiaWeb. (2021) AmphibiWeb. Retrieved 6 February 2021. http://amphibiaweb.org/amphibian/speciesnums.html

Armitage, J. J., Collier, J. S., Minshull, T. A., and Henstock, T. J. (2011) Thin oceanic crust and flood basalts: India-Seychelles breakup. *Geochemistry, Geophysics, Geosystems* 12(5). https://doi.org/10.1029/2010GC003316

Arnold, E. N. (1979) Indian Ocean giant tortoises: Their systematics and island adaptations. Philosophical Transactions of the Royal Society of London. *B, Biological Sciences* 286(1011): 127–145. https://doi.org/10.1098/rstb.1979.0022

Arteaga, A., Pyron, R. A., Penafiel, N. et al. (2016) Comparative phylogeography reveals cryptic diversity and repeated patterns of cladogenesis for amphibians and reptiles in Northwestern Ecuador. *PLoS One* 11(4), e0151746. https://doi.org/10.1371/journal.pone.0151746

Baker, J. G. (1877) *Flora of Mauritius and the Seychelles: A Description of the Flowering Plants and Ferns of Those Islands*: L. Reeve & Company, London.

Balinsky, J. B. (1981) Adaptation of nitrogen metabolism to hyperosmotic environment in Amphibia. *Journal of Experimental Zoology* 215(3): 335–350. https://doi.org/10.1002/jez.1402150311

Bauer, A. M., Parham, J. F., Brown, R. M. et al. (2011) Availability of new Bayesian-delimited gecko names and the importance of character-based species descriptions. *Proceedings of the Royal Society B: Biological Sciences* 278(1705): 490–492. https://doi.org/10.1098/rspb.2010.1330

Baxter-Gilbert, J., Riley, J. L., Wagener, C., Mohanty, N. P., and Measey, J. (2020) Shrinking before our isles: The rapid expression of insular dwarfism in two invasive populations of guttural toad (*Sclerophrys gutturalis*). *Biology Letters* 16(11), 20200651. https://doi.org/10.1098/rsbl.2020.0651

Benedict, B. (1984) The human population of the Seychelles. In: D. R. Stoddart (ed.) *Biogeography and Ecology of the Seychelles Islands*. W. Junk, The Hague; Boston: Hingham, MA, pp. 627–639.

Bickford, D., Lohman, D. J., Sodhi, N. S. et al. (2007) Cryptic species as a window on diversity and conservation. *Trends in Ecology and Evolution* 22(3): 148–155. https://doi.org/10.1016/j.tree.2006.11.004

Bickford, D., Howard, S. D., Ng, D. J. J., & Sheridan, J. A. (2010) Impacts of climate change on the amphibians and reptiles of Southeast Asia. *Biodiversity and Conservation* 19(4): 1043–1062. https://doi.org/10.1007/s10531-010-9782-4

Böhm, M., Collen, B., Baillie, J. E. M. et al. (2013) The conservation status of the world's reptiles. *Biological Conservation* 157: 372–385. https://doi.org/10.1016/j.biocon.2012.07.015

Bossuyt, F. and Milinkovitch, M. C. (2001) Amphibians as indicators of early tertiary "out-of-India" dispersal of vertebrates. *Science* 292(5514): 93–95. https://doi.org/10.1126/science.1058875

Boulenger, G. A. (1882) *Catalogue of the Batrachia Gradientia S. caudata and Batrachia apoda in the collection of the British Museum*: order of the Trustees.

Bour, R. (1984) Taxonomy, history and geography of Seychelles land tortoises and fresh-water turtles. In: D. R. Stoddart (ed.) *Biogeography and Ecology of the Seychelles Islands*. W. Junk, The Hague; Boston: Hingham, MA, pp. 281–308.

Bowler, J. (2006) *Wildlife of Seychelles*: WILDGuides, Old Basing.

Brandley, M. C., Schmitz, A., and Reeder, T. W. (2005) Partitioned Bayesian analyses, partition choice, and the phylogenetic relationships of scincid lizards. *Systematic Biology* 54(3): 373–390. https://doi.org/10.1080/10635150590946808

Braithwaite, C. J. R. (2016) The giant tortoise, *Aldabrachelys*, and its bearing on the biogeography and dispersal of terrestrial biota in the Western Indian Ocean. *Palaeogeography, Palaeoclimatology, Palaeoecology* 461: 449–459. https://doi.org/10.1016/j.palaeo.2016.08.010

Braithwaite, C. J. R., Taylor, J. D., and Kennedy, W. J. (1973) The evolution of an atoll: The depositional and erosional history of Aldabra. *Philosophical Transactions of the Royal Society of London. B, Biological Sciences* 266(878): 307–340. https://doi.org/10.1098/rstb.1973.0051

Bramble, D. M. and Hutchison, J. H. (1981) A reevaluation of plastral kinesis in African turtles of the genus *Pelusios*. *Herpetologica* 37(4): 205–212.

Brito, D. (2008) Amphibian conservation: Are we on the right track? *Biological Conservation* 141(11): 2912–2917. https://doi.org/10.1016/j.biocon.2008.08.016

Buitelaar, K. and de Pous, P. (2011) First record of *Pelusios castanoides intergularis* and rediscovery of *Pelusios subniger parietalis* on Cousin Island, Seychelles. *Herpetology Notes* 4: 9–10.

Bullock, D. (2009) Round Island – a tale of destruction. *Oryx* 14(1): 51–58. https://doi.org/10.1017/S0030605300014800

Caccone, A., Gentile, G., Gibbs, J. P. et al. (2002) Phylogeography and history of giant Galapagos tortoises. *Evolution* 56(10): 2052–2066. www.jstor.org/stable/3094648

Caccone, A., Karanth, K. P., Gerlach, J. et al. (2005) Native Seychelles tortoises or Aldabran imports? The importance of radiocarbon dating for ancient DNA studies. *Amphibia-Reptilia* 26(1): 116–121. https://doi.org/10.1163/1568538053693279

Camoin, G. F., Montaggioni, L. F., and Braithwaite, C. J. R. (2004) Late glacial to post glacial sea levels in the western Indian Ocean. *Marine Geology* 206(1–4): 119–146. https://doi.org/10.1016/j.margeo.2004.02.003

Chatterjee, S., Goswami, A., and Scotese, C. R. (2013) The longest voyage: Tectonic, magmatic, and paleoclimatic evolution of the Indian plate during its northward flight from Gondwana to Asia. *Gondwana Research* 23(1): 238–267. https://doi.org/10.1016/j.gr.2012.07.001

Cheke, A. (1982) *Phelsuma* Gray 1825 in the Seychelles and neighbouring islands: A re-appraisal of their taxonomy and description of two new forms (Reptilia: Sauria: Gekkonidae). *Senckenberg Biology* 62: 181–198.

(1984) Lizards of the Seychelles. In: D. R. Stoddart (ed.) *Biogeography and Ecology of the Seychelles Islands*. W. Junk, The Hague; Boston: Hingham, MA, pp. 245–258.

Cheke, A. and Hume, J. P. (2010) *Lost Land of the Dodo: An Ecological History of Mauritius, Réunion and Rodrigues*: Bloomsbury Publishing, London.

Cheke, A. S., Pedrono, M., Bour, R. et al. (2017) Giant tortoises spread to western Indian Ocean islands by sea drift in pre-Holocene times, not by later human agency: Response to Wilmé et al. (2016a). *Journal of Biogeography* 44(6): 1426–1429. https://doi.org/10.1111/jbi.12882

Coffin, M. F. and Rabinowitz, P. D. (1987) Reconstruction of Madagascar and Africa: Evidence from the Davie Fracture Zone and

Western Somali Basin. *Journal of Geophysical Research* 92(B9): 9385–9406. https://doi.org/10.1029/JB092iB09p09385

Cogan, B. H. (1984) Origins and affinities of Seychelles insect fauna. In: D. R. Stoddart (ed.) *Biogeography and Ecology of the Seychelles Islands*. W. Junk, The Hague; Boston: Hingham, MA, pp. 1–16.

Collier, J. S., Sansom, V., Ishizuka, O. et al. (2008) Age of Seychelles–India break-up. *Earth and Planetary Science Letters* 272(1–2): 264–277. https://doi.org/10.1016/j.epsl.2008.04.045

Colonna, M., Casanova, J., Dullo, W.-C., and Camoin, G. (1996) Sea-level changes and δ18O record for the Past 34,000 yr from Mayotte Reef, Indian Ocean. *Quaternary Research* 46(03): 335–339. https://doi.org/10.1006/qres.1996.0071

Courtillot, V., Besse, J., Vandamme, D. et al. H. (1986) Deccan flood basalts at the Cretaceous/Tertiary boundary? *Earth and Planetary Science Letters* 80(3–4): 361–374. https://doi.org/10.1016/0012-821X(86)90118-4

Courtillot, V., Féraud, G., Maluski, H. et al.(1988) Deccan flood basalts and the Cretaceous/Tertiary boundary. *Nature* 333(6176): 843–846. https://doi.org/10.1038/333843a0

Crottini, A., Madsen, O., Poux, C. et al. (2012) Vertebrate time-tree elucidates the biogeographic pattern of a major biotic change around the K-T boundary in Madagascar. *Proceedings of the National Academy of Sciences of the United States of America* 109(14): 5358–5363. https://doi.org/10.1073/pnas.1112487109

Cumberlidge, N. and Daniels, S. R. (2014) Recognition of two new species of freshwater crabs from the Seychelles based on molecular evidence (Potamoidea: Potamonautidae). *Invertebrate Systematics* 28(1): 17–31. https://doi.org/10.1071/IS13017

Daniels, S. R. (2011) Reconstructing the colonisation and diversification history of the endemic freshwater crab (*Seychellum alluaudi*) in the granitic and volcanic Seychelles Archipelago. *Molecular Phylogenetics and Evolution* 61(2): 534–542. https://doi.org/10.1016/j.ympev.2011.07.015

Deepak, V., Maddock, S. T., Williams, R. et al. (2021). Molecular phylogenetics of sub-Saharan African natricine snakes, and the biogeographic origins of the Seychelles endemic Lycognathophis seychellensis. *Molecular Phylogenetics and Evolution* 161: 107–152. https://doi.org/10.1016/j.ympev.2021.107152

Domingos, F. M., Bosque, R. J., Cassimiro, J. et al. (2014) Out of the deep: Cryptic speciation in a Neotropical gecko (Squamata, Phyllodactylidae) revealed by species delimitation methods. *Molecular Phylogenetics and Evolution* 80(0): 113–124. https://doi.org/10.1016/j.ympev.2014.07.022

Donihue, C. M., Daltry, J. C., Challenger, S., and Herrel, A. (2020a) Population increase and changes in behavior and morphology in the Critically Endangered Redonda ground lizard (*Pholidoscelis atratus*) following the successful removal of alien rats and goats. *Integrative Zoology*. https://doi.org/10.1111/1749-4877.12500

Donihue, C. M., Kowaleski, A. M., Losos, J. B. et al. (2020b) Hurricane effects on Neotropical lizards span geographic and phylogenetic scales. *Proceedings of the National Academy of Sciences of the United States of America* 117(19): 10429–10434. https://doi.org/10.1073/pnas.2000801117

Dowling, H. G. (1990) Taxonomic status and relationships of the genus *Lycognathophis*. *Herpetologica* 46(1): 60–66.

Dubois, A. (1981) Liste des genres et sous-genres nominaux de Ranoidea (Amphibiens Anoures) du monde, avec identification de leurs espéces-types: Conséquences nomenclaturales. *Monitore Zoologico Italiano: Supplemento* 15(1): 225–284. https://doi.org/10.1080/03749444.1981.10736637

Duméril, A. and Bibron, G. (1836) *Erpetologie Générale ou Histoire Naturelle Complete des Reptiles*. Vol. 3. Libr. Encyclopédique Roret, Paris.

(1841) *Erpétologie générale ou histoire naturelle complète des reptiles: Comprenant l'histoire génerale des batraciens, et la description des cinquante-deux genres et des cent soixante-trois espèces des deux premiers sous-ordres: les péroméles.* Vol. 8. Libr. Encyclopédique Roret, Paris.

Duncan, R. A. and Pyle, D. G. (1988) Rapid eruption of the Deccan flood basalts at the Cretaceous/Tertiary boundary. *Nature* 333 (6176): 841–843. https://doi.org/10.1038/333841a0

Feng, Y. J., Blackburn, D. C., Liang, D. et al. (2017) Phylogenomics reveals rapid, simultaneous diversification of three major clades of Gondwanan frogs at the Cretaceous-Paleogene boundary. *Proceedings of the National Academy of Sciences of the United States of America* 114 (29): E5864–E5870. https://doi.org/10.1073/pnas.1704632114

Florio, A. M., Ingram, C. M., Rakotondravony, H. A., Louis, E. E., and Raxworthy, C. J. (2012) Detecting cryptic speciation in the widespread and morphologically conservative carpet chameleon (*Furcifer lateralis*) of Madagascar. *Journal of Evolutionary Biology* 25(7): 1399–1414. https://doi.org/10.1111/j.1420-9101.2012.02528.x

Forlani, M. C., Tonini, J. F., Cruz, C. A., Zaher, H., and de Sa, R. O. (2017) Molecular and morphological data reveal three new cryptic species of *Chiasmocleis* (Mehely 1904) (Anura, Microhylidae) endemic to the Atlantic Forest, Brazil. *PeerJ* 5: e3005. https://doi.org/10.7717/peerj.3005

Frazão, A., da Silva, H. R., and Russo, C. A. (2015) The Gondwana breakup and the history of the Atlantic and Indian Oceans unveils two new clades for early neobatrachian diversification. *PLoS One* 10 (11): e0143926. https://doi.org/10.1371/journal.pone.0143926

Fritz, U., Branch, W. R., Gehring, P.-S. et al. (2012) Weak divergence among African, Malagasy and Seychellois hinged terrapins (*Pelusios castanoides, P. subniger*) and evidence for human-mediated oversea dispersal. *Organisms Diversity & Evolution* 13(2): 215–224. https://doi.org/10.1007/s13127-012-0113-3

Fritz, U., Branch, W. R., Hofmeyr, M. D.et al. (2010) Molecular phylogeny of African hinged and helmeted terrapins (Testudines: Pelomedusidae: *Pelusios* and *Pelomedusa*). *Zoologica Scripta* 40(2): 115–125. https://doi.org/10.1111/j.1463-6409.2010.00464.x

Frost, D. R. (2020) Amphibian Species of the World: An Online Reference. Version 6.1. Retrieved 29 June 2020, from American Museum of Natural History. https://doi.org/10.5531/db.vz.0001

Fujisawa, T. and Barraclough, T. G. (2013) Delimiting species using single-locus data and the Generalized Mixed Yule Coalescent approach: A revised method and evaluation on simulated data sets. *Systematic Biology* 62 (5): 707–724. https://doi.org/10.1093/sysbio/syt033

Funk, W. C., Caminer, M., and Ron, S. R. (2012) High levels of cryptic species diversity uncovered in Amazonian frogs. *Proceedings of the Royal Society B: Biological Sciences* 279 (1734): 1806–1814. https://doi.org/10.1098/rspb.2011.1653

Fusinatto, L. A., Alexandrino, J., Haddad, C. F. et al. (2013) Cryptic genetic diversity is paramount in small-bodied amphibians of the genus *Euparkerella* (Anura: Craugastoridae) endemic to the Brazilian Atlantic forest. *PLoS One* 8(11): e79504. https://doi.org/10.1371/journal.pone.0079504

Gamble, T., Greenbaum, E., Jackman, T. R., Russell, A. P., and Bauer, A. M. (2012) Repeated origin and loss of adhesive toepads in geckos. *PLoS One* 7(6): e39429. https://doi.org/10.1371/journal.pone.0039429

Ganerød, M., Torsvik, T. H., van Hinsbergen, D. J. J. et al. (2011) Palaeoposition of the Seychelles microcontinent in relation to the Deccan Traps and the Plume Generation Zone in Late Cretaceous-Early Palaeogene time. *Geological Society, London, Special*

Publications 357(1): 229–252. https://doi.org/10.1144/SP357.12

Gardner, A. S. (1984) The evolutionary ecology and population systematics of day geckos (*Phelsuma*) in the Seychelles. (PhD Doctoral dissertation), University of Aberdeen.

Gardner, A. S. (1986a) The biogeography of the lizards of the Seychelles Islands. *Journal of Biogeography* 13(3): 237–253. https://doi.org/10.2307/2844923

(1986b) Morphological evolution in the day gecko *Phelsuma sundbergi* in the Seychelles: A multivariate study. *Biological Journal of the Linnean Society* 29: 223–244. https://doi.org/10.1111/j.1095-8312.1986.tb01774.x

(1987) The systematics of the *Phelsuma madagascariensis* species group of day geckos (Reptilia: Gekkonidae) in the Seychelles. *Zoological Journal of the Linnean Society* 91: 93–105. https://doi.org/10.1111/j.1096-3642.1987.tb01724.x

Gardner, T. A., Barlow, J., and Peres, C. A. (2007) Paradox, presumption and pitfalls in conservation biology: The importance of habitat change for amphibians and reptiles. *Biological Conservation* 138(1–2): 166–179. https://doi.org/10.1016/j.biocon.2007.04.017

Gerlach, J. (1999) The origins of *Isometrus maculatus* and other scorpions on the smaller islands of the western Indian Ocean.

(2007) *Terrestrial and Freshwater Vertebrates of the Seychelles Islands*. Backhuys, Leiden.

(2008) Population and conservation status of the reptiles of the Seychelles islands. *Phelsuma* 16: 31–48.

Gerlach, J. and Canning, L. (1998) Taxonomy of Indian Ocean giant tortoises (*Dipsochelys*). *Chelonian Conservation and Biology* 3: 3–19.

Gerlach, J. and Ineich, I. (2006a) *Janetaescincus braueri* (Publication no. https://dx.doi.org/10.2305/IUCN.UK.2006.RLTS.T61437A12484604.en.). Retrieved 14 July 2020, from the IUCN Red List of Threatened Species 2006: e.T61437A12484604. www.iucnredlist.org/species/61437/12484604

(2006b) *Janetaescincus veseyfitzgeraldi* (Publication no. https://dx.doi.org/10.2305/IUCN.UK.2006.RLTS.T61438A12484713.en.). Retrieved 14 July 2020, from The IUCN Red List of Threatened Species 2006: e. T61438A12484713. www.iucnredlist.org/species/61438/12484713

(2006c) *Lycognathophis seychellensis*. (Publication no. https://dx.doi.org/10.2305/IUCN.UK.2006.RLTS.T61427A12481585.en). Retrieved 14 July 2020, from The IUCN Red List of Threatened Species 2006: e. T61427A12481585. www.iucnredlist.org/species/61427/12481585

(2006d) *Pamelaescincus gardineri* (Publication no. https://dx.doi.org/10.2305/IUCN.UK.2006.RLTS.T61439A12484813.en). Retrieved 14 July 2020, from The IUCN Red List of Threatened Species 2006: e. T61439A12484813. www.iucnredlist.org/species/61439/12484813

(2006e) *Urocotyledon inexpectata* (Publication no. https://dx.doi.org/10.2305/IUCN.UK.2006.RLTS.T61435A12484181.en). Retrieved 14 July 2020, from The IUCN Red List of Threatened Species 2006: e. T61435A12484181 www.iucnredlist.org/species/61435/12484181

Gerlach, J. and Willi, J. (2002) A new species of frog, genus *Sooglossus* (Anura, Sooglossidae) from Silhouette Island, *Seychelles. Amphibia-Reptilia* 23(4): 445–458. https://doi.org/10.1163/15685380260462356

Gerlach, J., Rhodin, A., Pritchard, P. et al. (2008a) Pelusios castanoides intergularis *Bour 1983-Seychelles Yellow-Bellied Mud Turtle, Seychelles Chestnut-Bellied Terrapin. Conservation Biology of Freshwater Turtles and Tortoises: A Compilation Project of the IUCN/SSC Tortoise and Freshwater Turtle Specialist Group*. Chelonian Research Foundation. Chelonian Research Monographs, Lunenburg, MA, 5, pp. 010.1–010.4.

(2008b) Pelusios *Subniger Parietalis* Bour 1983–Seychelles Black Mud Turtle. Conservation Biology of Freshwater Turtles and Tortoises: A Compilation Project of the IUCN/SSC Tortoise and Freshwater Turtle Specialist Group. Chelonian Research Foundation. Chelonian Research

Monographs, Lunenburg, MA, 5, pp. 016.01–016.4.

Gerlach, J., Rocamora, G., Gane, J., Jolliffe, K., & Vanherck, L. (2013) Giant tortoise distribution and abundance in the Seychelles Islands: Past, present, and future. *Chelonian Conservation and Biology* 12(1): 70–83. https://doi.org/10.2744/CCB-0902.1

Gillespie, R. (2004) Community assembly through adaptive radiation in Hawaiian spiders. *Science* 303(5656): 356–359. https://doi.org/10.1126/science.1091875

Gower, D. J., San Mauro, D., Giri, V. et al. (2011) Molecular systematics of caeciliid caecilians (Amphibia: Gymnophiona) of the Western Ghats, India. *Molecular Phylogenetics and Evolution* 59(3): 698–707. https://doi.org/10.1016/j.ympev.2011.03.002

Gower, D. J., Agarwal, I., Karanth, K. P., Datta-Roy, A., Giri, V. B., Wilkinson, M., et al. (2016) The role of wet-zone fragmentation in shaping biodiversity patterns in peninsular India: Insights from the caecilian amphibian *Gegeneophis*. *Journal of Biogeography* 43(6): 1091–1102. http://dx.doi.org/10.1111/jbi.12710

Grant, P. R. (1999) *Ecology and Evolution of Darwin's Finches*: Princeton University Press, Princeton, NJ.

Grant, P. R. and Grant, B. R. (2014) *40 Years of Evolution: Darwin's Finches on Daphne Major Island*: Princeton University Press, Princeton, NJ.

Hambler, C. (1994) Giant tortoise *Geochelone gigantea* translocation to Curieuse island (Seychelles): Success or failure? *Biological Conservation* 69(3): 293–299. https://doi.org/10.1016/0006-3207(94)90429-4

Hansen, D. M., Austin, J. J., Baxter, R. H. et al. (2017) Origins of endemic island tortoises in the western Indian Ocean: A critique of the human-translocation hypothesis. *Journal of Biogeography* 44(6): 1430–1435. https://doi.org/10.1111/jbi.12893

Harris, D. J., Perera, A., Valente, J., and Rocha, S. (2015) Deep genetic differentiation within *Janetaescincus* spp. (Squamata: Scincidae) from the Seychelles Islands. *Herpetological Journal* 25(4): 205–213.

Hartl, D. L. and Clark, A. G. (1997) *Principles of Population Genetics* (Vol. 116). Sinauer Associates, Sunderland, MA.

Hass, C. A., Nussbaum, R. A., and Maxson, L. R. (1993) Immunological insights into the evolutionary history of caecilians (Amphibia: Gymnophiona): Relationships of the Seychellean caecilians and a preliminary report on family-level relationships. *Herpetological Monographs* 7: 56–63. https://doi.org/10.2307/1466951

Hawlitschek, O., Ramirez Garrido, S., and Glaw, F. (2017) How marine currents influenced the widespread natural overseas dispersal of reptiles in the Western Indian Ocean region. *Journal of Biogeography* 44(6): 1435–1440. https://doi.org/10.1111/jbi.12940

Hedges, S. B., Nussbaum, R. A., and Maxson, L. R. (1993) Caecilian phylogeny and biogeography inferred from mitochondrial DNA sequences of the 12S rRNA and 16S rRNA genes (Amphibia: Gymnophiona). *Herpetological Monographs* 7. https://doi.org/10.2307/1466952

Hillis, D. M. (2019) Species delimitation in herpetology. *Journal of Herpetology* 53(1). https://doi.org/10.1670/18-123

Hime, P. M., Lemmon, A. R., Lemmon, E. C. M. et al. (2021) Phylogenomics reveals ancient gene tree discordance in the amphibian Tree of Life. *Systematic Biology* 70(1): 49–66. https://doi.org/10.1093/sysbio/syaa034

Hudson, M. A., Young, R. P., D'Urban Jackson, J. et al. (2016) Dynamics and genetics of a disease-driven species decline to near extinction: Lessons for conservation. *Scientific Reports* 6: 30772. https://doi.org/10.1038/srep30772

IUCN, SSC. (2013) *Guidelines for Reintroductions and other Conservation Translocations (Version 1.0)*. Gland, Switzerland.

Kamei, R. G., San Mauro, D., Gower, D. J. et al. (2012) Discovery of a new family of amphibians from northeast India with ancient links to Africa. *Proceedings of the Royal Society B: Biological Sciences* 279

(1737): 2396–2401. https://doi.org/10.1098/rspb.2012.0150

Kelly, C. M., Branch, W. R., Broadley, D. G., Barker, N. P., and Villet, M. H. (2011) Molecular systematics of the African snake family Lamprophiidae Fitzinger, 1843 (Serpentes: Elapoidea), with particular focus on the genera *Lamprophis* Fitzinger 1843 and *Mehelya* Csiki 1903. *Molecular Phylogenetics and Evolution* 58(3): 415–426. https://doi.org/10.1016/j.ympev.2010.11.010

Kindler, C., Moosig, M., Branch, W. R. et al. (2016) Comparative phylogeographies of six species of hinged terrapins (*Pelusios* spp.) reveal discordant patterns and unexpected differentiation in the *P. castaneus*/*P. chapini* complex and *P. rhodesianus*. *Biological Journal of the Linnean Society* 117(2): 305–321. https://doi.org/10.1111/bij.12647

Klaver, C. and Böhme, W. (1986) *Phylogeny and Classification of the Chamaeleonidae (Sauria) with Special Reference to Hemipenis Morphology*. Zoologisches Forschungsinstitut und Museum Alexander Koenig, Bonn.

Kolbe, J. J., Leal, M., Schoener, T. W., Spiller, D. A., and Losos, J. B. (2012) Founder effects persist despite adaptive differentiation: A field experiment with lizards. *Science* 335 (6072): 1086–1089. https://doi.org/10.1126/science.1209566.

Kopeczky, R. and Gemel, R. (2016) First record of *Lamprophis geometricus* (SCHLEGEL, 1837), on the Seychelles Island of La Digue (Indian Ocean). *Herpetozoa* 28(3/4): 178–180.

Kopp, R. E., Simons, F. J., Mitrovica, J. X., Maloof, A. C., and Oppenheimer, M. (2009) Probabilistic assessment of sea level during the last interglacial stage. *Nature* 462(7275): 863–867. https://doi.org/10.1038/nature08686

Kueffer, C., Beaver, K., and Mougal, J. (2013) Case study: Management of novel ecosystems in the Seychelles. In R. J. Hobbs, E. S. Higgs, and C. M. Hall (eds.) *Novel Ecosystems: Intervening in the New Ecological World Order*. John Wiley & Sons, Oxford, pp. 228–238.

Kueffer, C. and Kaiser-Bunbury, C. N. (2014) Reconciling conflicting perspectives for biodiversity conservation in the Anthropocene. *Frontiers in Ecology and the Environment* 12(2): 131–137. http://dx.doi.org/10.1890/120201

Kueffer, C., Schumacher, E., Dietz, H., Fleischmann, K., and Edwards, P. J. (2010) Managing successional trajectories in alien-dominated, novel ecosystems by facilitating seedling regeneration: A case study. *Biological Conservation* 143(7): 1792–1802. https://doi.org/10.1016/j.biocon.2010.04.031

Labisko, J., Maddock, S. T., Taylor, M. L. et al. (2015) Chytrid fungus (*Batrachochytrium dendrobatidis*) undetected in the two orders of Seychelles amphibians. *Herpetological Review* 46(1): 41–45.

Labisko, J., Griffiths, R. A., Chong-Seng, L., Bunbury, N., Maddock, S. T., Bradfield, K. S. et al. (2019) Endemic, endangered and evolutionarily significant: Cryptic lineages in Seychelles' frogs (Anura: Sooglossidae). *Biological Journal of the Linnean Society* 126 (3): 417–435. https://doi.org/10.1093/biolinnean/bly183

Lamichhaney, S., Berglund, J., Almen, M. S. et al. (2015) Evolution of Darwin's finches and their beaks revealed by genome sequencing. *Nature* 518(7539): 371–375. https://doi.org/10.1038/nature14181

Le, M. and Raxworthy, C. J. (2017) Human-mediated dispersals do not explain tortoise distribution on the Indian Ocean's islands. *Journal of Biogeography* 44(10): 2421–2424. https://doi.org/10.1111/jbi.13055

Le, M., Raxworthy, C. J., McCord, W. P., and Mertz, L (2006) A molecular phylogeny of tortoises (Testudines: Testudinidae) based on mitochondrial and nuclear genes. *Molecular Phylogenetics and Evolution* 40(2): 517–531. https://doi.org/10.1016/j.ympev.2006.03.003

Leache, A. D. and Fujita, M. K. (2010) Bayesian species delimitation in West African forest geckos (*Hemidactylus fasciatus*). *Proceedings of the Royal Society B: Biological Sciences* 277

(1697): 3071–3077. https://doi.org/10.1098/rspb.2010.0662

Lima, A., Harris, D. J., Rocha, S. et al. (2013) Phylogenetic relationships of *Trachylepis* skink species from Madagascar and the Seychelles (Squamata: Scincidae). *Molecular Phylogenetics and Evolution* 67(3): 615–620. https://doi.org/10.1016/j.ympev.2013.02.001

Liu, Z., Chen, G., Zhu, T. et al. (2018) Prevalence of cryptic species in morphologically uniform taxa – fast speciation and evolutionary radiation in Asian frogs. *Molecular Phylogenetics and Evolution* 127: 723–731. https://doi.org/10.1016/j.ympev.2018.06.020

Loader, S. P., Pisani, D., Cotton, J. A. et al. (2007) Relative time scales reveal multiple origins of parallel disjunct distributions of African caecilian amphibians. *Biology Letters* 3(5): 505–508. https://doi.org/10.1098/rsbl.2007.0266

López-Sepulcre, A., Doak, N., Norris, K., and Shah, N. J. (2008) Population trends of Seychelles magpie-robins *Copsychus sechellarum* following translocation to Cousin Island, Seychelles. *Conservation Evidence* 5: 33–37.

Losos, J. B. (1992) The evolution of convergent structure in Caribbean *Anolis* communities. *Systematic Biology* 41(4): 403–420. https://doi.org/10.1093/sysbio/41.4.403

MacArthur, R. H. and Wilson, E. O. (1967) *Theory of Island Biogeography*. Princeton University Press, Princeton, NJ.

Maddock, S. T., Wilkinson, M., and Gower, D. J. (2018) A new species of small, long-snouted *Hypogeophis* Peters, 1880 (Amphibia: Gymnophiona: Indotyphlidae) from the highest elevations of the Seychelles island of Mahé. *Zootaxa* 4450(3): 359–375. https://doi.org/10.11646/zootaxa.4450.3.3

Maddock, S. T., Day, J. J., Nussbaum, R. A., Wilkinson, M., and Gower, D. J. (2014) Evolutionary origins and genetic variation of the Seychelles treefrog, *Tachycnemis seychellensis* (Dumeril and Bibron, 1841) (Amphibia: Anura: Hyperoliidae). *Molecular Phylogenetics and Evolution* 75: 194–201. https://doi.org/10.1016/j.ympev.2014.02.004

Maddock, S. T., Briscoe, A. G., Wilkinson, M., Waeschenbach, A., San Mauro, D., Day, J. J. at al. (2016) Next-generation mitogenomics: A comparison of approaches applied to caecilian amphibian phylogeny. *PLoS One* 11 (6): e0156757. https://doi.org/10.1371/journal.pone.0156757

Maddock, S. T., Wilkinson, M., Nussbaum, R. A., & Gower, D. J. (2017) A new species of small and highly abbreviated caecilian (Gymnophiona: Indotyphlidae) from the Seychelles island of Praslin, and a recharacterization of *Hypogeophis brevis* Boulenger, 1911. *Zootaxa* 4329(4): 301. https://doi.org/10.11646/zootaxa.4329.4.1

Maddock, S. T., Nussbaum, R. A., Day, J. J., Latta, L., Miller, M., Fisk, D. L. at al. (2020) The roles of vicariance and isolation by distance in shaping biotic diversification across an ancient archipelago: Evidence from a Seychelles caecilian amphibian. *BMC Evolutionary Biology* 20(1): 110. https://doi.org/10.1186/s12862-020-01673-w

Makhubo, B. G., Tolley, K. A., and Bates, M. F. (2015) Molecular phylogeny of the *Afroedura nivaria* (Reptilia: Gekkonidae) species complex in South Africa provides insight on cryptic speciation. *Molecular Phylogenetics and Evolution* 82 Pt A(0): 31–42. https://doi.org/10.1016/j.ympev.2014.09.025

Malenda, H. F., Simpson, E. L., Szajna, M. J. et al. (2012). Taphonomy of lacustrine shoreline fish-part conglomerates in the Late Triassic age Lockatong Formation (Collegeville, Pennsylvania, USA): Toward the recognition of catastrophic fish kills in the rock record. *Palaeogeography, Palaeoclimatology, Palaeoecology* 313: 234–245. https://doi.org/10.1016/j.palaeo.2011.11.022

Malhotra, A. and Thorpe, R. S. (1991) Experimental detection of rapid evolutionary response in natural lizard populations. *Nature* 353(6342): 347–348. https://doi.org/10.1038/353347a0

Marshall, A. F., Bardua, C., Gower, D. J. et al. (2019) High-density three-dimensional morphometric analyses support conserved static (intraspecific) modularity in caecilian

(Amphibia: Gymnophiona) crania. *Biological Journal of the Linnean Society* 126(4): 721–742. https://doi.org/10.1093/biolinnean/blz001

Miller, K. G., Kominz, M. A., Browning, J. V. et al. (2005) The Phanerozoic record of global sea-level change. *Science* 310(5752): 1293–1298. https://doi.org/10.1126/science.1116412

Montaggioni, L. F. and Hoang, C. T. (1988) The last interglacial high sea level in the granitic Seychelles, Indian Ocean. *Palaeogeography, Palaeoclimatology, Palaeoecology* 64(1–2): 79–91. https://doi.org/10.1016/0031-0182(88)90144-7

Myers, G. (1953) Ability of amphibians to cross sea barriers with especial reference to Pacific zoogeography. *Proceedings of the 7th Pacific Science Congress, New Zealand* 4: 2.

Norder, S. J., Seijmonsbergen, A. C., Rughooputh, S. D. D. V. et al. (2017) Assessing temporal couplings in social-ecological island systems: historical deforestation and soil loss on Mauritius (Indian Ocean). *Ecology and Society* 22(1) https://doi.org/10.5751/ES-09073-220129

Nussbaum, R. A. (1984a) Amphibians of the Seychelles. In: D. R. Stoddart (ed.) *Biogeography and Ecology of the Seychelles Islands*. W. Junk, The Hague; Boston: Hingham, MA, pp. 379–415.

(1984b) Snakes of the Seychelles. In: D. R. Stoddart (ed.) *Biogeography and Ecology of the Seychelles Islands*. W. Junk, The Hague; Boston: Hingham, MA, pp. 361–377.

Nussbaum, R. A. and Wu, S. H. (1995) Distribution, variation, and systematics of the Seychelles treefrog, *Tachycnemis seychellensis* (Amphibia: Anura: Hyperoliidae). *Journal of Zoology* 236(3): 383–406. http://dx.doi.org/10.1111/j.1469-7998.1995.tb02720.x

O'Neill, E. M., Beard, K. H., and Pfrender, M. E. (2012) Cast adrift on an island: Introduced populations experience an altered balance between selection and drift. *Biology Letters* 8(5): 890–893. https://doi.org/10.1098/rsbl.2012.0312

Oliver, P. M., Adams, M., Lee, M. S., Hutchinson, M. N., and Doughty, P. (2009) Cryptic diversity in vertebrates: Molecular data double estimates of species diversity in a radiation of Australian lizards (*Diplodactylus, Gekkota*). *Proceedings of the Royal Society B: Biological Sciences* 276(1664): 2001–2007. https://doi.org/10.1098/rspb.2008.1881

Owen-Smith, T. M., Ashwal, L. D., Torsvik, T. H. et al. (2013) Seychelles alkaline suite records the culmination of Deccan Traps continental flood volcanism. *Lithos* 182: 33–47. https://doi.org/10.1016/j.lithos.2013.09.011

Padial, J. M., Miralles, A., De la Riva, I., and Vences, M. (2010) The integrative future of taxonomy. *Frontiers in Zoology* 7(1): 16. https://doi.org/10.1186/1742-9994-7-16

Palkovacs, E. P., Gerlach, J., and Caccone, A. (2002) The evolutionary origin of Indian Ocean tortoises (*Dipsochelys*). *Molecular Phylogenetics and Evolution* 24(2): 216–227. https://doi.org/10.1016/S1055-7903(02)00211-7

Palkovacs, E. P., Marschner, M., Ciofi, C., Gerlach, J., and Caccone, A. (2003) Are the native giant tortoises from the Seychelles really extinct? A genetic perspective based on mtDNA and microsatellite data. *Molecular Ecology* 12(6): 1403–1413. https://doi.org/10.1046/j.1365-294X.2003.01834.x

Papadopoulou, A. and Knowles, L. L. (2016) Toward a paradigm shift in comparative phylogeography driven by trait-based hypotheses. *Proceedings of the National Academy of Sciences of the United States of America* 113(29): 8018–8024. https://doi.org/10.1073/pnas.1601069113

Parker, H. W. (1941) I.—The Cæcilians of the Seychelles. *Annals and Magazine of Natural History* 7(37): 1–17. https://doi.org/10.1080/00222934108527137

(1958) Caecilians of the Seychelles Islands with description of a new subspecies. *Copeia* 1958(2): 71–76. https://doi.org/10.2307/1440543

Perez-Ponce de Leon, G. and Poulin, R. (2016) Taxonomic distribution of cryptic diversity among metazoans: Not so homogeneous after all. *Biology Letters* 12(8). https://doi.org/10.1098/rsbl.2016.0371

Peters, W. C. H. (1881) Über die verschiedenheit der lage der äusseren spalten der schallblasen als merkmal zur unterscheidung besonder Afrikanischer froscharten. *Sitzungsberichte der Gesellschaft Naturforschender Freunde zu Berlin* 1881: 162–163.

Priti, H., Roshmi, R. S., Ramya, B. et al. (2016) Integrative taxonomic approach for describing a new cryptic species of bush frog (*Raorchestes*: Anura: Rhacophoridae) from the Western Ghats, India. *PLoS One* 11(3): e0149382. https://doi.org/10.1371/journal.pone.0149382

Procter, J. (1984a) Floristics of the granitic islands of the Seychelles. In: D. R. Stoddart (ed.) *Biogeography and Ecology of the Seychelles Islands*. W. Junk, The Hague; Boston: Hingham, MA, pp. 209–220.

(1984b) Vegetation of the granitic islands of the Seychelles. In: D. R. Stoddart (ed.) *Biogeography and Ecology of the Seychelles Islands*. W. Junk, The Hague; Boston: Hingham, MA, pp. 193–208.

Pyron, R. A. (2014) Biogeographic analysis reveals ancient continental vicariance and recent oceanic dispersal in amphibians. *Systematic Biology* 63(5): 779–797. https://doi.org/10.1093/sysbio/syu042

Pyron, R. A., Burbrink, F. T., and Wiens, J. J. (2013) A phylogeny and revised classification of Squamata, including 4161 species of lizards and snakes. *BMC Evolutionary Biology* 13(1): 1–54. https://doi.org/10.1186/1471-2148-13-93

Radtkey, R. R. (1996) Adaptive radiation of day-geckos (*Phelsuma*) in the Seychelles Archipelago: A phylogenetic analysis. *Evolution* 50(2): 604–623. https://doi.org/10.1111/j.1558-5646.1996.tb03872.x

Raxworthy, C. J., Forstner, M. R., and Nussbaum, R. A. (2002) Chameleon radiation by oceanic dispersal. *Nature* 415(6873): 784–787. https://doi.org/10.1038/415784a

Raxworthy, C. J., Ingram, C. M., Rabibisoa, N., and Pearson, R. G. (2007) Applications of ecological niche modeling for species delimitation: A review and empirical evaluation using day geckos (*Phelsuma*) from Madagascar. *Systematic Biology* 56(6): 907–923. https://doi.org/10.1080/10635150701775111

Rendahl, H. (1939) Zur Herpetologie der Seychellen: I. Reptilien. Zoologische Jahrbücher. *Abteilung für Systematik, Geographie und Biologie der Tiere* 72: 255–328.

Rhodin, A. G., Iverson, J. B., Bour, R. et al. (2017) *Turtles of the World: Annotated Checklist and Atlas of Taxonomy, Synonomy, Distribution, and Conservation Status*. Chelonian Research Foundation and Turtle Conservancy, Lunenburg.

Rocamora, G. and Henriette, E. (2015) *Invasive Alien Species in Seychelles: Why and How to Eliminate Them? Identification and Management of Priority Species*. Biotope Éditions, Mèze.

Rocha, S., Harris, D. J., and Carretero, M. A. (2010a) Genetic diversity and phylogenetic relationships of *Mabuya* spp. (Squamata: Scincidae) from western Indian Ocean islands. *Amphibia-Reptilia* 31(3): 375–385. http://dx.doi.org/10.1163/156853810791769473

Rocha, S. et al. (2010b) Phylogenetic systematics of day geckos, genus *Phelsuma*, based on molecular and morphological data (Squamata: Gekkonidae). *Zootaxa* 28: 1–28. https://doi.org/10.11646/zootaxa.2429.1.1

Rocha, S., Harris, D. J., and Posada, D. (2011) Cryptic diversity within the endemic prehensile-tailed gecko *Urocotyledon inexpectata* across the Seychelles Islands: Patterns of phylogeographical structure and isolation at the multilocus level. *Biological Journal of the Linnean Society* 104: 177–191. https://doi.org/10.1111/j.1095-8312.2011.01710.x

Rocha, S., Posada, D., and Harris, D. J. (2013) Phylogeography and diversification history of the day-gecko genus *Phelsuma* in the

Seychelles islands. *BMC Evolutionary Biology* 13(1): 3. https://doi.org/10.1186/1471-2148-13-3

Rocha, S., Harris, D. J., Perera, A. et al. (2009) Recent data on the distribution of lizards and snakes of the Seychelles. *Herpetological Bulletin* 110: 20–32.

Rocha, S., Perera, A., Silva, A., Posada, D., and Harris, D. J. (2016) Evolutionary history of *Trachylepis* skinks in the Seychelles islands: Introgressive hybridization, morphological evolution and geographic structure. *Biological Journal of the Linnean Society* 119 (1): 15–36. http://dx.doi.org/10.1111/bij.12803

(2017) Speciation history and species-delimitation within the Seychelles bronze geckos, *Ailuronyx* spp.: Molecular and morphological evidence. *Biological Journal of the Linnean Society* 120(3): 518–538. https://doi.org/10.1111/bij.12895

Roelants, K., Gower, D. J., Wilkinson, M. et al. (2007) Global patterns of diversification in the history of modern amphibians. *Proceedings of the National Academy of Sciences of the United States of America* 104 (3): 887–892. https://doi.org/10.1073/pnas.0608378104

Roll, U., Feldman, A., Novosolov, M. et al. (2017) The global distribution of tetrapods reveals a need for targeted reptile conservation. *Nature Ecology and Evolution* 1(11): 1677–1682. https://doi.org/10.1038/s41559-017-0332-2

Ryan, W. B. F., Carbotte, S. M., Coplan, J. O. et al. (2009) Global multi-resolution topography synthesis. *Geochemistry, Geophysics, Geosystems* 10(3). https://doi.org/10.1029/2008GC002332

San Mauro, D., Gower, D. J., Muller, H. et al. (2014) Life-history evolution and mitogenomic phylogeny of caecilian amphibians. *Molecular Phylogenetics and Evolution* 73(0): 177–189. https://doi.org/10.1016/j.ympev.2014.01.009

Santamaria, C. A., Bluemel, J. K., Bunbury, N., and Curran, M. (2017) Cryptic biodiversity and phylogeographic patterns of Seychellois *Ligia* isopods. *PeerJ* 5: e3894. https://doi.org/10.7717/peerj.3894

Shapiro, B., Barlow, A., Heintzman, P. D. et al. (eds.) (2019) *Ancient DNA: Methods and Protocols* (Second ed.): Humana Press, New York.

Shellnutt, J. G., Yeh, M. W., Suga, K. et al. (2017) Temporal and structural evolution of the Early Palaeogene rocks of the Seychelles microcontinent. *Scientific Reports* 7(1): 179. https://doi.org/10.1038/s41598-017-00248-y

Silva, A., Rocha, S., Gerlach, J. et al. (2010) Assessment of mtDNA genetic diversity within the terrapins *Pelusios subniger* and *Pelusios castanoides* across the Seychelles islands. *Amphibia-Reptilia* 31(4): 583–588. https://doi.org/10.1163/017353710X524723

Simberloff, D. and Cox, J. (1987) Consequences and costs of conservation corridors. *Conservation Biology* 1(1): 63–71. https://doi.org/10.1111/j.1523-1739.1987.tb00010.x

Simo-Riudalbas, M., Metallinou, M., de Pous, P. et al. (2017) Cryptic diversity in *Ptyodactylus* (Reptilia: Gekkonidae) from the northern Hajar Mountains of Oman and the United Arab Emirates uncovered by an integrative taxonomic approach. *PLoS One* 12(8): e0180397. https://doi.org/10.1371/journal.pone.0180397

Smith, M. J., Cogger, H., Tiernan, B. et al. (2012) An oceanic island reptile community under threat: The decline of reptiles on Christmas Island, Indian Ocean. *Herpetological Conservation and Biology* 7 (2): 206–218.

Sodhi, N. S. and Ehrlich, P. R. (2010) *Conservation Biology for All.* Oxford University Press, Oxford.

Spurgin, L. G., Illera, J. C., Jorgensen, T. H., Dawson, D. A., and Richardson, D. S. (2014) Genetic and phenotypic divergence in an island bird: Isolation by distance, by colonization or by adaptation? *Molecular Ecology* 23(5): 1028–1039. https://doi.org/10.1111/mec.12672

Stoddart, D. R. (1984a) Impact of man in the Seychelles. In: D. R. Stoddart (ed.) *Biogeography and Ecology of the Seychelles Islands.* W. Junk, The Hague; Boston: Hingham, MA, pp. 641–654.

(1984b) Scientific studies in the Seychelles. In: D. R. Stoddart (ed.) *Biogeography and Ecology of the Seychelles Islands.* W. Junk, The Hague; Boston: Hingham, MA, pp. 1–16.

Stoddart, D. R. and Peake, J. F. (1997) Historical records of Indian Ocean giant tortoise populations. *Philosophical Transactions of the Royal Society of London. B, Biological Sciences* 286(1011): 147–161. https://doi.org/10.1098/rstb.1979.0023

Streicher, J. W., Sadler, R., and Loader, S. P. (2020) Amphibian taxonomy: Early 21st century case studies. *Journal of Natural History* 54(1–4): 1–13. https://doi.org/10.1080/00222933.2020.1777339

Stuart, B. L., Inger, R. F., and Voris, H. K. (2006) High level of cryptic species diversity revealed by sympatric lineages of Southeast Asian forest frogs. *Biology Letters* 2(3): 470–474. https://doi.org/10.1098/rsbl.2006.0505

Stuckas, H., Gemel, R., and Fritz, U. (2013) One extinct turtle species less: *Pelusios seychellensis* is not extinct, it never existed. *PLoS One* 8(4): e57116. https://doi.org/10.1371/journal.pone.0057116

Taylor, E. H. (1968) *The Caecilians of the World.* University of Kansas Press, Lawrence.

Taylor, J. D., Braithwaite, C. J. R., Peake, J. F., and Arnold, E. N. (1997) Terrestrial faunas and habitats of Aldabra during the Late Pleistocene. *Philosophical Transactions of the Royal Society of London. B, Biological Sciences* 286(1011): 47–66. https://doi.org/10.1098/rstb.1979.0015

Taylor, M. L., Bunbury, N., Chong-Seng, L. et al. (2012) Evidence for evolutionary distinctiveness of a newly discovered population of sooglossid frogs on Praslin Island, Seychelles. *Conservation Genetics* 13 (2): 557–566. https://doi.org/10.1007/s10592-011-0307-9

Thewissen, J. G. M. and McKenna, M. C. (1992) Paleobiogeography of Indo-Pakistan: A response to Briggs, Patterson, and Owen. *Systematic Biology* 41(2): 248–251. https://doi.org/10.2307/2992525

Thorpe, R. S., Reardon, J. T., and Malhotra, A. (2005) Common garden and natural selection experiments support ecotypic differentiation in the Dominican anole (*Anolis oculatus*). *The American Naturalist* 165(4): 495–504. https://doi.org/10.1086/428408

Thorpe, R. S., Barlow, A., Malhotra, A., and Surget-Groba, Y. (2015) Widespread parallel population adaptation to climate variation across a radiation: Implications for adaptation to climate change. *Molecular Ecology* 24(5): 1019–1030. https://doi.org/10.1111/mec.13093

Titley, M. A., Snaddon, J. L., and Turner, E. C. (2017) Scientific research on animal biodiversity is systematically biased towards vertebrates and temperate regions. *PLoS One* 12(12): e0189577. https://doi.org/10.1371/journal.pone.0189577

Tolley, K. A., Townsend, T. M., and Vences, M. (2013) Large-scale phylogeny of chameleons suggests African origins and Eocene diversification. *Proceedings of the Royal Society B: Biological Sciences* 280(1759): 20130184. https://doi.org/10.1098/rspb.2013.0184

Townsend, T. M., Tolley, K. A., Glaw, F., Bohme, W., and Vences, M. (2011) Eastward from Africa: Palaeocurrent-mediated chameleon dispersal to the Seychelles islands. *Biology Letters* 7(2): 225–228. https://doi.org/10.1098/rsbl.2010.0701

Trape, J.-F. and Mediannikov, O. (2016) Cinq serpents nouveaux du genre *Boaedon* Duméril, Bibron & Duméril, 1854 (Serpentes: Lamprophiidae) en Afrique centrale. *Bulletin de la Société Herpétologique de France* 159: 61–111.

Tschudi, J. J. v. (1838) *Classification der Batrachier mit Berucksichtigung der fossilen Thiere diese Abtheilung der Reptilien.* Petitpierre, Neuchatel.

Uetz, P., Hošek, J., and Hallermann, J. (2020) The Reptile Database. Retrieved 14 July 2020 www.reptile-database.org/

Valente, J., Rocha, S., and Harris, D. J. (2014) Differentiation within the endemic burrowing skink *Pamelaescincus gardineri*, across the Seychelles islands, assessed by mitochondrial and nuclear markers. *African Journal of Herpetology* 63(1): 25–33. http://dx.doi.org/10.1080/21564574.2013.856354

Velo-Antón, G., Zamudio, K. R., and Cordero-Rivera, A. (2012) Genetic drift and rapid evolution of viviparity in insular fire salamanders (*Salamandra salamandra*). *Heredity* 108(4): 410–418. https://doi.org/10.1038/hdy.2011.91

Vesey-Fitzgerald, D. (1940) On the Vegetation of Seychelles. *Journal of Ecology* 28(2): 465–483. https://doi.org/10.2307/2256241

Vieites, D. R., Wollenberg, K. C., Andreone, F. et al. (2009) Vast underestimation of Madagascar's biodiversity evidenced by an integrative amphibian inventory. *Proceedings of the National Academy of Sciences of the United States of America* 106(20): 8267–8272. https://doi.org/10.1073/pnas.0810821106

Wagner, P., Leaché, A. D., and Fujita, M. K. (2014) Description of four new West African forest geckos of the *Hemidactylus fasciatus* Gray, 1842 complex, revealed by coalescent species delimitation. *Bonn Zoological Bulletin* 63(1): 1–14

Wallace, A. R. (1880) *Island Life, or, the Phenomena and Causes of Insular Faunas and Floras: Including a Revision and Attempted Solution of the Problem of Geological Climates* (First ed.). Macmillan, London.

(1892) *Island Life, or, the Phenomena and Causes of Insular Faunas and Floras: Including a Revision and Attempted Solution of the Problem of Geological Climates* (Second ed.). Macmillan, London.

Wallach, V., Williams, K. L., and Boundy, J. (2014) *Snakes of the World: A Catalogue of Living and Extinct Species*. CRC press, London.

Warren, B. H., Strasberg, D., Bruggemann, J. H., Prys-Jones, R. P., and Thébaud, C. (2010) Why does the biota of the Madagascar region have such a strong Asiatic flavour? *Cladistics* 26(5): 526–538. https://doi.org/10.1111/j.1096-0031.2009.00300.x

Wegener, A. (1924) *The Origin of Continents and Oceans*. Methuen, New York.

Weinell, J. L., Branch, W. R., Colston, T. J. et al. (2019) A species-level phylogeny of *Trachylepis* (Scincidae: Mabuyinae) provides insight into their reproductive mode evolution. *Molecular Phylogenetics and Evolution* 136: 183–195. https://doi.org/10.1016/j.ympev.2019.04.002

Wilkinson, M. and Nussbaum, R. A. (2006) Caecilian phylogeny and classification. In: J.-M. Exbrayat (ed.) *Reproductive Biology and Phylogeny of Gymnophiona (Caecilians)* (Vol. 5). Science Publishers, Enfield, NH, pp. 39–78.

Winter, M., Fiedler, W., Hochachka, W. M. et al. (2016) Patterns and biases in climate change research on amphibians and reptiles: A systematic review. *Royal Society Open Science* 3(9): 160158. https://doi.org/10.1098/rsos.160158.

Woodroffe, S. A., Long, A. J., Milne, G. A., Bryant, C. L., and Thomas, A. L. (2015) New constraints on late Holocene eustatic sea-level changes from Mahé, Seychelles. *Quaternary Science Reviews* 115(0): 1–16. https://doi.org/10.1016/j.quascirev.2015.02.011

Wright, D. J., Shah, N. J., and Richardson, D. S. (2014) Translocation of the Seychelles warbler *Acrocephalus sechellensis* to establish a new population on Frégate Island, Seychelles. *Conservation Evidence* 11: 20–24.

Yan, F., Lü, J., Zhang, B. et al. (2018) The Chinese giant salamander exemplifies the hidden extinction of cryptic species. *Current Biology* 28(10): R590–R592. https://doi.org/10.1016/j.cub.2018.04.004

Yatheesh, V. (2020) Structure and tectonics of the continental margins of India and the adjacent deep ocean basins: Current status of knowledge and some unresolved problems. *Episodes* 43(1): 586–608. https://doi.org/10.18814/epiiugs/2020/020039

Zamudio, K. R., Bell, R. C., and Mason, N. A. (2016) Phenotypes in phylogeography: Species' traits, environmental variation, and vertebrate diversification. *Proceedings of the National Academy of Sciences of the United States of America* 113(29): 8041–8048. https://doi.org/10.1073/pnas.1602237113

Zhang, J., Kapli, P., Pavlidis, P., and Stamatakis, A. (2013) A general species delimitation method with applications to phylogenetic placements. *Bioinformatics* 29(22): 2869–2876. https://doi.org/10.1093/bioinformatics/btt499

11

Cryptic Diversity in European Terrestrial Flatworms of the Genus *Microplana* (Platyhelminthes, Tricladida, Geoplanidae)

Marta Álvarez-Presas, Eduardo Mateos, Ronald Sluys, and Marta Riutort

Introduction

What Are Cryptic Species?

Species constitute one of the main components of biodiversity but, nonetheless, the species concept has always been one of the most difficult to define. More than 24 different definitions of species have been proposed, which may yield different outcomes (Mayden 1997). However, it is essential to have a correct species delimitation when addressing various fields of research, such as taxonomy, ecology, conservation biology, and biogeography (Agapow et al. 2004). Nowadays, many researchers agree on a concept that considers species to be entities that during the speciation process emerge as segments of independently evolving metapopulation lineages (Mayden 1997; De Queiroz 1998; Bock 2004; Hey 2006; Fišer et al. 2018). Nevertheless, the existence of many different species concepts, together with the introduction of molecular data in the delimitation of species taxa, in recent years has resulted in a proliferation of species delimitation methods, which are based on different premises (Raxworthy et al. 2007; Fontaneto et al. 2015; Luo et al. 2018). The accuracy of these delimitation methods depends basically on the particular stage of speciation occupied by the species *in statu nascendi* (Dellicour and Flot 2015). Thanks to the application of these new methods, many cryptic species have been discovered in little-known taxonomic groups or even in those groups presumed to have been well analysed. But what exactly is a cryptic species? Bickford et al. (2007) gave an exhaustive and accurate explanation of the notion of cryptic species and their implications for biodiversity assessment. The authors consider two or more species to be *cryptic* if they are or have been 'classified as a single nominal species because they are at least morphologically indistinguishable in their external appearance'. This implies that several lines of evidence are necessary in the delimitation and recognition of a species. Thus, not only morphological or molecular characters should be used, but one should apply an integrative combination of morphological, anatomical, molecular-genetic, karyological, and behavioural data in order to arrive at a comprehensive delimitation of the species entities of a particular habitat or region. Each species, including the cryptic ones, exhibits a unique set of biological traits and, therefore, should be recognised as a unit of biodiversity (Janzen et al. 2017).

A final term remains to be defined, viz., *pseudo-cryptic* species, to which the criterion 'inadequacy of morphological analysis' also applies (Knowlton 1993). However, in the case of pseudo-cryptic species this inadequacy is not due to limitations of morphological methods but merely to insufficient depth of knowledge or to the criterion that has been applied during the description of the species. When genetic, ecological, karyological, and behavioural data are combined with a careful morphological analysis of a species that originally was considered to be cryptic, it is generally possible to find sufficient morphological features for a clear distinction, sometimes at different life-stages of the organism (Gomez et al. 2004; Will et al. 2005; Cardoso et al. 2009; Lajus et al. 2015).

The Importance of Detecting Hidden Diversity in a Conservation Context

Ignoring cryptic biodiversity complicates biodiversity conservation efforts, as it may result in an underestimation of the species richness of particular regions and at the same time may lead to an overestimation of the abundance of individual species (Casseta et al. 2019). Furthermore, insufficient determination of cryptic species may result in an overestimation of the geographical range of a species because there may be several geographically restricted species instead of one widespread species (Eme et al. 2018), while it may result also in an erroneous assessment of the generalist/specialist status of species (a single generalist species may be confounded with numerous specialised species) (Morard et al. 2016). Furthermore, connectivity patterns may be erroneously estimated when cryptic species are ignored (Pante et al. 2015). This could occur in cases where one takes samples from two sympatric cryptic species, but the collections from one area contain, by chance, only individuals of one of the species, while samples from other areas include only individuals of the other species. In such a case, our data will suggest a high genetic differentiation, despite the fact that individuals migrate extensively and randomly select reproduction partners from both regions.

11.1 The Case of Terrestrial Planarians in Europe

Terrestrial flatworms are small worms belonging to the phylum Platyhelminthes Claus, 1887 that are classified in the order Tricladida Lang, 1884 and in the family Geoplanidae Stimpson, 1857. They are soil-dwelling invertebrates that are top-predators of arthropods, annelids, snails, slugs, or even other flatworms. Generally, they exhibit nocturnal activity, when humidity conditions are optimal, because land planarians lack any mechanism to preserve water in their tissues, thus depending directly on the humidity of their preferred habitat, that is, wet forest soils. Owing to this lack of physiological water-saving adaptation, they are presumed to have low-dispersal capabilities (excepting human-induced transportation in potted ornamental plants; see Sluys 2016). Taxa in soil communities carry the genetic footprint of ancient climatic and geographical events that may have been lost in other organisms with higher dispersal capacities. The characteristics of these soil-dwelling flatworms make them optimal model organisms for the study of ancient events that have affected the evolution of biodiversity and at the same time make them excellent indicators in biodiversity and conservation studies (Sluys 1999; Álvarez-Presas et al. 2011; Álvarez-Presas et al. 2014; Álvarez-Presas et al. 2018; Sluys 2019). Therefore, land planarians form a paradigm of the cryptic edaphic fauna in humid forests.

The subfamily Microplaninae Pantin, 1953 represents the most abundant group of land planarians on the European continent, with most species belonging to the genus *Microplana* Vejdovsky, 1889 (Figure 11.1). Species of *Microplana* are small animals with dull colours that cannot be found easily because they inhabit the upper layer of the soil, hiding under leaf litter, stones, logs, and other debris in the forest, while they are not easily caught in traps. The highest number of species descriptions in this group was published in the first decade of the twentieth century.

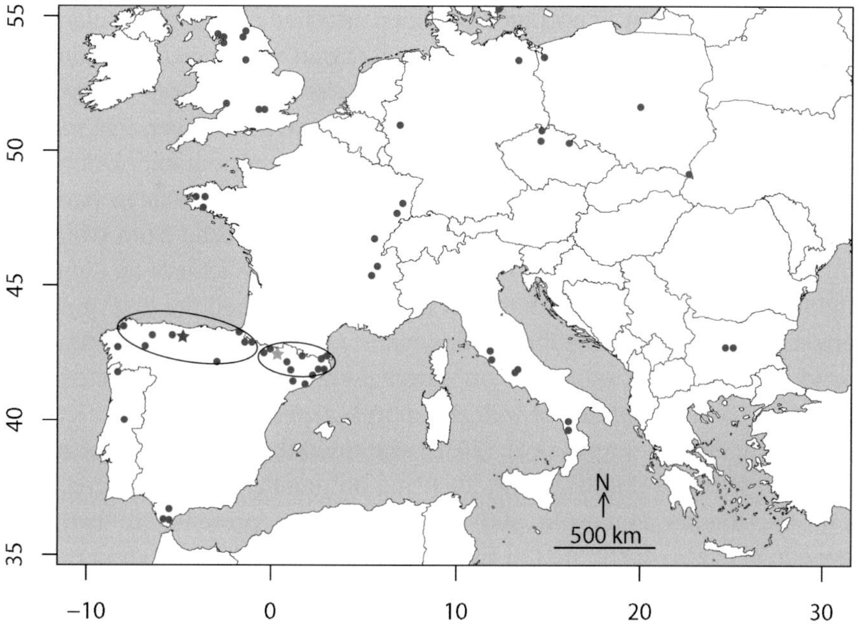

Figure 11.1 Map of European sampling localities (red dots) of species of *Microplana* visited during the years 2008–2017 (in many cases, localities in close proximity have been pooled). Blue asterisk: Picos de Europa National Park. Green asterisk: Ordesa y Monte Perdido National Park. Ellipses: two areas each with a molecularly distinct lineage of *Microplana terrestris*.

Up to 1998 only 12 species of *Microplana* had been described from Europe, mainly due to the fact that the study of European terrestrial planarians almost ceased after the Second World War. Until 2008 about 30 native land planarian species were known from Europe, of which 24 species belong to the genus *Microplana*. From 2008 onwards, however, this group of organisms received renewed attention, resulting in the description of several new species. Moreover, recent studies have demonstrated that our knowledge on the diversity of this group in temperate regions is still increasing, because the species accumulation curve has not yet reached an asymptote, as claimed by some researchers. On the contrary, the more the group is studied in detail, the more species are found (Sluys et al. 2016; Mateos et al. 2017).

There are still several constraints preventing a considerable increase in the number of described species, among which is the analytical, hypothesis-driven nature of taxonomic research (Sluys 2013). For example, it is rather difficult to describe species of land planarians. The first step is the collection of individuals, which in the case of terrestrial flatworms takes a long time due to the difficulty of finding these animals in their soil habitat. In addition, land planarians in Europe do not present a high population density, so that only few individuals can be found per sampling effort (Álvarez-Presas et al. 2018). In the next step it is necessary to study the anatomy of the worms, notably their reproductive apparatus, being the only complex structure making it possible to differentiate between species. This involves the time-consuming preparation and examination of histological sections. In addition, one must review the literature on already known and described species. Furthermore, in recent years great importance has been attached also to molecular information in the delimitation of new species. That is the reason why recently an effective system of identification by barcoding has been applied in European terrestrial flatworms by using a fragment of the mitochondrial gene cytochrome oxidase I. However, this requires the molecular sequencing of each of the collected individuals and subsequent phylogenetic analysis of these sequences, thus facilitating classification of the individuals in particular lineages. Next, from each lineage candidate specimens are then chosen from which the morphology and anatomy will be studied in detail, in order to finally achieve an identification or the description of a new species. Mateos et al. (2009) published the first molecular approach on terrestrial planarians of the Iberian Peninsula, by collecting 35 individuals and assigning these to ten morphotypes, based on external morphological characters. The molecular results were basically consistent with the morphotypes.

In a subsequent study, Álvarez-Presas et al. (2012) examined the most abundant species in Europe, *Microplana terrestris* (Müller, 1773), focusing on populations distributed in the northern part of the Iberian Peninsula and highlighting the presence of two well-differentiated genetic lineages, distributed in the western part and the eastern part of the range. From these two groups, the western one presented more structured and genetically divergent populations, while the eastern group was characterised by a higher uniformity in the genetic content across populations. It was also detected that some populations from the United Kingdom (UK) were genetically similar to the eastern Iberian populations. The hypothesis was formulated that, in view of the preference of these animals for humid forests, populations of *M. terrestris* may have followed the expansions and contractions of the forests during and after the glaciations in Europe, thus colonising European territories and reaching the UK from putative refugia in the northern section of the Iberian Peninsula.

In order to test that hypothesis, it was necessary to collect samples of *Microplana terrestris* from other countries and regions, especially those that are known to have harboured glacial refugia, such as the Italian peninsula and the Balkan area. Strategic countries, in this respect, with previous records of species of *Microplana* were visited. Mateos et al. (2017) published some of the results from these sampling campaigns, which started in 2013. The species *Microplana terrestris* was present in England, France, Spain, Germany, Sweden, Poland, and the Czech Republic. However, it was not found in Portugal, Italy, and Bulgaria (the latter representing two of the presumed glacial refugia), where instead other *Microplana* species were found. This does not necessarily imply that *M. terrestris* does not occur in these three countries, as it is difficult to confirm an absence, but at least it gives an idea of its abundance, since where it is present, *M. terrestris* is a frequently encountered species. The conclusion was drawn that the species that was believed to be the most abundant in Europe, *M. terrestris*, was not present in every country, and that individuals with a similar external appearance in fact belonged to other species.

The application of molecular tools was key to detecting the presence of different species with similar external appearance (Figure 11.2) and it was the first time that cryptic species of European terrestrial planarians received detailed study. In their paper, Mateos et al. (2017) described no less than seven new species with an external appearance similar to that of *M. terrestris*. However, we may consider these other species to be pseudo-cryptic instead of cryptic, because the specimens can be assigned their proper taxonomic species status on the basis of their copulatory apparatus and also because there is molecular evidence that clearly places the various species in different evolutionary lineages. From these sampling campaigns there is still material that remains to be analysed, while some of the putative new species had to remain as unconfirmed candidate species, being based only on molecular evidence because there was no material available for anatomical studies. In conclusion, the number of pseudo-cryptic species of land planarians in Europe is rather high and may be expected to increase even further when subjected to detailed studies.

11.2 The Case History of the Northern Iberian Peninsula

As mentioned in Section 11.1, terrestrial planarians are excellent indicators of both the state of their local habitat as well as environmental change. Therefore, it is interesting and useful to analyse the biodiversity of these animals in protected areas, in an attempt to discover which factors influence the distribution and abundance of these animals, as well as to gain insight into the status and functioning of their communities. However, no community-level study of land planarians in Europe was available until Álvarez-Presas et al. (2018) published their results on a study in two national parks in northern Spain, viz., Picos de Europa (Asturias, León, and Cantabria) and Ordesa y Monte Perdido (Huesca). The study focused on the distribution patterns of the genetic diversity of *Microplana terrestris* by intensively sampling, in two consecutive years, two different types of forest (beech and pine) in Ordesa

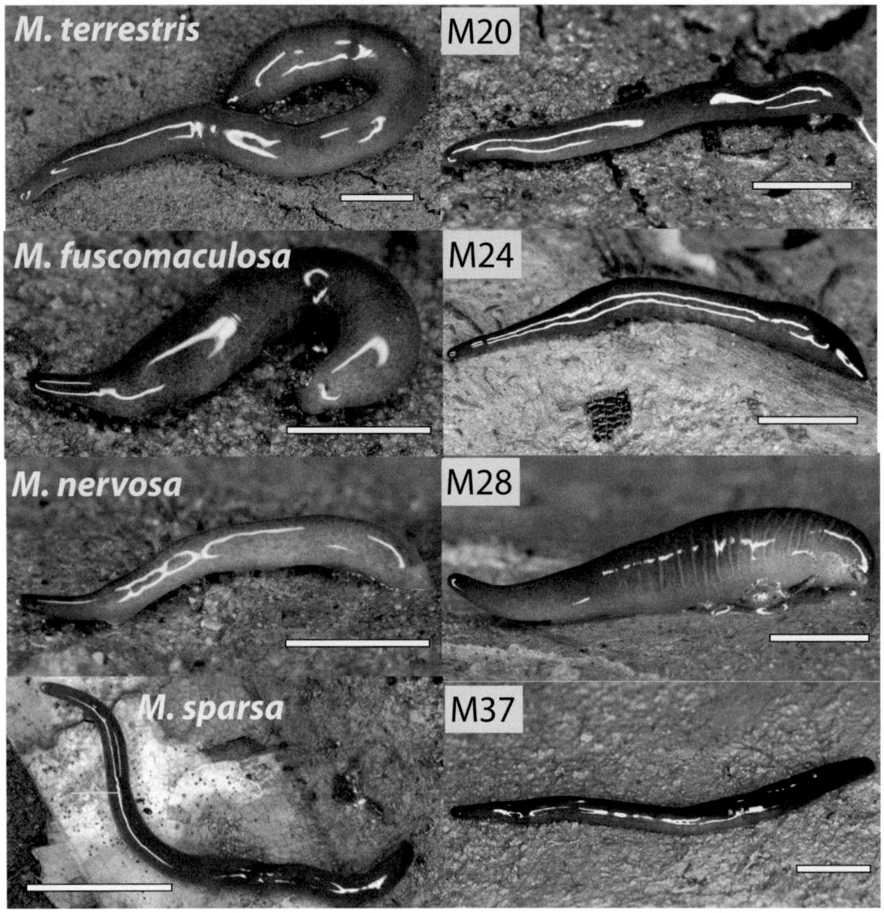

Figure 11.2 Pseudo-cryptic species in the *Microplana terrestris* complex. On the left, nominal species *M. terrestris* and three pseudo-cryptic species recently described as different species. On the right, four specimens molecularly different from *M. terrestris* but externally indistinguishable, awaiting anatomical investigations. Scale bar = 5 mm. Pictures: Eduardo Mateos.

y Monte Perdido during spring and autumn in all four valleys in the area (Añisclo, Escuaín, Pineta, and Ordesa). The same procedure was followed for Picos de Europa, where flatworm samples were collected in three different types of forest (beech, oak, and mixed forest), also in spring and autumn, which are the wettest periods of the year when there is, thus, a greater probability of finding terrestrial flatworms.

In the study of Álvarez-Presas et al. (2018) the survey unit consisted of a forest plot (approximately 50 × 50 m) where two persons searched for terrestrial flatworms under logs and stones for a period of one hour. For each of the plots, information was obtained on temperature, humidity, and pH, with the main objective of determining whether these environmental parameters differed between the various national parks and types of forest and, subsequently, whether these variables correlated with the presence/absence of terrestrial planarians. The results of these analyses suggested that none of these abiotic factors had any effect on the presence/absence and the local abundance of planarians in these two national parks. Only in the park Picos de Europa was there a significant correlation between pH and the abundance of the worms, similar to the situation found in Brazilian terrestrial planarians (Fick et al. 2006).

In terms of the number of species, as identified by external appearance, most of the individuals seemed to belong to the *M. terrestris* morphotype. However, detailed analyses revealed that a good number of pseudo-cryptic species were subsumed under this morphotype. Of the 17 Molecular Operational Taxonomic Units (MOTUs) found in the two parks, one could be attributed to *M. terrestris*, while seven MOTUs corresponded to new species, albeit that their external features were similar to *M. terrestris*.

It was discovered that there is also a second morphotype, concerning small worms with a pale colour, corresponding to a number of cryptic species. Although these worms match the description of *Microplana aixandrei* Vila-Farré, Mateos, Sluys & Romero, 2008, it appears that there are actually four MOTUs involved, three of which correspond to new pseudo-cryptic species (Figure 11.3).

The results of this study revealed the presence of significant differences between the national parks, in terms of species composition and richness of terrestrial planarians. In view of the phylogenetic relationships among species and the genetic structure of species shared between parks, these differences may result from historical events experienced by the forests during glacial periods. Nonetheless, the basal position of the genus *Microplana* in the phylogenetic tree of the land planarians suggests that the diversification of the genus predates the glaciations. This is similar to the situation in some Neotropical planarians, where ancient geological events may have underlain the early diversification, while the

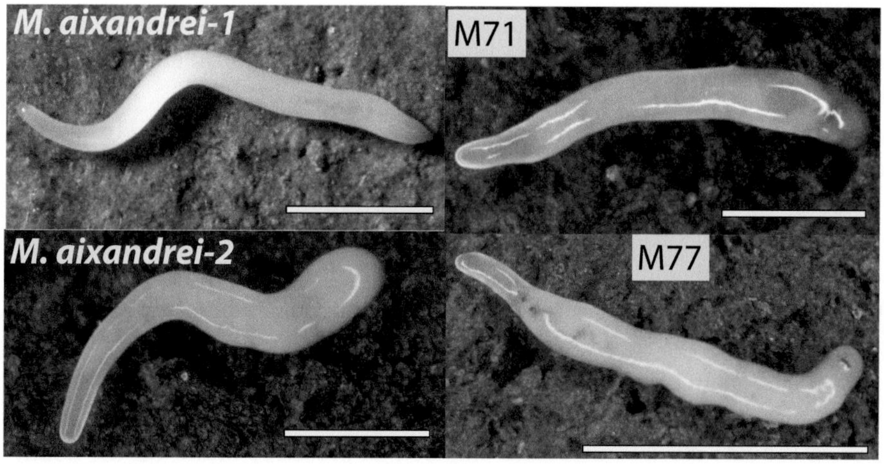

Figure 11.3 Pseudo-cryptic species in the *Microplana aixandrei* complex. On the left, specimens of two cryptic species, molecularly distinguishable but with similar external and internal (copulatory apparatus) morphology. On the right, two specimens molecularly different from *M. aixandrei* but externally indistinguishable, awaiting anatomical investigations. Scale bar = 2 mm. Pictures: Eduardo Mateos.

more recent Pleistocene vicissitudes presumably influenced the present-day distribution of the worms (Álvarez-Presas et al. 2011, 2014).

11.3 Intra- and Interspecific Genetic Diversity in Terrestrial Planarians

For the conservation of biodiversity and proper functioning of ecosystems, it is desirable that the latter host a species complex that is representative of their original biodiversity (Araújo et al. 2007), including levels of genetic diversity necessary for the maintenance of their populations that is representative of the diversity in the region. Therefore, knowledge of this genetic diversity is important, both at intraspecific and interspecific levels. Our study on the two Spanish national parks mentioned previously was the first in which we were able to analyse both intra- and interspecific genetic diversity in European terrestrial planarians (Sluys et al. 2016; Mateos et al. 2017). The wider extension of Picos de Europa and its higher diversity of forests, as compared with Ordesa y Monte Perdido, may be of influence on the distribution of the genetic diversity between and within the parks, resulting in higher diversity both at the species level and at the genetic level in the first-mentioned park (Álvarez-Presas et al. 2018). Nevertheless, both parks harbour species with a high genetic diversity that occupy basal positions in the phylogenetic trees and, therefore, may represent ancient lineages that before the Last Glacial Maximum (LGM) had a wide distribution throughout Europe.

At the intraspecific level, the genetic structure of the land planarian populations seems to reflect the evolutionary history of their habitat, which consisted of several isolated refugia during the Last Glaciation. Since these animals depend on the presence of moist forest soils, the existence and location of their supposed refugia can be reconstructed from information on the presence of wet forests in the northern part of the Iberian Peninsula before, during, and after the glacial period (Álvarez-Presas et al. 2012).

Extensive studies based on fossil wood, pollen records, and genetic variation have been published mainly for beech and oak forests in Europe (Ramil-Rego et al., 2000; Petit et al., 2003; Magri, 2008), showing that during the last glacial period beech forests were reduced to a few refugia of very small size. During the late Holocene, after the LGM, these forests slowly expanded, but probably did not invade the region north of the Pyrenees mountains, with the result that beeches of the Iberian Peninsula did not contribute to the recolonisation of this type of forest in northern Europe.

Studies of chloroplast sequences of oaks in the Iberian Peninsula show that haplotypes present in the Cantabrian and Basque regions probably persisted in a refugium on the west coast, while during the late glacial interstadial haplotypes presently found on the south-eastern slopes of the Pyrenees and in Catalonia could have arrived from Italy or even from the Balkans, via France (Petit et al. 2002). Thus, *Fagus* and *Quercus* agree in showing a clear differentiation between eastern and western populations in the north of the Iberian Peninsula, and in the fact that the populations in the east probably have an East European origin.

The mitochondrial genetic differentiation between the eastern and western clades of *Microplana*, their marked genetic structure within the western region, and the low genetic divergence in the eastern region all may have resulted from this complex history of their preferred habitat. At the beginning of the glacial period, the populations of planarians may have been restricted to a few refugia, that is, remnants of humid forests located mainly in the western section of the Iberian Peninsula, the worms thus being isolated in several distant localities. After the LGM, western populations may have expanded their range from their respective refugia and thus established secondary contacts, probably after the northward expansion of the oak forests. However, populations from Ordesa y Monte Perdido park exhibit also some genetic structure, albeit less pronounced. This suggests that despite the reduction of their habitat during the glacial period, the orography of the Pyrenees still facilitated retention of the genetic structure, a situation that is mirrored by beech forests in that region (Magri et al. 2006, 2008).

11.4 Diversity and Crypsis in the Neotropical Region and Europe

Generally, biodiversity in the tropics is higher than in temperate zones, such as the European continent. The same holds true for terrestrial planarians, for which higher species richness in the tropics, especially in areas originally covered by the Brazilian Atlantic rainforest, is well documented (Sluys 1999; Carbayo et al. 2002; Fick et al. 2006). Although there is an evident collector's bias in the currently known global distributional data of the land planarians, it is also true that tropical diversity is higher than in Europe. Nevertheless, in the latter the number of species has increased proportionally to the amount of work dedicated to their study, without yet having reached an asymptote in the species accumulation curve (Mateos et al. 2017).

Neotropical land planarians are generally much larger than their European counterparts, and much more brightly coloured. Despite their colourful external appearance, Neotropical land planarians also exhibit a good number of cryptic and pseudo-cryptic species. Some of these cases of crypsis have been detected in an integrative manner by using both morphological and molecular data (Lemos et al. 2014; Álvarez-Presas et al. 2015; Amaral et al. 2018; Carbayo et al. 2018), but others have been described only on the basis of morphological features (Lemos et al. 2008; Negrete et al. 2019), which complicates their detection.

One may wonder what the underlying cause is of the many externally similar species of terrestrial flatworms. In the case of European species, it simply may be the result of convergence because there is not much variation possible, given their small size and dull colours. In the case of Neotropical species, however, colourful pigmentation may signal danger to potential predators, as a potent neurotoxin has been detected in two species of the land planarian genus *Bipalium* that could be used both to subdue prey items and as a defence mechanism (Stokes et al. 2014). Thus, in Neotropical species crypsis may not only result from random convergence, as is presumably the case in European species, but may involve both Batesian and Müllerian mimicry as a way to ensure protection from predators.

Our past and ongoing studies on the European fauna of microplanid land flatworms illustrate the positive effect of an increased taxonomic knowledge based on our understanding of the actual biodiversity and distribution of groups of organisms that previously were considered to be obscure and poorly represented.

References

Agapow, P. M., Bininda-Emonds, O. R. P., Crandall, K. A. et al. (2004) The impact of species concept on biodiversity studies. *Quarterly Review of Biology* 79(2): 161–179. https://doi.org/10.1086/383542.

Álvarez-Presas, M., Mateos, E., and Riutort, M. (2018) Hidden diversity in forest soils: Characterization and comparison of terrestrial flatworm's communities in two national parks in Spain. *Ecology and Evolution* 8: 7386–7400. https://doi.org/10.1002/ece3.4178.

Álvarez-Presas, M., Amaral, S. V., Carbayo, F. et al. (2015) Focus on the details: Morphological evidence supports new cryptic land flatworm (Platyhelminthes) species revealed with molecules. *Organisms Diversity and Evolution* 15: 379–403. https://doi.org/10.1007/s13127-014-0197-z.

Álvarez-Presas, M., Carbayo, F., Rozas, J., and Riutort, M. (2011) Land planarians (Platyhelminthes) as a model organism for fine-scale phylogeographic studies: Understanding patterns of biodiversity in the Brazilian Atlantic Forest hotspot. *Journal of Evolutionary Biology* 24: 887–896. https://doi.org/10.1111/j.1420-9101.2010.02220.x

Álvarez-Presas, M., Mateos, E., Vila-Farré, M., Sluys, R., and Riutort, M. (2012) Evidence for the persistence of the land planarian species *Microplana terrestris* (Müller, 1774) (Platyhelminthes, Tricladida) in microrefugia during the Last Glacial Maximum in the northern section of the Iberian Peninsula. *Molecular Phylogenetics and Evolution* 64: 491–499. https://doi.org/10.1016/j.ympev.2012.05.001.

Alvarez-Presas, M., Sánchez-Gracia, A., Carbayo, F., Rozas, J., and Riutort, M. (2014) Insights into the origin and distribution of biodiversity in the Brazilian Atlantic forest hot spot: A statistical phylogeographic study using a low-dispersal organism. Heredity *(Edinb)* 112: 656–665. https://doi.org/10.1038/hdy.2014.3

Amaral, S. V., Ribeiro, G. G., Valiati, V. H., and Leal-Zanchet, A. M. (2018) Body doubles: An integrative taxonomic approach reveals new sibling species of land planarians. *Invertebrate Systematics* 32: 533–550. https://doi.org/10.1071/IS17046.

Araújo, M. B., Lobo, J. M., and Moreno, J. C. (2007) The effectiveness of Iberian protected areas in conserving terrestrial biodiversity. *Conservation Biology* 21: 1423–1432. https://doi.org/10.1111/j.1523-1739.2007.00827.x.

Bickford, D., Lohman, D. J., Sodhi, N. S. et al. (2007) Cryptic species as a window on diversity and conservation. *Trends in Ecology & Evolution* 22(3): 148–155. https://doi.org/10.1016/j.tree.2006.11.004.

Bock, W. J. (2004) Species: The concept, category and taxon. *Journal of Zoological Systematics and Evolutionary Research* 42(3): 178–190. https://doi.org/10.1111/j.1439-0469.2004.00276.x.

Carbayo, F., Leal-Zanchet, A. M., and Vieira, E. M. (2002) Terrestrial flatworm (Platyhelminthes: Tricladida: Terricola) diversity versus man-induced disturbance in an ombrophilous forest in southern Brazil. *Biodiversity and Conservation* 11: 1091–1104. https://doi.org/10.1023/A:1015865005604.

Carbayo, F., Silva, M. S., Riutort, M., and Álvarez-Presas, M. (2018) Rolling into the deep of the land planarian genus *Choeradoplana* (Tricladida, Continenticola, Geoplanidae) taxonomy. *Organisms Diversity and Evolution* 18: 187–210. https://doi.org/10.1007/s13127-017-0352-4.

Cardoso, A., Serrano, A., and Vogler, A. P. (2009) Morphological and molecular variation in tiger beetles of the *Cicindela hybrida* complex: Is an “integrative taxonomy” possible? *Molecular Ecology* 18: 648–664. https://doi.org/10.1111/j.1365-294X.2008.04048.x.

Casetta, E., da Silva, J. M., and Vecchi, D. (eds.) (2019) From Assessing to Conserving Biodiversity. Springer Open, https://doi.org/10.1007/978-3-030-10991-2.

Dellicour, S. and Flot, J. F. (2015) Delimiting species-poor data sets using single molecular markers: A study of barcode gaps, haplowebs and GMYC. *Systematic Biology* 64 (6): 900–908. https://doi.org/10.1093/sysbio/syu130.

De Queiroz, K. (1998) The general lineage concept of species, species criteria, and the process of speciation and terminological recommendations. In: D. J. Howard and S. H. Berlocher (eds.) *Endless Forms: Species and Speciation*. Oxford University Press, Oxford, pp. 57–75.

Eme, D., Zagmajster, M., Delic, T. et al. (2018) Do cryptic species matter in macroecology? Sequencing European groundwater crustaceans yields smaller ranges but does not challenge biodiversity determinants. *Ecography* 41: 424–436. https://doi.org/10.1111/ecog.02683.

Fick, I. A., Leal-Zanchet, A. M., and Vieira, E. M. (2006) Community structure of land flatworms (Platyhelminthes, Terricola): Comparisons between *Araucaria* and Atlantic forest in Southern Brazil. *Invertebrate Biology* 125: 306–313. https://doi.org/10.1111/j.1744-7410.2006.00062.x.

Fišer, C., Robinson, C. T., and Malard, F. (2018) Cryptic species as a window into the paradigm shift of the species concept. *Molecular Ecology* 27(3): 613–635. https://doi.org/10.1111/mec.14486.

Fontaneto, D., Flot, J.-F., and Tang, C. Q. (2015) Guidelines for DNA taxonomy, with a focus on the meiofauna. *Marine Biodiversity* 45(3): 433–451. https://doi.org/10.1007/s12526-015-0319-7.

Gómez, S., Fleeger, J. W., Rocha-Olivares, A., and Foltz, D. (2004) Four new species of *Cletocamptus* Schmankewitsch, 1875, closely related to *Cletocamptus deitersi* (Richard, 1897) (Copepoda: Harpacticoida). *Journal of Natural History* 38: 2669–2732. https://doi.org/10.1080/0022293031000156240.

Hey, J. (2006) On the failure of modern species concepts. *Trends in Ecology and Evolution* 21 (8): 447–450. https://doi.org/10.1016/j.tree.2006.05.011.

Janzen, D. H., Burns, J. M., Cong, Q. et al. (2017) Nuclear genomes distinguish cryptic species suggested by their DNA barcodes and ecology. *Proceedings of the National Academy of Sciences of the United States of America* 114: 8313–8318. https://doi.org/10.1073/pnas.1621504114.

Knowlton, N. (1993). Sibling species in the sea. *Annual Review on Ecology and Systematics* 24: 189–216. https://doi.org/10.1146/annurev.es.24.110193.001201.

Lajus, D., Sukhikh, N., and Alekseev, V. (2015) Cryptic or pseudocryptic: Can morphological methods inform copepod taxonomy? An analysis of publications and a case study of the *Eurytemora affinis* species complex. *Ecology and Evolution* 5: 2374–2385. https://doi.org/10.1002/ece3.1521.

Lemos, V. S. and Leal-Zanchet, A. M. (2008) Two new species of *Notogynaphallia* Ogren & Kawakatsu (Platyhelminthes: Tricladida: Terricola) from Southern Brazil. *Zootaxa* 1907(1): 28–46. htpps://doi.org/ 10.11646/zootaxa.1907.1.2.

Lemos, V. S. A., Cauduro, G. P. B., Valiati, V. H. B., and Leal-Zanchet, A. M. (2014) Phylogenetic relationships within the flatworm genus *Choeradoplana* Graff (Platyhelminthes: Tricladida) inferred from molecular data with the description of two new sympatric species from *Araucaria* moist forests. *Invertebrate Systematics* 28: 605–627. https://doi.org/http://dx.doi.org/10.1071/IS14003.

Luo, A., Ling, C., Ho, S. Y. W., and Zhu, C.-D. (2018) Comparison of methods for molecular species delimitation across a range of speciation scenarios. *Systematic Biology* 67(5): 830–846. https://doi.org/10.1093/sysbio/syy011.

Magri, D. (2008) Patterns of post-glacial spread and the extent of glacial refugia of European beech (*Fagus sylvatica*). *Journal of Biogeography* 35: 450–463. https://doi.org/10.1111/j.1365-2699.2007.01803.x.

Magri, D., Vendramin, G. G., Comps, B. et al. (2006) A new scenario for the Quaternary history of European beech populations: Palaeobotanical evidence and genetic consequences. *New Phytologist* 171: 199–221. https://doi.org/10.1111/j.1469-8137.2006.01740.x.

Mateos, E., Cabrera, C., Carranza, S., and Riutort, M. (2009) Molecular analysis of the diversity of terrestrial planarians (Platyhelminthes, Tricladida, Continenticola) in the Iberian Peninsula. *Zoologica Scripta* 38: 637–649. https://doi.org/10.1111/j.1463-6409.2009.00398.x.

Mateos, E., Sluys, R., Riutort, M., and Álvarez-Presas, M. (2017) Species richness in the genus *Microplana* (Platyhelminthes, Tricladida, Microplaninae) in Europe: As yet no asymptote in sight. *Invertebrate Systematics* 31: 269–301. http://dx.doi.org/10.1071/IS16038.

Mayden, R. L. (1997) A hierarchy of species concepts: The denouement in the saga of the species problem. In: M. F. Claridge, H. A. Dawah, and M. R. Wilson (ed.) *Species: The Units of Biodiversity*. Chapman and Hall, London, pp. 381–423.

Morard, R., Escarguel, G., Weiner, A. K. M. et al. (2016) Nomenclature for the nameless: A proposal for an integrative molecular taxonomy of cryptic diversity exemplified by planktonic foraminifera. *Systematic Biology* 65: 925–940. https://doi.org/10.1093/sysbio/syw031.

Negrete, L., Gira, R. D., and Brusa, F. (2019) Two new species of land planarians (Platyhelminthes, Tricladida, Geoplanidae) from protected areas in the southern extreme of the Paranaense Rainforest, Argentina. *Zoologischer Anzeiger* 279: 38–51. https://doi.org/10.1016/j.jcz.2019.01.002.

Pante, E., Puillandre, N., Viricel, A. et al. (2015) Species are hypotheses: Avoid connectivity assessments based on pillars of sand. *Molecular Ecology* 24, 525–544. https://doi.org/10.1111/mec.13048.

Petit, R. J., Aguinagalde, I., de Beaulieu, J.-L. et al.(2003) Glacial refugia: Hotspots but not melting pots of genetic diversity. *Science* 300: 1563–1565. https://doi.org/10.1126/science.1083264

Petit, R. J., Brewer, S., Bordacs, S. et al. (2002) Identification of refugia and post-glacial colonisation routes of European white oaks based on chloroplast DNA and fossil pollen evidence. *Forest Ecology and Management* 156: 49–74. https://doi.org/10.1016/S0378-1127(01)00634-X.

Ramil-Rego, P., Guitian, M. A. R., Sobrino, C. M., and Gomez-Orellana, L. (2000) Some considerations about the postglacial history and recent distribution of *Fagus sylvatica* in the NW Iberian Peninsula. *Folia Geobotanica* 35: 241–271. https://doi.org/10.1007/BF02803118.

Raxworthy, C. J., Ingram, C. M., Rabibisoa, N., and Pearson, R. G (2007) Applications of ecological niche modeling for species delimitation: A review and empirical evaluation using day geckos (*Phelsuma*) from Madagascar. *Systematic Biology* 56(6): 907–923. https://doi.org/10.1080/10635150701775111.

Sluys, R. (1999) Global diversity of land planarians (Platyhelminthes, Tricladida, Terricola): A new indicator-taxon in biodiversity and conservation studies. *Biodiversity & Conservation* 8: 1663–1681. https://doi.org/10.1023/A:1008994925673.

(2013) The unappreciated, fundamentally analytical nature of taxonomy and the implications for the inventory of biodiversity. *Biodiversity & Conservation* 22: 1095–1105.

(2016) Invasion of the flatworms. *American Scientist* 104: 288–295.

(2019) The evolutionary terrestrialization of planarian flatworms (Platyhelminthes,

Tricladida, Geoplanidae): A review and research programme. *Zoosystematics and Evolution* 95: 543–556.

Sluys, R., Mateos, E., Riutort, M., and Álvarez-Presas, M. (2016) Towards a comprehensive, integrative analysis of the diversity of European microplaninid land flatworms (Platyhelminthes, Tricladida, Microplaninae), with the description of two peculiar new species. *Systematics and Biodiversity* 14: 9–31. https://doi.org/10.1080/14772000.2015.1103323.

Stokes, A. N., Ducey, P. K., Neuman-Lee, L. et al. (2014) Confirmation and distribution of tetrodotoxin for the first time in terrestrial invertebrates: Two terrestrial flatworm species (*Bipalium adventitium* and *Bipalium kewense*). *PLoS One* 9: e100718. https://doi.org/10.1371/journal.pone.0100718

Will, K. W., Mishler, B. D., and Wheeler, Q. D. (2005) The perils of DNA barcoding and the need for integrative taxonomy. *Systematic Biology* 54: 844–851. https://doi.org/10.1080/10635150500354878.

Index

Systematics Association Special Volumes

1. The New Systematics (1940)[a]
   Edited by J. S. Huxley (reprinted 1971)
2. Chemotaxonomy and Serotaxonomy (1968)*
   Edited by J. C. Hawkes
3. Data Processing in Biology and Geology (1971)*
   Edited by J. L. Cutbill
4. Scanning Electron Microscopy (1971)*
   Edited by V. H. Heywood
5. Taxonomy and Ecology (1973)*
   Edited by V. H. Heywood
6. The Changing Flora and Fauna of Britain (1974)*
   Edited by D. L. Hawksworth
7. Biological Identification with Computers (1975)*
   Edited by R. J. Pankhurst
8. Lichenology: Progress and Problems (1976)*
   Edited by D. H. Brown, D. L. Hawksworth and R. H. Bailey
9. Key Works to the Fauna and Flora of the British Isles and Northwestern Europe, fourth edition (1978)*
   Edited by G. J. Kerrich, D. L. Hawksworth and R. W. Sims
10. Modern Approaches to the Taxonomy of Red and Brown Algae (1978)*
    Edited by D. E. G. Irvine and J. H. Price
11. Biology and Systematics of Colonial Organisms (1979)*
    Edited by C. Larwood and B. R. Rosen
12. The Origin of Major Invertebrate Groups (1979)*
    Edited by M. R. House
13. Advances in Bryozoology (1979)*
    Edited by G. P. Larwood and M. B. Abbott
14. Bryophyte Systematics (1979)*
    Edited by G. C. S. Clarke and J. G. Duckett
15. The Terrestrial Environment and the Origin of Land Vertebrates (1980)*
    Edited by A. L. Panchen
16. Chemosystematics: Principles and Practice (1980)*
    Edited by F. A. Bisby, J. G. Vaughan and C. A. Wright
17. The Shore Environment: Methods and Ecosystems (two volumes) (1980)*
    Edited by J. H. Price, D. E. C. Irvine and W. F. Farnham

18. The Ammonoidea (1981)*
Edited by M. R. House and J. R. Senior

19. Biosystematics of Social Insects (1981)*
Edited by P. E. House and J.-L. Clement

20. Genome Evolution (1982)*
Edited by G. A. Dover and R. B. Flavell

21. Problems of Phylogenetic Reconstruction (1982)*
Edited by K. A. Joysey and A. E. Friday

22. Concepts in Nematode Systematics (1983)*
Edited by A. R. Stone, H. M. Platt and L. F. Khalil

23. Evolution, Time and Space: The Emergence of the Biosphere (1983)*
Edited by R. W. Sims, J. H. Price and P. E. S. Whalley

24. Protein Polymorphism: Adaptive and Taxonomic Significance (1983)*
Edited by G. S. Oxford and D. Rollinson

25. Current Concepts in Plant Taxonomy (1983)*
Edited by V. H. Heywood and D. M. Moore

26. Databases in Systematics (1984)*
Edited by R. Allkin and F. A. Bisby

27. Systematics of the Green Algae (1984)*
Edited by D. E. G. Irvine and D. M. John

28. The Origins and Relationships of Lower Invertebrates (1985)‡
Edited by S. Conway Morris, J. D. George, R. Gibson and H. M. Platt

29. Infraspecific Classification of Wild and Cultivated Plants (1986)‡
Edited by B. T. Styles

30. Biomineralization in Lower Plants and Animals (1986)‡
Edited by B. S. C. Leadbeater and R. Riding

31. Systematic and Taxonomic Approaches in Palaeobotany (1986)‡
Edited by R. A. Spicer and B. A. Thomas

32. Coevolution and Systematics (1986)‡
Edited by A. R. Stone and D. L. Hawksworth

33. Key Works to the Fauna and Flora of the British Isles and Northwestern Europe, fifth edition (1988)‡
Edited by R. W. Sims, P. Freeman and D. L. Hawksworth

34. Extinction and Survival in the Fossil Record (1988)‡
Edited by G. P. Larwood

35. The Phylogeny and Classification of the Tetrapods (two volumes) (1988)‡
Edited by M. J. Benton

36. Prospects in Systematics (1988)‡
Edited by J. L. Hawksworth

37. Biosystematics of Haematophagous Insects (1988)‡
Edited by M. W. Service

38. The Chromophyte Algae: Problems and Perspective (1989)‡
Edited by J. C. Green, B. S. C. Leadbeater and W. L. Diver

39. Electrophoretic Studies on Agricultural Pests (1989)[‡]
Edited by H. D. Loxdale and J. den Hollander

40. Evolution, Systematics and Fossil History of the Hamamelidae (two volumes) (1989)[‡]
Edited by P. R. Crane and S. Blackmore

41. Scanning Electron Microscopy in Taxonomy and Functional Morphology (1990)[‡]
Edited by D. Claugher

42. Major Evolutionary Radiations (1990)[‡]
Edited by P. D. Taylor and G. P. Larwood

43. Tropical Lichens: Their Systematics, Conservation and Ecology (1991)[‡]
Edited by G. J. Galloway

44. Pollen and Spores: Patterns and Diversification (1991)[‡]
Edited by S. Blackmore and S. H. Barnes

45. The Biology of Free-Living Heterotrophic Flagellates (1991)[‡]
Edited by D. J. Patterson and J. Larsen

46. Plant–Animal Interactions in the Marine Benthos (1992)[‡]
Edited by D. M. John, S. J. Hawkins and J. H. Price

47. The Ammonoidea: Environment, Ecology and Evolutionary Change (1993)[‡]
Edited by M. R. House

48. Designs for a Global Plant Species Information System (1993)[‡]
Edited by F. A. Bisby, G. F. Russell and R. J. Pankhurst

49. Plant Galls: Organisms, Interactions, Populations (1994)[‡]
Edited by M. A. J. Williams

50. Systematics and Conservation Evaluation (1994)[‡]
Edited by P. L. Forey, C. J. Humphries and R. I. Vane-Wright

51. The Haptophyte Algae (1994)[‡]
Edited by J. C. Green and B. S. C. Leadbeater

52. Models in Phylogeny Reconstruction (1994)[‡]
Edited by R. Scotland, D. I. Siebert and D. M. Williams

53. The Ecology of Agricultural Pests: Biochemical Approaches (1996)[**]
Edited by W. O. C. Symondson and J. E. Liddell

54. Species: The Units of Diversity (1997)[**]
Edited by M. F. Claridge, H. A. Dawah and M. R. Wilson

55. Arthropod Relationships (1998)[**]
Edited by R. A. Fortey and R. H. Thomas

56. Evolutionary Relationships among Protozoa (1998)[**]
Edited by G. H. Coombs, K. Vickerman, M. A. Sleigh and A. Warren

57. Molecular Systematics and Plant Evolution (1999)[‡‡]
Edited by P. M. Hollingsworth, R. M. Bateman and R. J. Gornall

58. Homology and Systematics (2000)[‡‡]
Edited by R. Scotland and R. T. Pennington

59. The Flagellates: Unity, Diversity and Evolution (2000)[‡‡]
Edited by B. S. C. Leadbeater and J. C. Green

60. Interrelationships of the Platyhelminthes (2001)[‡‡]
Edited by D. T. J. Littlewood and R. A. Bray

61. Major Events in Early Vertebrate Evolution (2001)[‡‡]
Edited by P. E. Ahlberg

62. The Changing Wildlife of Great Britain and Ireland (2001)[‡‡]
Edited by D. L. Hawksworth

63. Brachiopods Past and Present (2001)[‡‡]
Edited by H. Brunton, L. R. M. Cocks and S. L. Long

64. Morphology, Shape and Phylogeny (2002)[‡‡]
Edited by N. MacLeod and P. L. Forey

65. Developmental Genetics and Plant Evolution (2002)[‡‡]
Edited by Q. C. B. Cronk, R. M. Bateman and J. A. Hawkins

66. Telling the Evolutionary Time: Molecular Clocks and the Fossil Record (2003)[‡‡]
Edited by P. C. J. Donoghue and M. P. Smith

67. Milestones in Systematics (2004)[‡‡]
Edited by D. M. Williams and P. L. Forey

68. Organelles, Genomes and Eukaryote Phylogeny (2004)[‡‡]
Edited by R. P. Hirt and D. S. Horner

69. Neotropical Savannas and Seasonally Dry Forests: Plant Diversity, Biogeography and Conservation (2006)[‡‡]
Edited by R. T. Pennington, G. P. Lewis and J. A. Rattan

70. Biogeography in a Changing World (2006)[‡‡]
Edited by M. C. Ebach and R. S. Tangney

71. Pleurocarpous Mosses: Systematics and Evolution (2006)[‡‡]
Edited by A. E. Newton and R. S. Tangney

72. Reconstructing the Tree of Life: Taxonomy and Systematics of Species Rich Taxa (2006)[‡‡]
Edited by T. R. Hodkinson and J. A. N. Parnell

73. Biodiversity Databases: Techniques, Politics, and Applications (2007)[‡‡]
Edited by G. B. Curry and C. J. Humphries

74. Automated Taxon Identification in Systematics: Theory, Approaches and Applications (2007)[‡‡]
Edited by N. MacLeod

75. Unravelling the Algae: The Past, Present, and Future of Algal Systematics (2008)[‡‡]
Edited by J. Brodie and J. Lewis

76. The New Taxonomy (2008)[‡‡]
Edited by Q. D. Wheeler

77. Palaeogeography and Palaeobiogeography: Biodiversity in Space and Time (2011)[‡‡]
Edited by P. Upchurch, A. McGowan and C. Slater

---

[a] Published by Clarendon Press for the Systematics Association
[*] Published by Academic Press for the Systematics Association
[‡] Published by Oxford University Press for the Systematics Association
[**] Published by Chapman & Hall for the Systematics Association
[‡‡] Published by CRC Press for the Systematics Association